Formulas for Computing Volumes and Surface Areas

Rectangular Solid: The length is l, the width w, and the height h.

Volume: $V = lwh$

Surface Area: $S = 2(lw + lh + wh)$

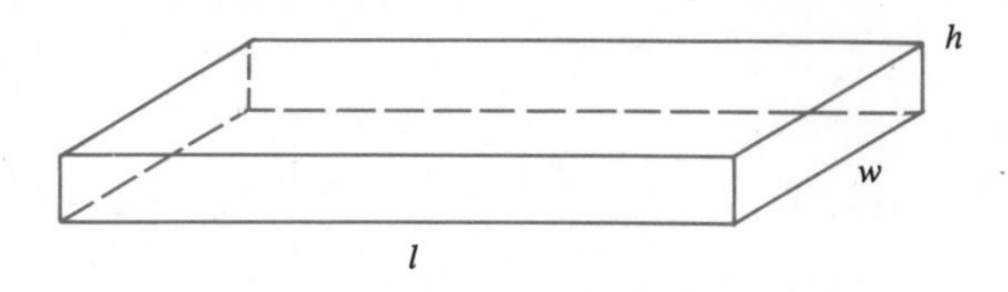

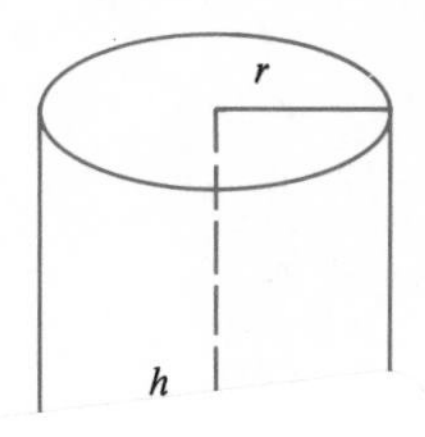

Right Circular Cylinder: The height is h. The radius of each circular base is r.

Volume: $V = \pi r^2 h$

Total Surface Area: $S = 2\pi r^2 + 2\pi rh$

Lateral Surface Area: $L = 2\pi rh$

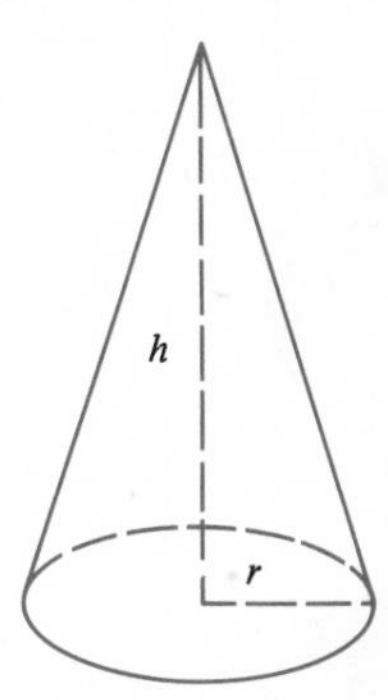

Right Circular Cone: The height is h. The radius of the circular base is r.

Volume: $V = \frac{1}{3}\pi r^2 h$

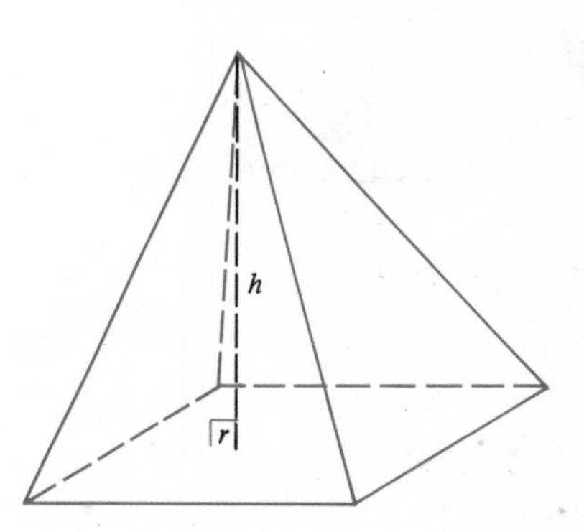

Right Pyramid: The height is h. The area of the base is B.

Volume: $V = \frac{1}{3}Bh$

Other Formulas

Distance: If t is time, r is speed, and d is distance, then

$$d = rt$$

Simple Interest: If t is time, r is the rate of interest (usually expressed as a percent), p is the principal, and i is the interest earned, then

$$i = prt$$

ELEMENTARY ALGEBRA

ELEMENTARY ALGEBRA

Andrew Demetropoulos
Montclair State College

Kenneth C. Wolff
Montclair State College

MACMILLAN PUBLISHING COMPANY
NEW YORK

Collier Macmillan Publishers
London

Cover photo: Immigrant Pass, Death Valley, California. Photo by William Garnett.

Printed in the United States of America

Macmillan Publishing Company
866 Third Avenue, New York, New York 10022

Collier Macmillan Canada, Inc.

Library of Congress Cataloging in Publication Data

Demetropoulos, Andrew.
Elementary algebra.

Includes index.
1. Algebra. I. Wolff, Kenneth C. II. Title.
QA152.2.D45 1983 512.9 82-25911
ISBN 0-02-328580-X

Printing: 1 2 3 4 5 6 7 8 Year: 4 5 6 7 8 9 0 1

ISBN 0-02-328580-X

PREFACE

Elementary Algebra is designed for students with no previous background in algebra and for those needing a review. Specifically, we have in mind students in the following categories:

1. Traditional and second career students who need a course in beginning algebra before taking other courses in the quantitative areas of business, the social sciences, psychology, or the traditional sciences.
2. Students who need a review of algebra to satisfy general education requirements in their curriculum.

The scope of the text ranges from the material covered in most elementary algebra courses to selected topics in intermediate algebra. The text may be used in either a one-semester or a one-quarter course at two-year and four-year colleges. We have built sufficient flexibility into the text to enable instructors to emphasize material suitable to the needs of their students. For example, for those instructors who prefer an early introduction to graphing in the Cartesian coordinate system, Chapter 6 may be presented immediately after Chapter 2. Also, for those instructors who are able to present more material than is covered in the standard elementary algebra course, we have included Chapter 9. This chapter contains topics more appropriate to intermediate algebra. The sections of Chapter 9 are independent of each other with the exception of 9.3, which depends on 9.2.

Mastery of the topics in this text will provide the student with practical algebraic skills and problem solving ability *together* with an understanding of the processes involved. Students who learn elementary algebra with understanding rather than by rote are better equipped to deal with subsequent quantitatively related courses. To this end, we have done the following in writing this text:

1. Written the text using language that is clear and readable. (Readability is an advantage in large classes.)
2. Presented the contents of the text in an organized and coherent manner.
3. Been as precise and mathematically correct as possible at this elementary level.
4. Provided many real-world applications of elementary algebra.

In accordance with the general characteristics of the text given above, we list below under appropriate headings some of the specific features of our text.

Pedagogy

- Each chapter contains a brief introduction to the topics that are presented in the chapter.
- Each new rule or concept presented is introduced by appropriate mathematical motivation and then illustrated by numerous examples.
- Examples have step-by-step explanations, usually with supporting reasons in the margin.
- Summaries are provided of the techniques employed to solve various types of problems, such as solving equations, graphing, applied problems, and factoring.
- Verbal (or word) problems are incorporated early in the text and used consistently in the exercises, where appropriate, in a nontrival way.
- Applied problems that a student might encounter in daily activities are included. Also, some of the answers to the applied problems require a ''commonsense'' interpretation by the student.
- Key concepts covered in each chapter are listed at the end of the chapter.

Exercises

- Approximately 3600 exercises. Exercises at the end of each section are graduated in difficulty.
- The majority of exercises are provided with references to worked-out examples in the text.
- In general, an even-numbered exercise corresponds to the preceding odd-numbered exercise.
- Almost all chapters contain numerous verbal exercises.
- An abundance of exercises dealing with applications.
- Each chapter has a chapter test, which can also be used as a review.
- Answers to odd-numbered exercises and all answers to chapter test questions are given at the end of the text.

Other Features

- A slowly developed and extensive treatment of fundamental algebraic concepts is presented at the beginning of the text. In particular, there is a careful development of the real number system that may be used by the instructor to the extent desired.
- The language of sets is used in the text proper only when it is useful in achieving economy of language and thought.
- There are examples that contain fractions, decimals, and percents, thus enabling the student to deal with any difficulties involving arithmetic skills. These are usually found at the end of the exercise sets.
- The inside cover contains common formulas for finding perimeters, areas, volumes, and so on. A table of square roots is provided at the end of the book.

Accompanying the text is an instructor's manual containing additional sample tests for each chapter, sample final examinations, and answers to all even-numbered exercises, sample tests, and sample final examinations.

We have used the text in regular 15-week semesters meeting 3 times per week for 50 minutes and also in an accelerated 6-week summer session. We found that the text was exceptionally well received by students. The students commented especially favorably on the readability of the text.

We wish to thank the following reviewers for their helpful suggestions for improving the text: Ignacio Bello, Hillsborough Community College; Harold Oxsen, Diablo Valley College; Ronald E. Harrell, Allegheny College; Arthur Goodman, Queens College; A. H. Tellez, Glendale Community College; David Petrie, Cypress College; Roger Pitasky, Marietta College; Lynne Tooley, Bellevue Community College; John Pace, Essex Community College; Betty L. Shores, Tidewater Community College; and Michael Karelius, American River College.

We also thank our Mathematics Editor, Susan Saltrick, for her kindly and thoughtful assistance; Elaine Wetterau, our Production Supervisor, for her gracious and expert guidance of the manuscript through the production process, and Linda Lodzinski, for checking all the solutions. Our families were also supportive despite long periods of inattentiveness. Finally, we thank Pat Demetropoulos for an excellent job of typing both the manuscript and the instructor's manual.

Andrew Demetropoulos
Kenneth C. Wolff

CONTENTS

ELEMENTARY ALGEBRA

CHAPTER 1

Fundamentals

Positive and negative numbers are used in algebra. These numbers are also used in weather forecasting: for example, when a weather forecaster states that on a cold winter day the temperature is 12° but the wind-chill factor makes it feel like −7°.

Source: *U.S. Department of Commerce, National Oceanic and Atmospheric Administration*

CHAPTER 1
CHAPTER 1
CHAPTER 1
CHAPTER 1
CHAPTER 1
CHAPTER 1
CHAPTER 1
CHAPTER 1
CHAPTER 1
CHAPTER 1
CHAPTER 1
CHAPTER 1
CHAPTER 1
CHAPTER 1
CHAPTER 1
CHAPTER 1
CHAPTER 1
CHAPTER 1
CHAPTER 1
CHAPTER 1
CHAPTER 1
CHAPTER 1
CHAPTER 1
CHAPTER 1
CHAPTER 1
CHAPTER 1
CHAPTER 1
CHAPTER 1
CHAPTER 1
CHAPTER 1
CHAPTER 1
CHAPTER 1
CHAPTER 1

In this chapter we introduce some of the fundamental concepts that are used throughout our study of elementary algebra. One of these basic concepts is the idea of using letters of the alphabet to represent numbers.

We shall become familiar with different types of numbers, such as integers, rational numbers, and irrational numbers.

The use of letters to represent numbers in algebra enables us to recognize and state certain properties that these numbers share, such as the commutative, associative, and distributive properties. We discuss these and other properties because they are important to an understanding of the correct use of many of the rules of elementary algebra. Finally, we shall also practice translating simple word problems into the language of algebra.

1.1 Basic Algebraic Symbols

In algebra we use the letters of the alphabet to represent numbers. Letters at the end of the alphabet, such as x, y, and z, are usually used to represent *any* number of a specified set of numbers. They are, therefore, called *variables*.

The letters at the beginning of the alphabet, such as a, b, and c, are usually used to represent a number that remains *fixed* in a mathematical expression. Therefore, they are called *constants*. Specific numbers such as 2 or 5 appearing in a mathematical expression are also referred to as constants.

The fundamental operations involving numbers in elementary arithmetic are addition (+), subtraction (−), multiplication (×), and division (÷). In algebra the basic operations (addition, subtraction, multiplication, and division) are performed on variables and constants. We list in Table 1.1 some basic algebraic expressions involving operation symbols and variables, together with a verbal statement of their meaning. Notice the additional ways of representing multiplication and division in algebra.

In Table 1.1, let x and y represent numbers.

Table 1.1

Operation	Basic Algebraic Expression	Verbal Statement
Addition	$x + y$	The *sum* of x and y.
Subtraction	$x - y$	The *difference* of x and y.
Multiplication	$x \cdot y$, xy, $x(y)$, $(x)(y)$	The *product* of x and y; x and y are called *factors* of the product.
Division	$x \div y$, $\frac{x}{y}$, x/y	The *quotient* of x by y; y is called the *divisor* of x.

Note that the "times" symbol, $\times$, is not generally used in algebra to represent multiplication because it can be easily confused with the letter x, which is frequently used to represent numbers.

In the following examples of algebraic expressions, we use a familiar symbol, $=$, the equals sign. The symbol $=$ means "is the same as," "is," or "is equal to."

In Examples 1 and 2 we present statements of equality involving the basic algebraic expressions listed above. In Example 1 the variables have been replaced by numerical constants. The important thing to notice in these examples is the variety of ways of writing the product and quotient of two numbers.

Example 1. Translate each statement of equality into a verbal statement.

(a) $7 + 2 = 9$. (b) $5 - 3 = 2$.

(c) $(2)(5) = 2 \cdot 5 = (2)5 = 10$. (d) $10 \div 2 = \dfrac{10}{2} = 10/2 = 5$.

Solution

(a) $7 + 2 = 9$. the sum of 7 and 2 *equals* 9

(b) $5 - 3 = 2$. the difference of 5 and 3 *is* 2

(c) $(2)(5) = 2 \cdot 5 = (2)5 = 10$. the product of 2 and 5 *is* 10 (2 and 5 are called *factors* of 10)

(d) $10 \div 2 = \dfrac{10}{2} = 10/2 = 5$. the quotient of 10 by 2 *is* 5

Example 2. Translate each statement of equality into a verbal statement.

(a) $2x = 8$. (b) $y - a = b$. (c) $\dfrac{y}{3} = 9$.

Solution

(a) $2x = 8$. the product of 2 and x is 8

(b) $y - a = b$. the difference of y and a is b

(c) $\dfrac{y}{3} = 9$ the quotient of y by 3 is 9

We are aware that when we compare two quantities they are not always equal. To compare quantities that are not equal, we require symbols of inequality.

We can negate most mathematical symbols by placing a slash, /, through the symbol. Thus $\neq$ means "not equal to." For example,

$6 \neq 9$. **is read "6 does not equal 9"**

If two quantities are not equal, then either one quantity is *less than* or possibly *greater than* the other quantity.

We use the symbol $<$ to represent the relation ''less than.'' Thus, when comparing 6 to 9, we write

$$6 < 9. \qquad \textbf{read ``6 is less than 9''}$$

The symbol $>$ means ''greater than.'' Thus, when comparing 9 to 6, we write

$$9 > 6. \qquad \textbf{read ``9 is greater than 6''}$$

Notice that the symbols $<$ and $>$ always point to the smaller quantity.

We also use the symbol $\leq$ to mean ''less than or equal to.'' The inequality $x \leq y$ is true if either statement ''x less than y'' or ''x equal to y'' is true. A similar statement can be made for the symbol $\geq$, which represents the relation ''greater than or equal to.''

Example 3. State why each inequality is true.
(a) $7 \leq 9$. (b) $8 \geq 8$.

Solution
(a) $7 \leq 9$ is true because $7 < 9$ is true.
(b) $8 \geq 8$ is true because $8 = 8$ is true.

In the following examples we translate simple statements of equality and inequality involving basic algebraic expressions.

Example 4. Express each basic algebraic statement verbally.
(a) $6 - 2 \neq 3$. (b) $x + 3 < 7$. (c) $2y \geq 6$.

Solution
(a) $6 - 2 \neq 3$. the difference of 6 and 2 is not equal to 3
(b) $x + 3 < 7$. the sum of x and 3 is less than 7
(c) $2y \geq 6$. the product of 2 and y is greater than or equal to 6

Example 5. Translate each verbal statement into mathematical symbols.
(a) The sum of twice x and 5 is 25.
(b) The difference of z and 3 is less than 12.
(c) The quotient of 10 by y is greater than or equal to 5.

Solution
(a) $2x + 5 = 25$.
(b) $z - 3 < 12$.
(c) $\dfrac{10}{y} \geq 5$.

A summary of the various symbols of comparison follows.

Let x and y represent numbers or mathematical expressions.

Equality:	$x = y$	x is the same as y; x equals y
Inequality:	$x \neq y$	x does not equal y
	$x < y$	x is less than y
	$x > y$	x is greater than y
	$x \leq y$	x is less than or equal to y
	$x \geq y$	x is greater than or equal to y

EXERCISES 1.1

For Exercises 1–20, evaluate the expression.

1. $5 + 7$ **2.** $6 + 2$ **3.** $\frac{10}{5}$ **4.** $\frac{26}{2}$ **5.** $7 \cdot 2$

6. $3 \cdot 12$ **7.** $4(8)$ **8.** $6(5)$ **9.** $7 - 2$ **10.** $10 - 3$

11. $(3)(9)$ **12.** $(7)(3)$ **13.** $12 \div 2$ **14.** $30 \div 10$ **15.** $9(\frac{1}{3})$

16. $8(\frac{1}{2})$ **17.** $\frac{3}{2} \cdot \frac{2}{3}$ **18.** $\frac{5}{4} \cdot \frac{4}{5}$ **19.** $8/4$ **20.** $12/3$

For Exercises 21–34, determine whether the statement of comparison is true or false. See Example 3.

21. $3 - 1 \neq 3$ **22.** $5 + 2 = 8$ **23.** $7 - 1 > 5$ **24.** $9 - 2 < 4$

25. $\frac{10}{3} < \frac{10}{3}$ **26.** $\frac{30}{5} \geq 6$ **27.** $5 + 2 = 10 - 3$ **28.** $8 - 4 \neq 6 - 2$

29. $(2)(3) \leq 6(1)$ **30.** $\frac{75}{5} \neq \frac{85}{5}$ **31.** $2 > 2$ **32.** $3 > 2$

33. $5 \geq 1$ **34.** $2 \leq 1$

For Exercises 35–54, write a verbal expression, using the words sum, difference, product, *and* quotient, *which is equivalent to the mathematical statement of comparison. See Examples 1, 2, and 4.*

35. $3 + 6 = 9$ **36.** $12 + 3 = 15$ **37.** $9 - 3 = 6$ **38.** $15 - 12 = 3$

39. $4 \cdot 5 = 20$ **40.** $6 \cdot 2 = 12$ **41.** $\frac{20}{5} = 4$ **42.** $\frac{12}{2} = 6$

43. $x + 2 = 3$ **44.** $y + 1 = 5$ **45.** $2x < 10$ **46.** $\dfrac{x}{2} < 5$

47. $x - y = 25$ **48.** $w - x = 50$ **49.** $x + 2 > 3$ **50.** $rs + 1 > 3$

51. $xy \leq 1$ **52.** $rs \leq 2$ **53.** $\dfrac{x}{5} + 1 \geq 2$ **54.** $3y + 1 \geq 4$

For Exercises 55–60, use the symbols, $<$, $>$, $\leq$, or $\geq$ to fill in the blanks to obtain a true statement of comparison. See Example 3.

55. 30 ____ 50 **56.** 2 ____ 2 **57.** 8 ____ 5

58. $3 + 2$ ____ 4 **59.** $\frac{2}{8}$ ____ $\frac{3}{8}$ **60.** $\frac{2}{3}$ ____ $\frac{1}{3}$

For Exercises 61–74, translate the verbal statement into mathematical symbols. See Examples 1, 2, 4, and 5.

61. The sum of 12 and 13 equals 25.

62. The difference of 8 and 5 is 3.

63. The product of 5 and 11 is 55.

64. The quotient of 100 by 20 equals 5.

65. The product of 2 and x equals 16.

66. The sum of y and 5 is 18.

67. The quotient of z by 3 equals 12.

68. The difference of x and y equals z.

69. The sum of 3 with the product of 4 and y equals 15.

70. The sum of z and 2 does not equal z.

71. The product of w and x is greater than or equal to 20.

72. The product of 5 and x is less than or equal to 20.

73. The sum of w and 3 is greater than 12.

74. The quotient of r by 2 is less than 4.

1.2 Evaluating Basic Mathematical Expressions

The symbol 5^4 represents the product $5 \cdot 5 \cdot 5 \cdot 5$; also, 6^5 stands for $6 \cdot 6 \cdot 6 \cdot 6 \cdot 6$. Thus, if b is a number, then an expression such as b^{20} means the following:

$$b^{20} = \underbrace{b \cdot b \cdot b \cdots b}_{\substack{b \text{ is used} \\ 20 \text{ times}}}.$$

the three dots mean "and so on"; they represent the b's not written

The number b is called the *base,* and 20 is called the *exponent*. We say that b^{20} represents *b raised to the 20th power*. Therefore, 5^4 is read as "5 raised to the 4th power." Furthermore, since

$$5^4 = 5 \cdot 5 \cdot 5 \cdot 5 = 625,$$

we say that the *value* of 5^4 is 625.

Special powers such as b^2 and b^3 are read as "*b squared*" and "*b cubed,*" respectively. Note that b^1 is the same as b.

Example 1. Write a brief verbal statement describing each power and its value. (a) 5^2. (b) 4^3. (c) 2^5.

Solution

(a) $5^2 = 5 \cdot 5$. 5 squared is 25

(b) $4^3 = 4 \cdot 4 \cdot 4 = 64$. 4 cubed equals 64

(c) $2^5 = 2 \cdot 2 \cdot 2 \cdot 2 \cdot 2 = 32$. 2 to the 5th power equals 32

We will discuss exponents and rules of exponents more fully in Chapter 3.

In mathematical expressions involving more than one operation, we often use grouping symbols such as parentheses, ( ), to indicate which operations are to be performed first. Thus

$$2 \cdot (5 + 3)$$

means do the addition within the parentheses first, then multiply. Therefore,

$$\begin{aligned} 2 \cdot (5 + 3) &= 2 \cdot (8) \\ &= 16. \end{aligned}$$

If additional grouping symbols are required, we use brackets, [ ], and braces, { }, in that order. We usually do the work in the innermost grouping symbols first and then work outward. Thus, to evaluate

$$4 \cdot [(5 + 3) - 2],$$

we proceed as follows:

$4 \cdot [(5 + 3) - 2] = 4 \cdot [8 - 2]$	evaluate the expression in the innermost grouping symbols, parentheses ( )
$= 4 \cdot [6]$	evaluate the expression in the "next" grouping symbol, brackets, [ ]
$= 24.$	multiplication

We now consider the problem of evaluating a mathematical expression involving more than one operation, which does not contain grouping symbols as a guide. For example, suppose that we are to evaluate the arithmetic expression

$$3 \cdot 2 + 5.$$

If we proceed by multiplying 3 and 2 first, we obtain

$$\begin{aligned} \underbrace{3 \cdot 2}_{6} + 5 &= 6 + 5 \\ &= 11. \end{aligned}$$

But if we choose to do the addition first, we find that we get a different answer. That is,

$$\begin{aligned} 3 \cdot \underbrace{2 + 5}_{7} &= 3 \cdot 7 \\ &= 21. \end{aligned}$$

Clearly, this is undesirable! We would like to evaluate a numerical expression so that we obtain one and only one answer. Therefore, the following rules are agreed upon in mathematics for the order in which arithmetic operations are to be performed.

Order of Operations Rule

When evaluating arithmetic expressions involving exponents, addition, subtraction, multiplication, and division:

1. Without grouping symbols—do the operations in the following order:
   (a) Proceed from left to right evaluating all powers.
   (b) Again proceed from left to right and perform only multiplication and division in the order in which they occur.
   (c) Again proceed from left to right and perform addition and subtraction in the order in which they occur.
2. With grouping symbols—work from within the innermost grouping symbols to the outermost. Within each grouping symbol use the order of operations rule (1).

Example 2. Evaluate each expression using the order of operations rule.
(a) $5 \cdot 6 + 7$. (b) $2^3 + 7 \cdot 6 - 3$. (c) $10 \div 2 \cdot 6 + 6 \div 2$. (d) $2^3 \cdot 0 \cdot 5$.

Solution

(a) $5 \cdot 6 + 7 = 30 + 7$ multiplication first
$= 37.$ addition

(b) $2^3 + 7 \cdot 6 - 3 = 8 + 7 \cdot 6 - 3$ powers first
$= 8 + 42 - 3$ multiplication
$= 50 - 3$ addition
$= 47.$ subtraction

(c) $10 \div 2 \cdot 6 + 6 \div 2 = 5 \cdot 6 + 6 \div 2$ leftmost division occurs first
$= 30 + 6 \div 2$ multiplication next
$= 30 + 3$ division
$= 33.$ addition

(d) $2^3 \cdot 0 \cdot 5 = 8 \cdot 0 \cdot 5$ evaluate powers first
$= 0 \cdot 5$ multiplication next
$= 0.$ multiplication

Notice in part (d) that if 0 appears in a product, the value of the product is automatically 0.

Example 3. Evaluate each expression.
(a) $2 \cdot (3 + 2 \cdot 4)$. (b) $3 \cdot [(4 \div 2 - 1) + 7]$.

Solution

(a) $2 \cdot (3 + 2 \cdot 4) = 2 \cdot (3 + 8)$ use the order of operations rule inside the parentheses first
$= 2 \cdot (11)$
$= 22.$ multiplication

(b) $3 \cdot [(4 \div 2 - 1) + 7] = 3 \cdot [(2 - 1) + 7]$ — do operations in the innermost grouping symbols, the parentheses

$= 3 \cdot [1 + 7]$

$= 3 \cdot [8]$ — perform operations in the next grouping symbols, the brackets

$= 24.$ — multiplication

The following example is a little more complicated.

Example 4. Evaluate the quotient $\dfrac{20 + 2^3}{4 \cdot 2 - 3 \cdot 2}$.

Solution: Because this is read as $(20 + 2^3) \div (4 \cdot 2 - 3 \cdot 2)$, we first evaluate the "numerator" and "denominator" separately using the order of operations rule.

Numerator: $20 + 2^3 = 20 + 8 = 28$ — powers first, then addition

Denominator: $4 \cdot 2 - 3 \cdot 2 = 8 - 6 = 2$ — products first, then subtraction

Evaluating the quotient, we find that

$$\frac{20 + 2^3}{4 \cdot 2 - 3 \cdot 2} = \frac{28}{2} = 14.$$

Expressions involving symbols of operation, variables, and constants, such as

$$2xy, \qquad 4xy + xy, \qquad 6x^2 + y,$$

are called **algebraic expressions.**

When we replace a variable in an algebraic expression with a number, we say that we are assigning a *value* to the variable. An algebraic expression assumes numerical values when values are assigned to the variables appearing in the expression. We say that the algebraic expression has been *evaluated* for those values. For example, if $x = 2$ (x is 2) and $y = 3$ (y is 3) in the expression

$$4x + xy,$$

then $4x + xy$ has a *value* of 14 because

$$4(2) + (2)(3) = 14.$$

Example 5. Evaluate each algebraic expression when $x = 2$ and $y = 3$.
(a) $2x + y$. (b) $5x + 2y^2$.

Solution

(a) If we substitute 2 for x and 3 for y in the expression $2x + y$, we obtain $2(2) + 3 = 4 + 3 = 7$. Thus the expression $2x + y$ has the value 7 when $x = 2$ and $y = 3$.

(b) Substituting 2 for x and 3 for y in the expression $5x + 2y^2$ results in the numerical expression

$$\begin{aligned} 5(2) + 2(3)^2 &= 5(2) + 2(9) \\ &= 10 + 18 \\ &= 28. \end{aligned}$$

Therefore, $5x + 2y^2$ has value 28 when $x = 2$ and $y = 3$.

EXERCISES 1.2

For Exercises 1–10, evaluate the expression. See Example 1.

1. 2^4 **2.** 3^2 **3.** 2^3 **4.** 6^2 **5.** 1^7

6. 2^1 **7.** 10^4 **8.** 3^5 **9.** 8^2 **10.** 4^4

For Exercises 11–36, evaluate the expression. See Examples 2–4.

11. $5 + 2 \cdot 5$ **12.** $7 \cdot 2 + 5$ **13.** $3 \cdot 6 + 4 \cdot 2$

14. $4 \cdot 2 + 3 \cdot 5$ **15.** $2^3 \cdot 4 + 3$ **16.** $3^2 \cdot 2 + 7$

17. $6 \cdot 3 - 2 \cdot 5$ **18.** $10 \cdot 4 - 2 \cdot 15$ **19.** $2 \cdot 6 \div 3 + 1$

20. $5 + 4 \cdot 8 \div 2$ **21.** $5 \cdot 2 + \frac{10}{2} - 2 \cdot 3$ **22.** $\frac{30}{2} - 2 \cdot 3 + 11$

23. $12/3 - 2 + 3^2$ **24.** $15/5 - 2 + 5^2$ **25.** $(6 + 2) \cdot 3$

26. $4 \cdot (5 + 2)$ **27.** $2 + 3 \cdot [(3^2 - 2) + 3]$

28. $4 + 5 \cdot [(2^2 + 6) - 8]$ **29.** $5 \cdot [(2 + 3) - (5 - 1)]$

30. $2 + [(6 - 2) - (10 - 8)]$ **31.** $3 \cdot [3 \cdot (3 - 1) + (4 - 2)]$

32. $6 \cdot [2 \cdot (4 + 3) - (9 - 2)]$ **33.** $\dfrac{6 + 4}{3 + 2}$

34. $\dfrac{20 - 8}{5 - 4}$ **35.** $\dfrac{2^3 + 4(3 + 1)}{2 + 2 \cdot 5}$

36. $\dfrac{3^2 + 8(2)}{13 - 4(2)}$

For Exercises 37–50, find the value of the algebraic expression when $x = 3$ and $y = 2$. See Example 5.

37. $x - y$ **38.** $x + y$ **39.** $x(x + 4)$ **40.** $y(y - 1)$

41. $2xy + 1$ **42.** $7yx + 6$ **43.** $x + y + xy + 1$ **44.** $xy - x + y + 4$

45. $x^2 - y^2$ **46.** $x^2 + y^2$ **47.** $(x + y)(x - y)$ **48.** $xy + y^2$

49. $y/(x + y)$ **50.** $(x + y)/x$

For Exercises 51–61, if x and y represent numbers, what algebraic expression containing x and y will represent the phrase indicated?

51. The product of x and the sum of x and y.

52. Five times the difference of x and y.

53. Six minus the product of x and y.

54. The sum of squares of x and y.

55. The difference between x squared and y squared.

56. The quotient of x by the sum of y and 2.

57. Two less than the quotient of x by y.

58. y cubed less the product of 6 and x.

59. The product of the sum and difference of x and y.

60. The difference between x cubed and y cubed.

61. The sum of the cubes of x and y.

Real Numbers 1.3

The group of numbers most familiar to us is the set of *natural numbers,* also called the *counting numbers*. If we denote this set by N, then

$$N = \{1, 2, 3, 4, 5, \ldots\}.$$

The number 0 added to the set N forms the set of *whole numbers, W,* where

$$W = \{0, 1, 2, 3, 4, 5, \ldots\}.$$

It is useful in the study of algebra to display the whole numbers geometrically by constructing a number line. In doing so, we will introduce the concepts of *positive and negative numbers*. These numbers are also called *signed numbers*.

To construct a number line, we select any point on the line and label it 0. This zero point is called the *origin*. Next we mark off equal segments of unit length to the right of 0. To the endpoint of each segment we assign, in order, a whole number. Thus we obtain the "geometric picture" of the set of whole numbers, W, a part of which is shown in Figure 1.1.

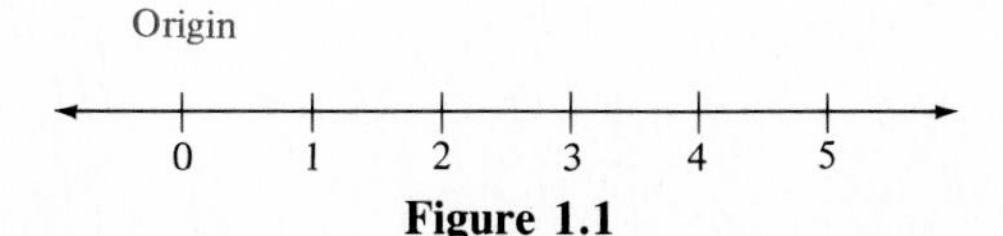

Figure 1.1

Since a point is like a "picture" of a number, the words *point* and *number* can be used interchangeably. Thus, with the number line in mind, we use the phrases "point 2" and "number 2" to mean the same thing.

The concept of a signed number may be introduced by considering the direction we move from one point on the number line to another point. When we move from *left to right* on the number line, we say that we are moving in a *positive direction*. For example, moving from 0 to 2, or 3 to 5 is represented by +2, the distance moved to the right. We use arrows to represent these movements, as shown in Figure 1.2.

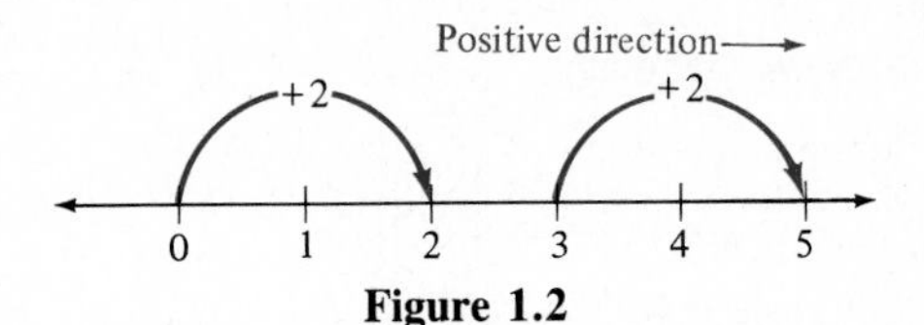

Figure 1.2

The set of movements in a positive direction, $\{+1, +2, +3, \ldots\}$, are called *positive whole numbers*. Thus +2 is read as "positive two." We consider the natural numbers $\{1, 2, 3, \ldots\}$ to be the same numbers as the positive whole numbers $\{+1, +2, +3, \ldots\}$. Thus 2 and +2 represent the same number. Therefore, we call either set of numbers

$$\{+1, +2, +3, +4, \ldots\} \qquad \text{or} \qquad \{1, 2, 3, 4, \ldots\}$$

the *positive whole numbers*.

If 2 or +2 represents moving 2 units in a positive direction on the number line, then moving 2 units in the *opposite direction,* say from 2 to 0, or from 5 to 3, can be thought of as moving 2 units in the *negative direction*. We symbolize this movement in the negative direction by −2, read as "negative two." Again, we use arrows to represent these movements in a negative direction along the number line as shown in Figure 1.3.

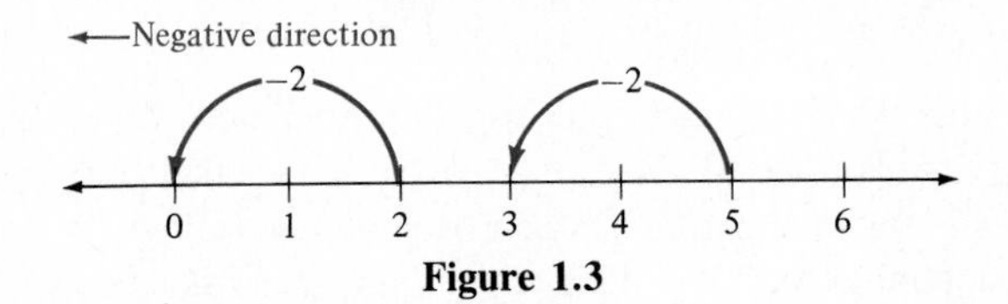

Figure 1.3

The set of negative whole numbers may be displayed on the number line by marking off equal unit segments to the *left* of the origin. The positive and negative whole numbers, as well as 0, are displayed on the number line as shown in Figure 1.4. Note that the number 0 is neither positive nor negative.

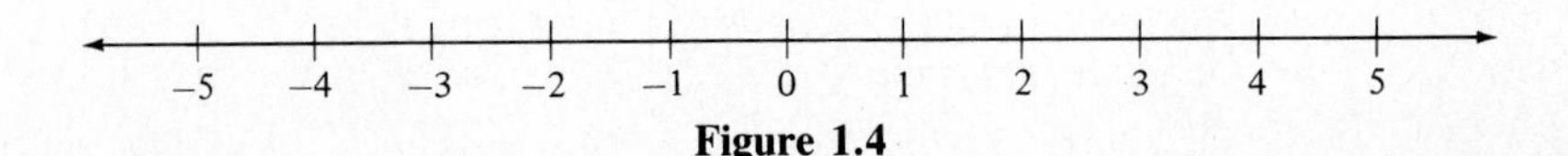

Figure 1.4

The set consisting of all positive whole numbers, all negative whole numbers, and 0 is called the set of *integers*. Because of the positive (+) and negative signs (−), these numbers are also referred to as *signed numbers*.

Here Z stands for the set of integers.

The Set of Integers

$$Z = \{\ldots, -3, -2, -1, 0, 1, 2, 3, \ldots\}.$$

Since signed numbers represent measure in which direction is considered, they have many practical applications. For example, we indicate *temperature readings* above zero with a positive number and below zero with a negative number. Thus a temperature of −8°F on a cold winter day represents 8 degrees below zero. We also use signed numbers to indicate the *distance of a point above or below sea level*. The sign indicates the direction in which altitude is measured from sea level or zero. Thus the altitude *in feet* of the lowest location in Louisiana, New Orleans, is −5 (5 feet below sea level). The highest point in Louisiana is Driskill Mountain, with an altitude of +535 feet (535 feet above sea level). Other applications include *profit* and *loss,* and *increases* and *decreases*.

It is necessary in algebra to extend the set of signed numbers to include *positive* and *negative* fractions such as $-\frac{4}{3}, -\frac{1}{2}, \frac{1}{2}, \frac{4}{3}$, and so on. Each such fraction (with a whole number in the numerator and a nonzero whole number in the denominator) may be assigned to a point on the number line. The positive fractions are located to the right of the origin and the negative fractions to the left of the origin. We now indicate how to locate these fractions on the number line.

For example, to locate the fraction $\frac{4}{3}$, we divide each unit segment into three equal segments. Then $\frac{4}{3}$ is associated with the endpoint of the fourth segment to the right of zero. The fraction $-\frac{4}{3}$ is located in a similar manner to the left of zero as shown in Figure 1.5. Can you locate $-\frac{5}{3}$ on the number line? It is the endpoint of the fifth segment to the left of zero.

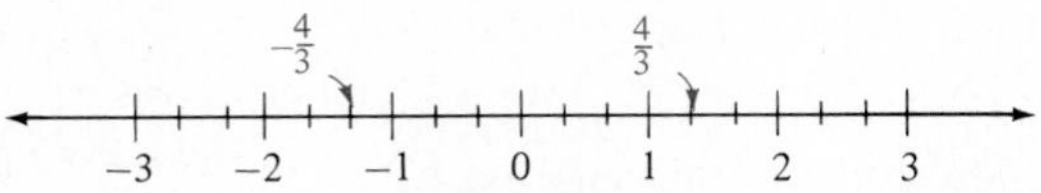

$-\frac{4}{3}$ $\frac{4}{3}$

−3 −2 −1 0 1 2 3

Figure 1.5

If we proceed in the manner just outlined, each positive and negative fraction can be assigned to a point on the number line. The points on the number line associated with some fractions are shown in Figure 1.6.

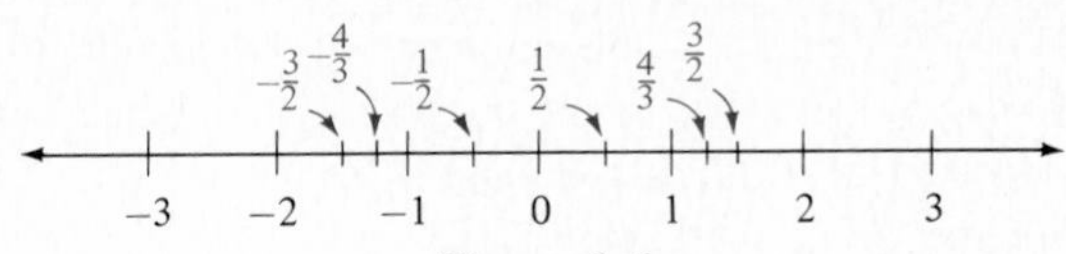

Figure 1.6

The entire collection of numbers consisting of the positive and negative integers, zero, and positive and negative fractions is called the set of *rational numbers*. We can also define a rational number simply as any number that can be written as the ratio of two integers (with denominator not 0). If we let Q represent the set of rational numbers, we may express this symbolically as follows.

The Set of Rational Numbers

$$Q = \left\{ q \;\middle|\; q = \frac{a}{b}, \text{ where } a \text{ and } b \text{ are integers, } b \neq 0 \right\}.$$

This is read as "the set of numbers q such that $q = \frac{a}{b}$, where a and b are integers, $b \neq 0$."

Thus

$$\frac{2}{3}, \quad \frac{-1}{3}, \quad \frac{+4}{-5}, \quad \frac{+3}{1}$$

are examples of rational numbers. Notice that integers such as $+3$ and -2 can be written as

$$+3 = \frac{+3}{1} \quad \text{and} \quad -2 = \frac{-2}{1}.$$

Therefore, with this "second" definition of a rational number, we see that each integer is also a rational number.

Example 1. Use set notation to list the numbers in the set

$$\left\{-2, \frac{-3}{2}, 0, \frac{1}{2}, 5, \frac{7}{1}\right\}$$

that are

(a) Integers.
(b) Rational numbers but are not integers.
(c) Rational numbers.

Solution

(a) $\left\{-2, 0, 5, \frac{7}{1}\right\}$. note that $\frac{7}{1} = 7$

(b) $\left\{\frac{-3}{2}, \frac{1}{2}\right\}$.

(c) $\left\{-2, \frac{-3}{2}, 0, \frac{1}{2}, 5, \frac{7}{1}\right\}$.

The point on the number line associated with a number is called the *graph* of the number and the number is called the *coordinate* of the point. We indicate the graph of a number on the number line by putting a dot on the point associated with the number as shown in the following example.

Example 2. Graph the set $\left\{-3, -1, -\frac{1}{3}, 2, 4\right\}$ on the number line.

Solution: The graph is shown in Figure 1.7.

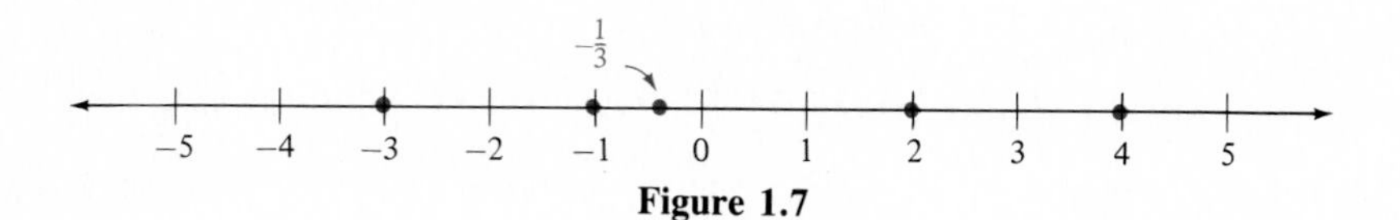

Figure 1.7

If we were to graph all the numbers considered so far (the set of rational numbers), we would discover that there were still points on the number line not assigned a number. These remaining points are associated with numbers that cannot be expressed as the ratio of two integers (i.e., as fractions). They are called the set of *irrational numbers*. Irrational numbers, therefore, are numbers that cannot be expressed as the ratio of two integers. We denote the set of irrational numbers by the letter I. The following discussion describes two irrational numbers that are frequently encountered in mathematics.

The ratio of the circumference of a circle to its diameter (Figure 1.8) is represented by the Greek letter π (read pi). The number π is an irrational number. It cannot be represented as the ratio of two integers. This means that c and d cannot both be integers. However, the rational number $\frac{22}{7}$ is often used as a rough estimate of π for computational purposes.

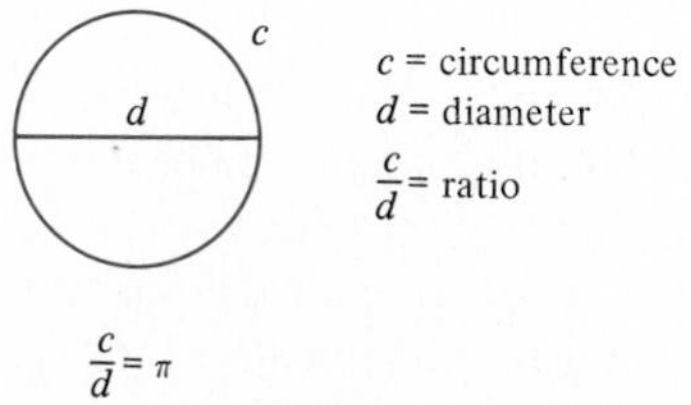

c d

c = circumference
d = diameter
$\frac{c}{d}$ = ratio

$\frac{c}{d} = \pi$

Figure 1.8

The symbol $\sqrt{2}$ is read "the square root of 2." It represents a number that when multiplied by itself yields 2; thus $\sqrt{2} \cdot \sqrt{2} = 2$. We will discuss square roots later in the book. The number $\sqrt{2}$ is an example of an irrational number. That is, it cannot be represented as a fraction.

It can be shown that the rational numbers are numbers that have decimal representation that either terminate (have a last nonzero digit) or repeat. For example,

$$\frac{1}{2} = .5, \qquad \frac{5}{8} = .625, \qquad \frac{1}{6} = .1666\ldots, \qquad \frac{35}{99} = .3535\ldots.$$

On the other hand, any decimal that terminates or repeats can be written as a fraction (or quotient of two integers). Thus

$$.75 = \frac{3}{4}, \qquad .666\ldots = \frac{2}{3}, \qquad .4545\ldots = \frac{5}{11}.$$

are all rational numbers.

Therefore, the set of rational numbers may be defined as the set of fractions (positive and negative) and zero or alternatively as

the set of terminating or nonterminating repeating decimals.

Thus numbers that are not rational, that is, irrational numbers, are those numbers whose decimal expansion does not terminate or repeat. For example,

$$.01001000100001\ldots$$

is an irrational number, since it does not repeat, because if the pattern continues there is one more 0 after each 1. We note that the irrational number $\sqrt{2}$ can be approximated with the rational number 1.414.

Example 3. Use set notation to list the numbers in the set

$$\left\{-3, \frac{-1}{3}, .666\ldots, .1313313331\ldots, \sqrt{2}, \pi, 2, \frac{22}{7}, 1.414, 3.14\right\}$$

that are (a) rational numbers or (b) irrational numbers.

Solution

(a) $\left\{-3, \frac{-1}{3}, .666\ldots, 2, \frac{22}{7}, 1.414, 3.14\right\}$.

(b) $\{.1313313331\ldots, \sqrt{2}, \pi\}$.

As we have noted, the points on our number line that do not have rational numbers as coordinates are associated with irrational numbers. Our number line is now complete in the sense that each point on the number line has either a rational or an irrational number as a coordinate. Also, given any rational or irrational number, there is exactly one point on the number line that is its graph.

The collection of all rational numbers, Q, and irrational numbers, I, is called the set of *real numbers*. The symbol $\mathbb{R}$ stands for the set of real numbers. Thus, in set notation,

$$\mathbb{R} = Q \cup I \quad \text{and} \quad Q \cap I = \varnothing,$$

where $\varnothing$ designates the empty set.

The corresponding number line is called the *real number line*. Figure 1.9 shows the location of a few real numbers on the real number line.

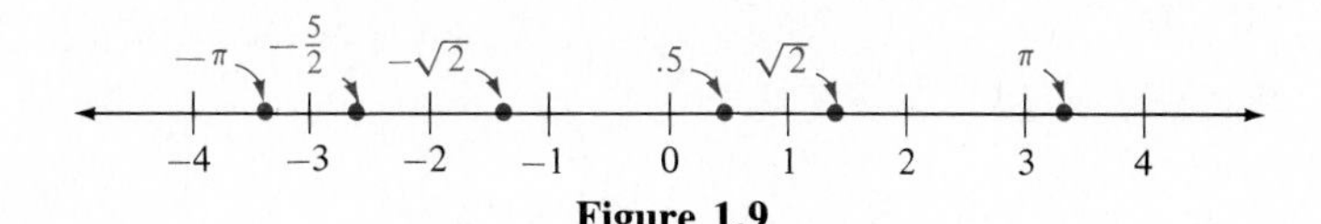

Figure 1.9

An outline of the relation between the real numbers and its subsets is shown in Figure 1.10.

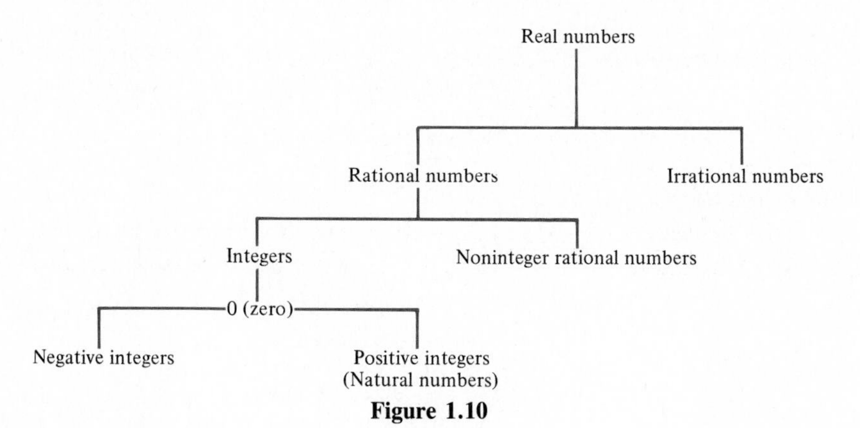

Figure 1.10

EXERCISES 1.3

1. Use a signed number to represent the following temperatures.
(a) 4 degrees below zero Fahrenheit.
(b) 10 degrees below zero Celsius.

2. Use a signed number to represent the following altitudes.
(a) The surface of the Dead Sea (Middle East) is 1292 feet below sea level.
(b) The peak of Mt. Kilimanjaro (Africa) is 19,340 feet above sea level.

3. Use a signed number to represent the following gains or losses in football.
(a) A loss of 15 yards on the first down.

(b) The net yardage gained after a loss of 5 yards on the first down followed by a gain of 7 yards on the second down.

4. Use a signed number to represent the balance in the checking account of

(a) A person who overdraws \$30 on the account.

(b) A person who opens an account with \$100 and then writes a check for \$150 without adding any more money to the account.

For Exercises 5–8, see Example 2.

5. Graph the set $\{-4, -1, 0, 1, 4\}$ on the number line.

6. Graph the set $\{-3, -2, 0, +2, +3\}$ on the number line.

7. Graph the set $\{-\frac{5}{2}, -2, 0, \frac{7}{2}, 3\}$ on the number line.

8. Graph the set $\{-\frac{7}{3}, -\frac{5}{3}, 0, \frac{2}{3}, \frac{8}{3}\}$ on the number line.

For Exercises 9–14, see Examples 1 and 3.

9. Use set notation to list the members in the set $\{-5, -3, .5, \frac{3}{2}, \sqrt{2}, 2, \pi\}$ that are

(a) Positive integers **(b)** Integers
(c) Rational numbers **(d)** Irrational numbers
(e) Real numbers

10. Use set notation to list the members in the set $\{-\frac{7}{2}, -\sqrt{2}, -8, \frac{1}{3}, \pi, 4, 6.2\}$ that are

(a) Negative integers **(b)** Integers
(c) Rational numbers **(d)** Irrational numbers
(e) Real numbers

11. State whether the real number is rational or irrational.

(a) 0.3 **(b)** .72772777277772 . . .
(c) $\frac{22}{7}$ **(d)** -2
(e) .010110111011110 . . . **(f)** .21353535 . . .

12. State whether the real number is rational or irrational.

(a) -0.8 **(b)** 2/3
(c) .1515515551 . . . **(d)** 3.14
(e) .2020020002 . . . **(f)** .20202020 . . .

13. Identify the real number as integer, rational, or irrational.

(a) 6 **(b)** $\frac{5}{3}$ **(c)** $-\frac{7}{2}$
(d) .121212 . . . **(e)** .121221222122221 . . . **(f)** $-\pi$
(g) $\frac{40}{4}$ **(h)** -7 **(i)** $-\sqrt{2}$

14. Identify the real number as integer, rational, or irrational.

(a) -12 **(b)** $\frac{11}{3}$ **(c)** .157157157 . . .
(d) .157157715777 . . . **(e)** .6666 . . . **(f)** $\sqrt{2}$
(g) $\frac{10}{3}$ **(h)** -130 **(i)** π

15. Determine whether the statement is true or false.

(a) The set of whole numbers is a subset of the integers.

(b) Some numbers can be both rational and irrational.

(c) Rational numbers are numbers whose decimal representations repeat periodically or terminate.
(d) The real numbers that are not rational numbers are called irrational numbers.
(e) The set of integers is a subset of the rational numbers.

16. Determine whether the statement is true or false.
(a) The set of integers is a subset of the natural numbers.
(b) The set of integers is a subset of the set of rational numbers.
(c) A rational number can be expressed as the ratio of two integers.
(d) Irrational numbers cannot be expressed as the ratio of two integers.
(e) Every point on the real number line is associated with a rational number.

Absolute Value of a Real Number 1.4

We can use the real number line to introduce **the absolute value of a real number** and **the additive inverse of a real number.**

We will also use the number line to give us a rule for determining when one real number is *less than* or *greater than* another real number.

If we look at the number line and locate the numbers $+3$ and -3, we see that both of those numbers are the same number of units (distance) from the origin, as shown in Figure 1.11.

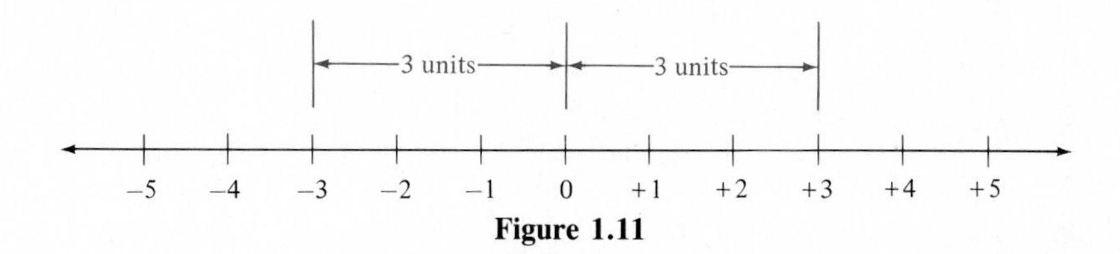

Figure 1.11

The distance that a real number is from the origin (0 point) without reference to a positive or negative direction is called the *absolute value* of the number. Thus the absolute value of both $+3$ and -3 is 3. Symbolically, we indicate the absolute value of a number by vertical bars on either side of the number. We write

$|+3| = 3$ **read "absolute value of +3 is 3"**

$|-3| = 3.$ **read "absolute value of −3 is 3"**

The absolute value of a number is never negative because it measures distance and not direction. However, because $+3$ and 3 stand for the same number, we may write $|+3| = +3$ and $|-3| = +3$.

Example 1. Find the absolute value of

(a) $+5$. (b) -3.5. (c) $-\dfrac{3}{2}$.

Solution

(a) $|+5| = 5$.
(b) $|-3.5| = 3.5$.
(c) $\left|-\frac{3}{2}\right| = \frac{3}{2}$.

Real numbers with opposite signs such as -3 and $+3$ are called additive inverses of each other. *Additive inverses* have the same absolute value but have opposite signs. Therefore, we often use the word *opposite* in place of "additive inverse." The number 0 is its own additive inverse, since -0 is the same as 0. Thus, to obtain the additive inverse (opposite) of a real number, we change its sign from either positive (+) to negative (−), or negative (−) to positive (+).

Example 2. Find the additive inverse of
(a) $+2$. (b) $-\frac{2}{3}$.

Solution

	Number	*Additive Inverse*
(a)	$+2$	-2
(b)	$-\frac{2}{3}$	$+\frac{2}{3}$ or $\frac{2}{3}$

To indicate the additive inverse of a real number represented by a variable such as x, we use the negative sign, −, and write

$-x$. **read "additive inverse of x" or "the opposite of x"**

If x is a positive number, then its opposite, $-x$, is a negative number. If x is a negative number, then its opposite, $-x$, is positive. The following example illustrates what was just stated.

Example 3. Find $-x$, the additive inverse (or opposite) of x, when x has the value (a) -2. (b) $+4$.

Solution

	x	$-x$ *(Additive Inverse of* x*)*
(a)	-2	$-(-2) = +2$
(b)	$+4$	$-(+4) = -4$

We now discuss statements of inequality involving real numbers. The real numbers include negative as well as positive numbers. How do we compare, for example, -3 and -5? Is -3 "less than" -5 or is it "greater than" -5? The same question arises when we compare positive and negative numbers such as 2 and -3. The number line can be used to provide us with a rule that will determine whether one signed number is less than (or greater than) another signed number.

Let x and y be *real numbers*.

x is less than y, written $x < y$, if x lies to the left of y on the real number line, as shown in Figure 1.12.

y is *greater than* x, written $y > x$, if y lies to the right of x on the real number line, as shown in Figure 1.12.

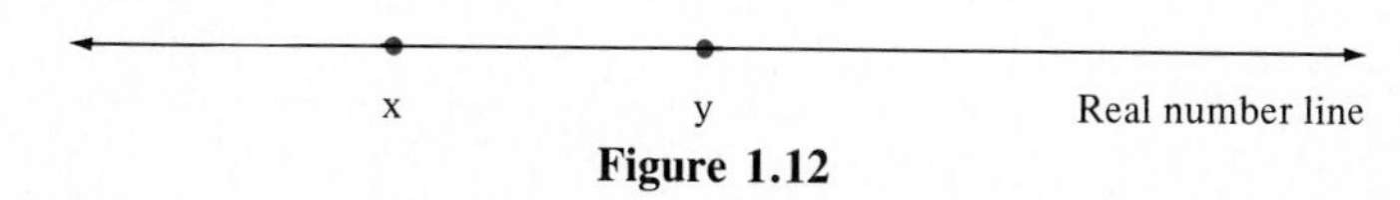

Figure 1.12

The following examples illustrate the rule for using the number line to compare signed numbers.

Example 4. Fill the blank spaces between each pair of numbers with the correct comparison symbol $<$ or $>$.
(a) 3 0. (b) -2 3. (c) -2 -5. (d) -5 0.

Solution: First we graph the set $\{2, -2, 3, -5, 0\}$, consisting of all the different numbers that we are comparing (Figure 1.13). Next we observe that numbers to the left of a number are less than that number, and numbers to the right of a number are greater than that number. Thus we have
(a) $3 > 0$. (b) $-2 < 3$. (c) $-2 > -5$. (d) $-5 < 0$.

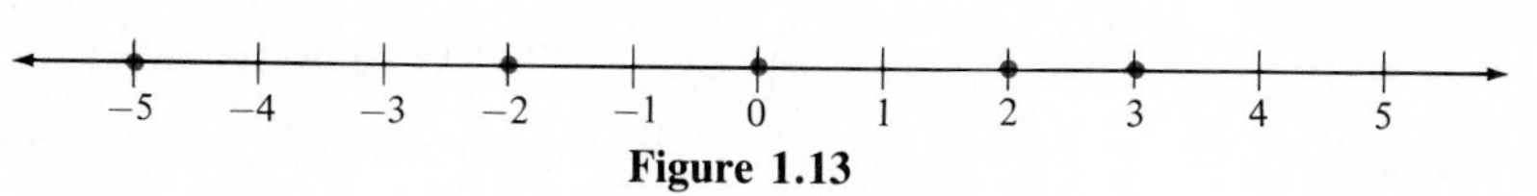

Figure 1.13

Note that a negative number is always *less than* 0, since it lies to the left of 0 on the number line. Thus $x < 0$ is often used to symbolize that *x is a negative number*. Similarly, every positive number is *greater than* 0, since it lies to the right of 0 on the number line. Therefore, $x > 0$ is often used to symbolize that *x is a positive number*.

The inequality symbols $\leq$ (less than or equal to) and $\geq$ (greater than or equal to), introduced in Section 1.1, are also used to compare real numbers.

Example 5. Write a verbal statement describing each inequality statement.
(a) $x \leq -5$. (b) $x \leq 0$. (c) $x \geq 0$.

Solution

(a) $x \leq -5$. x is less than or equal to -5
(b) $x \leq 0$. x is negative or equal to 0
(c) $x \geq 0$. x is positive or equal to 0, or x is *nonnegative*

Example 6. Which statements of inequality are true and which are false?
(a) $-2 \leq -1$. (b) $|-2| < |-1|$. (c) $-10 \leq |-20|$.

Solution

(a) True, since $-2 < -1$ is a true statement.
(b) False. Note that $|-2| = 2$ and $|-1| = 1$. But $2 < 1$ is a false statement.
(c) True. Note that $|-20| = 20$. Therefore, $-10 \leq |-20|$, which is the same as $-10 \leq 20$, is true.

EXERCISES 1.4

For Exercises 1–10, graph the pair of numbers on the number line and determine their absolute values.

1. $-1, +1$ **2.** $-2\frac{1}{2}, +2\frac{1}{2}$ **3.** $-2, +2$ **4.** $-6, +6$ **5.** $-4.5, +4.5$
6. $-8, +8$ **7.** $-3\frac{1}{4}, +3\frac{1}{4}$ **8.** $-5, +5$ **9.** $-7, +7$ **10.** $-\frac{5}{3}, +\frac{5}{3}$

For Exercises 11–22, find the additive inverse of the number. See Example 2.

11. -6 **12.** 5 **13.** $+8.5$ **14.** 1000 **15.** $-3\frac{1}{4}$ **16.** -85
17. $-\sqrt{2}$ **18.** 0 **19.** $8\frac{1}{2}$ **20.** -27 **21.** -8 **22.** π

For Exercises 23–34, simplify the expression by writing the number with one sign. See Example 3.

23. $-(-81)$ **24.** $-(+81)$ **25.** $-(+7)$ **26.** $-(-6)$ **27.** $-(-\pi)$
28. $-(-3\frac{1}{2})$ **29.** $-(+2\frac{1}{2})$ **30.** $-(-(-3))$ **31.** $-(|-32|)$ **32.** $-(|+32|)$
33. $-(-(+2))$ **34.** $-(+(-3))$

For Exercises 35–46, which number is the smaller, or smallest, in the set of numbers? See Example 4.

35. $\{2, 3\}$ **36.** $\{-5, 3\}$ **37.** $\{-8, -2\}$ **38.** $\{-15, -3\}$
39. $\{-15, +2\}$ **40.** $\{-18, 3\}$ **41.** $\{-5, -3\}$ **42.** $\{-|-3|, -|-2|\}$
43. $\{-1, -2, 5\}$ **44.** $\{-5, -2, -1\}$ **45.** $\{-5, -2, 1\}$ **46.** $\{|-6|, |-3|, |-2|\}$

For Exercises 47–52, write a verbal statement that describes the inequality statement. See Example 5.

47. $x \leq -3$ **48.** $x \leq -1$ **49.** $x > 0$ **50.** $x > 7$ **51.** $x < 0$ **52.** $x \leq 4$

For Exercises 53–66, state whether the statement of inequality is true or false. See Example 6.

53. $-5 < 1$ **54.** $5 < 1$ **55.** $-10 < -20$ **56.** $-100 < 0$

57. $|-10| < |-20|$ **58.** $7 > -(-3)$ **59.** $-|-8| < -(+10)$ **60.** $0 \geq -(-7)$

61. $-7 \geq -10$ **62.** $-12 \geq -12$ **63.** $-12 > -16$ **64.** $70 > 0$

65. $-(-8) > 0$ **66.** $0 \leq -13$

Addition of Real Numbers 1.5

In this section we answer a simple question: How do we add signed numbers? For example, what is the sum of +3 and −5? To answer this question, we use the real number line to obtain appropriate rules.

First, recall that signed numbers, such as +3 or −3, represent two ideas. They are

1. The *distance* moved along the number line.
2. The *direction* in which the movement takes place.

Thus a signed number represents a *directed distance* along the number line.

In the following examples we illustrate the procedure for adding two signed numbers, thinking of them as directed distances along the number line. These examples will suggest some simple rules for adding signed numbers without having to use the number line.

Example 1. Add +2 and +3 on the number line.

Solution: We write the sum as $+2 + (+3)$. To add +2 and +3, we first draw an arrow from the origin to +2, as shown in Figure 1.14. The arrow represents a directed distance of 2 units along the number line in the positive direction (+).

Next, since we are adding +3 to +2, we draw a second arrow in the positive direction beginning at 2 and ending at +5 as shown in Figure 1.14. Thus

$$+2 + (+3) = +5.$$

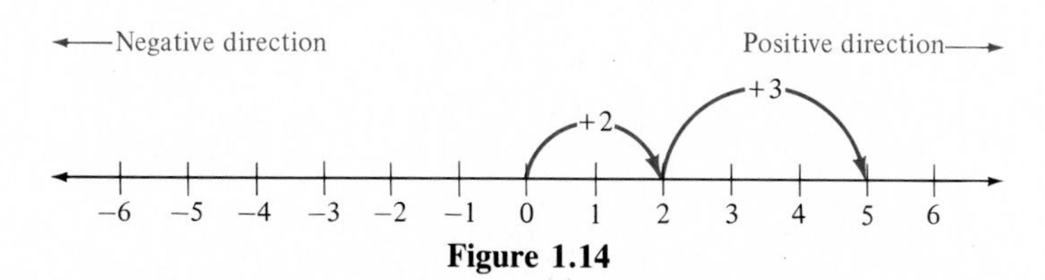

Figure 1.14

The next example deals with adding two negative numbers.

Example 2. Use the number line to find the sum

$$-2 + (-3).$$

Solution: First we draw an arrow from 0 to −2 on the number line as shown in Figure 1.15. Since we are adding −3 to −2, we draw a second arrow from −2 to a point 3 units to the left of -2 (negative direction). We find that we reach the point labeled -5. Thus

$$-2 + (-3) = -5.$$

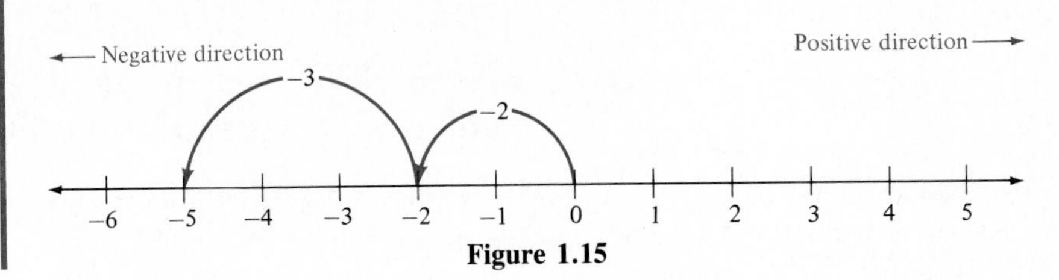

Figure 1.15

Note that in both Examples 1 and 2, we added the distances represented by the numbers (absolute values) and then used the common sign in our answer. These examples suggest the following rule for adding two numbers having the *same* sign, without having to use the number line.

Addition Rule for Adding Numbers with Like Signs

When adding two signed numbers with *like signs,* omit the signs of the two numbers (this gives the absolute values of the numbers) and add them as in ordinary arithmetic. Then use the common sign in your answer.

Example 3. Find the sum using the rule for adding signed numbers with like signs. (a) $+4 + (+18)$. (b) $-5 + (-4)$.

Solution

(a) We "drop" the plus signs on $+4$ and $+18$. Then we add 4 and 18 to obtain 22. The common sign of the two numbers is $+$; therefore,

$$+4 + (+18) = +22.$$

(b) Omitting the negative signs, we add 5 and 4 to obtain the sum 9. Using the common negative sign, we find that the answer to our problem is -9. Therefore,

$$-5 + (-4) = -9.$$

We now use the number line to obtain the sum of two real numbers with *unlike* signs. The following examples illustrate the procedure for finding their sum and also give us a rule for adding numbers with different signs.

Example 4. Find $-3 + (+5)$ using the number line.

Solution: We first draw an arrow from 0 to -3 on the number line, as shown in Figure 1.16. From the point -3 we draw an arrow 5 units in the positive (+) direction to $+2$. Thus we see that

$$-3 + (+5) = +2.$$

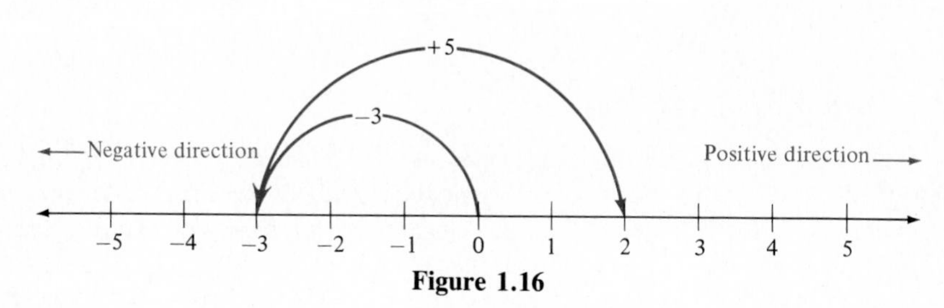

Figure 1.16

Example 5. Find $+3 + (-5)$ using the number line.

Solution: Starting at 0, we draw an arrow to a point 3 units to the right (positive direction), to the point 3. From that point we draw an arrow, in the negative direction, to a point 5 units to the left of 3. Thus we end up at the point -2 as shown in Figure 1.17. As a result we see that

$$(+3) + (-5) = -2.$$

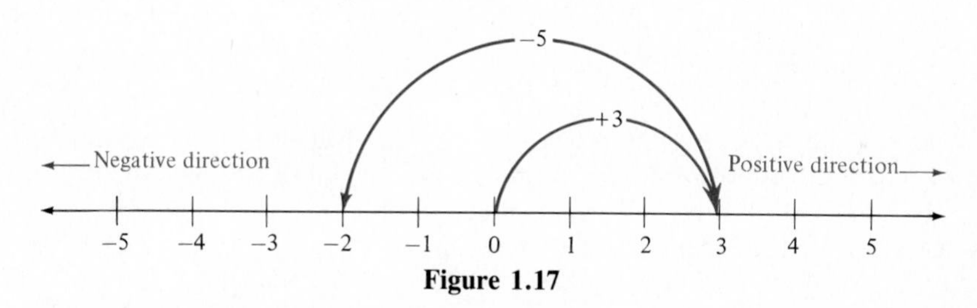

Figure 1.17

Examples 4 and 5 suggest the following rule for adding two signed numbers with unlike signs without having to use the number line.

Addition Rule for Adding Numbers with Unlike Signs

> Consider the numbers without the signs (the absolute values of the numbers). Subtract the smaller of the "unsigned numbers" from the larger one. To this result attach the original sign of the larger of the two unsigned numbers.

Example 6. Find the following sum using the rule for adding numbers with unlike signs:

$$-37 + (+12).$$

Solution: If we "drop" the signs of -37 and $+12$, we obtain their absolute values 37 and 12. Next we subtract 12 from 37 to get 25. The number that is larger when the signs are dropped is 37. Since the original sign of 37 was $-$, we use the negative sign in our answer. As a result,

$$-37 + (+12) = -25.$$

Example 7. Find $13 + (-10)$.

Solution: The number 13 is the same as $+13$. The difference between the "unsigned" numbers 13 and 10 is 3. We use the sign of 13 (which is understood to be $+$) because it is the larger of the two "unsigned" numbers. Our answer, therefore, is $+3$. We write

$$13 + (-10) = +3 \qquad \text{or} \qquad 13 + (-10) = 3.$$

Example 8. Evaluate $[+3 + (-5)] + (-7)$.

Solution

$\underbrace{[+3 + (-5)]}_{-2} + (-7) = (-2) + (-7)$ — rule for adding numbers with unlike signs

$= -9.$ — rule for adding numbers with like signs

Example 9. A man visiting an Atlantic City Casino decides to try his luck and play at the roulette wheel twice. The first time he played he lost $50, but on his second play he won $30.

(a) Use a signed number to represent the amount he won or lost each time he played.

(b) Compute his total gain or loss.

Solution

(a) The first time he played he lost $50, so he was out $50 or -50 dollars. The second time he won $30, that is, $+30$ dollars.

(b) Since $-50 + (+30) = -20$, the man lost $20 in his two plays at roulette.

EXERCISES 1.5

For Exercises 1–10, find the following sums using either the number line or the rules for addition of signed numbers. See Examples 1–8.

1. $+4 + (+3)$ **2.** $+1 + (+2)$ **3.** $-5 + (-1)$ **4.** $-4 + (-2)$ **5.** $+3 + (-1)$

6. $+7 + (-5)$ **7.** $-5 + (+3)$ **8.** $-2 + (-2)$ **9.** $4 + (-3)$ **10.** $-6 + (3)$

For Exercises 11–40, find the following sum using the rules for addition of signed numbers. See Examples 3 and 6–8.

11. $+8 + (+5)$
12. $+6 + (+9)$
13. $-6 + (-4)$
14. $-3 + (-14)$
15. $-10 + (+2)$
16. $-15 + (-4)$
17. $15 + (-5)$
18. $20 + (-12)$
19. $7 + [+2 + (-8)]$
20. $12 + [+6 + (-3)]$
21. $[-8 + (-3)] + (+16)$
22. $[-4 + (-6)] + (+18)$
23. $-\frac{1}{3} + (-\frac{1}{3})$
24. $-\frac{3}{4} + (-\frac{1}{4})$
25. $-\frac{1}{2} + (+\frac{1}{2})$
26. $-\frac{3}{4} + (+\frac{1}{4})$
27. $-.25 + (.75)$
28. $-1.25 + (.50)$
29. $7 + (-2\frac{1}{2})$
30. $19 + (-4\frac{1}{2})$
31. $+3\frac{1}{2} + 4$
32. $2\frac{1}{3} + 6$
33. $-3.5 + 3.0$
34. $-6.2 + 6.2$
35. $\pi + (-\pi)$
36. $\sqrt{2} + (-\sqrt{2})$
37. $+7.2 + [-3.5 + (-.5)]$
38. $6.8 + [-3.2 + (-.8)]$
39. $[+7.2 + (-3.5)] + (-.5)$
40. $[6.8 + (-3.2)] + (-.8)$

For Exercises 41–50, indicate whether the statement is true or false.

41. $-6 + (+2) = -4$
42. $-8 + 4 = -4$
43. $+2 + (-6) = -4$
44. $|-6 + 4| = -6 + 4$
45. $|-8| + |-2| = -8 + (-2)$
46. $-3 + (-2) = (|-3| + |-2|)$
47. $|-12 + (+3)| = -12 + (+3)$
48. $[(-2) + (-3)] + 7 = -2 + [-3 + 7]$
49. $-10 + 0 = -10$
50. $-8 + 2 = 2 + (-8)$

Solve Exercises 51–55. See Example 9.

51. The temperature at 5 A.M. is −5°F. Five hours later the temperature has increased 15°F.
(a) Use a signed number to represent the increase in temperature.
(b) What is the temperature at 10 A.M.?

52. The temperature at 12 P.M. is 10°F above 0. Seven hours later it has dropped 13°F.
(a) Use a signed number to represent the drop (decrease) in temperature.
(b) What is the temperature at 7 P.M.?

53. When the stock market opened on Monday, a certain share of stock had a value of $48 per share. When the stock market closed that day, the stock had dropped in value $2 per share.
(a) Use a signed number to represent the loss in value of each share of the stock for that day.
(b) On Tuesday the stock dropped in value $4 per share. Use a signed number to represent the total loss in value of each share of stock over the two days.

54. A man owes his friend $130. He pays him $65.
(a) Use signed numbers to represent the amount of money he originally owed his friend and the amount that he paid his friend.
(b) Add the signed numbers in part (a) to find the signed number that represents the amount of money he still owes his friend.

55. A football team has the football for four plays. They make gains of 3 and 6 yards on the first two downs, and lose 5 yards on the third down.
(a) Use signed numbers to represent the yards gained or lost on each down.
(b) Add the signed numbers in part (a) to determine the total gain or loss in yardage on the first three downs.
(c) How many yards does the team need on the fourth down in order to gain a total of 10 yards?

1.6 Subtraction of Real Numbers

In this section we state the rule for subtracting real numbers. To indicate that a real number y is to be subtracted from a real number x, we write

$x - y.$ read "subtract y from x," or "x minus y"

Let us consider a type of subtraction problem that we often encounter in algebra. Suppose that a man has \$60 in a checking account. He writes a check for \$100. We would like to be able to evaluate the expression

$60 - 100,$ 60 dollars minus 100 dollars

which represents his balance in the checking account before the bank draws on his cash reserve. We shall use signed numbers so that it "makes sense" to subtract a larger number from a smaller number. To this end we let

$+60$ represent the amount in the checking account

and

$-100.$ represent the check amount to be removed from the checking account

Then the subtraction problem $60 - 100$ can be written as an *addition* problem involving signed numbers as follows:

$$60 - 100 = \underset{\left(\begin{array}{c}\text{in checking}\\\text{account}\end{array}\right)}{60} + \underset{\left(\begin{array}{c}\text{amount taken}\\\text{from checking}\end{array}\right)}{(-100)} = \underset{\left(\begin{array}{c}\text{balance in the}\\\text{checking account}\end{array}\right)}{-40.}$$

Thus -40 represents his current balance. Since it is negative, he owes the bank \$40.

Notice from the previous discussion that with the use of signed numbers, we are able to write that $60 - 100 = -40$.

The example above motivates the following rule for subtracting one real number from another real number.

Subtraction Rule

> Let x and y be real numbers; then
>
> $$x - y = x + (-y).$$

This rule states that *to subtract y from x, add to x the additive inverse (opposite) of y.*

Simply stated, this rule says that to subtract one real number y from another real number x, change the sign of y to the opposite sign and proceed as in addition.

Example 1. Find $+3 - (+10)$.

Solution: Using the subtraction rule, we write

$$+3 \underset{\substack{\uparrow \\ \text{subtraction}}}{-} (+10) = 3 + (\underset{\substack{\uparrow \\ \text{additive inverse of } +10}}{-}10).$$

Then, using our addition rule,

$$\begin{aligned} +3 - (+10) &= 3 + (-10) \\ &= -7. \end{aligned}$$

Example 2. Find $7 - (-3)$.

Solution: Using our rule, we change the sign of -3 to $+3$. Next, we add 7 and $+3$; thus

$$7 - (-3) = 7 + (+3) = 10.$$

Example 3. Find $-3 - (-9)$.

Solution: Since the opposite of -9 is $+9$,

$$\begin{aligned} -3 - (-9) &= -3 + (+9) \\ &= +6. \end{aligned}$$

Example 4. Find $-8 - (-2)$.

Solution: Proceeding as in Example 3, we have

$$\begin{aligned} -8 - (-2) &= -8 + (+2) \\ &= -6. \end{aligned}$$

Example 5. The temperature at 1 A.M. is 3°F and by 5 A.M. has decreased by 10°. What is the temperature at 5 A.M.?

Solution: To determine the temperature at 5 A.M., we subtract the decrease in temperature (10°F) from the temperature at 1 A.M. (3°F). Using the rule for subtraction, we find that

$$\begin{aligned} 3 - 10 &= 3 + (-10) \\ &= -7. \end{aligned}$$

Therefore, the temperature at 5 A.M. is −7°F (or 7° below zero).

EXERCISES 1.6

For Exercises 1–24, evaluate the expression. See Examples 1–4.

1. $3 - 8$ **2.** $2 - 9$ **3.** $11 - 3$ **4.** $15 - 2$

5. $-5 - (+8)$ **6.** $-3 - (+10)$ **7.** $-4 - 8$ **8.** $-3 - 12$

9. $13 - (-2)$ **10.** $14 - (-3)$ **11.** $-12 - (-3)$ **12.** $-18 - (-4)$

13. $-4 - (-16)$ **14.** $-5 - (-20)$ **15.** $0 - 7$ **16.** $0 - (-3)$

17. $-9 - (-9)$ **18.** $-4 - (-4)$ **19.** $-12 - 12$ **20.** $-7 - 7$

21. $3.1 - (-2.4)$ **22.** $2.7 - (-5.3)$ **23.** $-7.25 - 3.05$ **24.** $-5.75 - (+3.20)$

For Exercises 25–34, use the subtraction rule to evaluate the expression.

25. $(5 - 12) + 5$ **26.** $(2 - 10) + 6$ **27.** $(2 - 6) - (-6)$ **28.** $(3 - 8) - (+8)$

29. $-8 - (2 - 7)$ **30.** $-4 - (7 - 3)$ **31.** $(-7 - 2) - 5$ **32.** $(7 - 11) - 13$

33. $-11 + (5 - 3)$ **34.** $-3 + [(+5) - 10]$

For Exercises 35–44, perform the operation indicated.

35. Subtract -3 from -8.

36. Subtract -1 from -7.

37. Subtract -5 from 7.

38. Subtract -4 from -4.

39. What number is 4 less than -9?

40. What number is 5 less than -7?

41. How much greater is 9 than -5?

42. How much greater is 10 than -12?

43. From -24, subtract -20.

44. From -32, subtract -10.

Solve Exercises 45–49. See Example 5.

45. The temperature at 12 A.M. is 5°F. It drops 10° by 5 A.M. What is the temperature at 5 A.M.?

46. The temperature at 5 P.M. is 4°F. If it is falling 1° per hour, what is the temperature at 11 P.M.?

47. A certain share of stock at the beginning of the day on the stock market has a value of $47 per share. When the stock market closes for that day, it has a value of $45 per share. The net change for the day is either positive or a negative number, representing a gain or loss in value per share. What is the net change in the value of the stock per share for that day?

48. A man has assets of $25,000. His brother has liabilities (debts) represented by −$17,000. The brother subtracts −17,000 from 25,000 to determine how far apart they are in financial worth. What is the result of the brother's computation?

49. A missile is launched in a vertical direction from a submarine 7520 feet below sea level. It hits a target directly above it that is 6325 feet above sea level. If sea level is taken to be 0:

(a) Use a signed number to represent the depth of the submarine below sea level.

(b) Use a signed number to represent the altitude of the target above sea level.

(c) Find how many feet the missile traveled in a vertical direction by subtracting the answer in part (a) from the answer to part (b).

Properties of Real Numbers 1.7

In order to motivate the rules for multiplying and dividing signed numbers, we discuss in this section certain properties of real numbers related to addition and multiplication. These properties are also important because they are the basis for many of the rules used in elementary algebra.

We learned in arithmetic that adding 2 to 3 yields the same result as adding 3 to 2; also, that 2 times 3 is the same as 3 times 2. Actually, the order in which we add or multiply *any* two real numbers does not affect the final result. This characteristic of addition and multiplication of real numbers is called the *commutative property of addition and multiplication*.

Commutative Properties

If x and y are real numbers, then

$$x + y = y + x$$

and

$$x \cdot y = y \cdot x.$$

The significance of the commutative property is best understood by considering an operation for which the property does not hold. We note that the operations of subtraction and division *are not* commutative. For instance,

$$7 - 3 \neq 3 - 7 \qquad \text{and} \qquad 8 \div 2 \neq 2 \div 8$$
$$4 \neq -4 \qquad\qquad\qquad 4 \neq \frac{2}{8}.$$

Therefore, we see that it does make a difference in which order we subtract or divide numbers.

The following statements illustrate the commutative property.

$$(-3) + 5 = 5 + (-3) \qquad \text{and} \qquad 2x = x(2).$$

We often find it necessary to add more than two numbers together. Does the way in which we add the numbers together affect the sum? For example, suppose that we wish to find the sum $2 + 3 + 7$. One way to proceed would be to evaluate this sum as $(2 + 3) + 7 = 5 + 7$, which is 12. The sum $2 + 3 + 7$ can also be found by evaluating $2 + (3 + 7) = 2 + 10$, which is again 12. Since both ways of adding the three numbers yield the same answer, we have the following statement of equality:

$$(2 + 3) + 7 = 2 + (3 + 7).$$

This last statement of equality illustrates that the way we group or associate numbers together when we add them does not affect the final answer. This property of addition is called the *associative property of addition*. There is also an *associative property of multiplication*.

Associative Properties

If x, y, and z are real numbers, then

$$(x + y) + z = x + (y + z)$$

and

$$(xy)z = x(yz).$$

Example 1. Use the associative property to rewrite the following expressions: (a) $3 + (7 + x)$ (b) $6(3y)$

Solution

(a) By the associative property for addition, we have that

$$\begin{aligned} 3 + (7 + x) &= (3 + 7) + x \\ &= 10 + x. \end{aligned}$$

(b) Using the associative property for multiplication, we see that

$$\begin{aligned} 6(3y) &= (6 \cdot 3)y \\ &= 18y. \end{aligned}$$

We can make clear the significance of the associative property of addition by showing that the associative property is not valid for subtraction. The following example shows that if we assume subtraction is associative, we are led to a false statement.

$$\begin{aligned} (2 - 5) - 3 &= 2 - (5 - 3) \\ (-3) - 3 &= 2 - (2) \\ -6 &= 0 \quad \text{a } \textit{false statement}. \end{aligned}$$

We note that division of real numbers is another example of an operation that is *not* associative. Readers should convince themselves of this by trying a few examples.

Subtraction problems such as

$$2 - 5 - 3$$

can be evaluated correctly by rewriting subtraction in terms of addition. Remember that to subtract one number from another number we change the sign of the number being subtracted and proceed as in addition. Thus we evaluate $2 - 5 - 3$ as follows:

$$\begin{aligned} 2 - 5 - 3 &= 2 + (-5) + (-3). \\ &= -3 + (-3) \quad \text{associative property of addition} \\ &= -6. \end{aligned}$$

Problems involving subtraction of real numbers of the type already discussed can be done quickly by supplying the addition operation mentally. This is illustrated in the following example.

Example 2. Evaluate each expression.

(a) $-3 - 7 - 2$. (b) $-9 - 2 - (-5)$.

Solution

(a) Note that $-3 - 7 - 2$ means $(-3) + (-7) + (-2)$. As a result, to evaluate the expression $-3 - 7 - 2$, we add -3, -7, and -2 to obtain -12. Thus

$$-3 - 7 - 2 = -12.$$

(b) The expression $-9 - 2 - (-5)$ means

$$-9 + (-2) + (+5) \quad \text{or} \quad -9 + (-2) + 5 \qquad (+5 = 5)$$

Therefore, to evaluate $-9 - 2 - (-5)$, we first add -9 and -2 to obtain the sum -11. Then to the sum -11 we add 5 to obtain -6. Hence

$$-9 - 2 - (-5) = -6.$$

There are two special real numbers, 0 for addition and 1 for multiplication, called *identity elements*. They are called identity elements because when 0 is added to any real number, the result is the identical number. Similarly, any number multiplied by 1 results in the identical number as the product.

Identity Properties

For any real number x:

$$x + 0 = x \quad \text{and} \quad 0 + x = x$$
$$x \cdot 1 = x \quad \text{and} \quad 1 \cdot x = x.$$

The following statements illustrate the identity properties of 0 and 1:

$$0 + (-3) = -3 \quad \text{and} \quad 1 \cdot (-2) = -2.$$

Next we discuss additive and multiplicative inverses of real numbers. The sum of a number and its additive inverse is 0. Similarly, the product of the *multiplicative inverse* of a number and the number is 1. The multiplicative inverse of a number is also called the *reciprocal* of that number.

Inverse Properties

For any real number x, the additive inverse of x, symbolized by $-x$, is the unique real number such that

$$x + (-x) = 0 \quad \text{or} \quad (-x) + x = 0.$$

For any real number x (except 0), the multiplicative inverse (reciprocal) of x, symbolized by $\frac{1}{x}$, is the unique real number such that

$$x \cdot \left(\frac{1}{x}\right) = 1 \quad \text{or} \quad \left(\frac{1}{x}\right) \cdot x = 1 \qquad \text{provided that } x \neq 0.$$

Note that 0 does not have a reciprocal. The reason for this is that it can be shown that the following property of zero is valid.

Zero Property

For any real number x,

$$x \cdot 0 = 0 \quad \text{or} \quad 0 \cdot x = 0$$

(the product of 0 and any number is 0).

Therefore, there does not exist a real number x such that x times 0 equals 1. In other words, 0 *does not* have a multiplicative inverse.

The following statements illustrate the inverse properties:

$$3 + (-3) = 0, \qquad \left(-\frac{3}{4}\right) + \frac{3}{4} = 0, \qquad \frac{1}{2} \cdot 2 = 1.$$

The final property of real numbers that we discuss in this section relates multiplication to addition. It is called the *distributive property* of multiplication over addition. This property is the basis for many rules used for manipulating algebraic symbols.

Distributive Properties

For any real numbers x, y, and z,

$$x \cdot (y + z) = x \cdot y + x \cdot z$$

and

$$(y + z) \cdot x = y \cdot x + z \cdot x.$$

It can be shown, using the distributive property together with the other properties of real numbers, that the following additional properties for x, y, and z real numbers are valid:

$$x \cdot (y - z) = x \cdot y - x \cdot z \quad \text{and} \quad (y - z) \cdot x = y \cdot x - z \cdot x.$$

We also refer to these properties as distributive properties.

Example 3. Use the distributive property and the results of this property to rewrite each algebraic expression.
(a) $3(x + 2)$. (b) $3x + 3y$. (c) $7(x - 2)$.

Solution

(a) $3(x + 2) = 3 \cdot x + 3 \cdot 2$
$= 3x + 6$.

(b) $3x + 3y = 3(x + y)$.

(c) $7(x - 2) = 7 \cdot x - 7 \cdot 2$
$= 7x - 14$.

For future reference, we list in convenient summary form the properties of the real number system.

In the following statements, x, y, and z, represent real numbers.

Commutative Properties: $x + y = y + x$; $xy = yx$

Associative Properties: $(x + y) + z = x + (y + z)$; $(xy)z = x(yz)$

Identity Properties: $x + 0 = x$; $0 + x = x$; $x \cdot 1 = x$; $1 \cdot x = x$

Inverse Properties:

$x + (-x) = 0$; $(-x) + x = 0$; $x \cdot \frac{1}{x} = 1$; $\frac{1}{x} \cdot x = 1$ $(x \neq 0)$

Distributive Properties:

$$x \cdot (y + z) = x \cdot y + x \cdot z; \quad (y + z) \cdot x = y \cdot x + z \cdot x$$
$$x \cdot (y - z) = x \cdot y - x \cdot z; \quad (y - z) \cdot x = y \cdot x - z \cdot x$$

Zero Property: $x \cdot 0 = 0$; $0 \cdot x = 0$

EXERCISES 1.7

For Exercises 1–24, state the property or properties that justify the statement of equality.

1. $x + 3 = 3 + x$
2. $y + 7 = 7 + y$
3. $(3 + r) + s = 3 + (r + s)$
4. $(x + 2) + y = x + (2 + y)$
5. $3 \cdot x = x \cdot 3$
6. $5 \cdot r = r \cdot 5$
7. $3 \cdot (5x) = 15 \cdot x$
8. $3 + (5 + y) = 8 + y$
9. $x + 1 = 1 + x$
10. $0 \cdot x = x \cdot 0$
11. $+5 + (-5) = 0$
12. $-1 + (+1) = 0$
13. $0 \cdot x = 0$
14. $1 \cdot a = a \cdot 1$
15. $2 \cdot (3x) = 6x$
16. $0 + a = a$
17. $(-3) \cdot \frac{1}{-3} = 1$
18. $(-2) \cdot \frac{1}{-2} = 1$
19. $-2 + 0 = -2$
20. $-8 + 0 = -8$
21. $1 \cdot (x + 3) = 1 \cdot x + 1 \cdot 3$
22. $1 \cdot (x + y) = (x + y)$
23. $3 \cdot a + 4 \cdot a = (3 + 4) \cdot a$
24. $5 \cdot y + 8 \cdot y = (5 + 8) \cdot y$

For Exercises 25–42, indicate whether the statement is true or false.

25. $(2 + 1)x = 2x + 1x$
26. $(5 + 1)y = 5 \cdot y + 1 \cdot y$
27. $3(x + 2) = 3x + 2$
28. $2(x + 1) = 2x + 1$
29. $x \cdot 0 = x$
30. $3 \cdot 1 = \frac{1}{3}$
31. $0 + 0 = 0$
32. $1 + 0 = 1$

33. $3 \cdot (8x) = (3 \cdot 8) \cdot x$

34. $a \cdot 1 = a$

35. $1 \cdot 0 = 1$

36. $(5 \cdot 4) \cdot y = 5 \cdot (4 \cdot y)$

37. $x + 0 = 0 + x$

38. $5 + 0 = 0 + 5$

39. $1 \cdot \frac{1}{x} = \frac{1}{x} \quad (x \neq 0)$

40. $\frac{1}{z} \cdot 1 = \frac{1}{z} \quad (z \neq 0)$

41. $[5 + (-5)] + 7 = 5 + [(-5) + 7]$

42. $(3 - 7) - 2 = 3 - (7 - 2)$

For Exercises 43–52, evaluate the expression. See Example 2.

43. $3 - 5 + 4$ **44.** $2 - 3 + 6$ **45.** $10 - 5 - 20$ **46.** $5 - 10 - 8$

47. $4 + 7 - (-3)$ **48.** $8 + 6 - (-2)$ **49.** $-4 - 6 - 8$ **50.** $-2 - 8 - 12$

51. $-3 - 4 - (-5)$ **52.** $-10 + 2 - 10$

For Exercises 53–66, use the distributive property to write an expression without parentheses equal to the expression given. See Example 3.

53. $5(x + 2)$ **54.** $6(y + 3)$ **55.** $2(x - 2)$ **56.** $7(x - 2)$

57. $(r + 2) \cdot 3$ **58.** $(s + 5) \cdot 2$ **59.** $(xy)(z + w)$ **60.** $(ab)(c + d)$

61. $4(x - y)$ **62.** $\pi(a - b)$ **63.** $(x - 3) \cdot 2$ **64.** $(y - 4) \cdot 5$

65. $(4 - y) \cdot 3$ **66.** $(10 - z) \cdot 4$

1.8 Multiplication of Real Numbers

In this section we motivate and state the rules for multiplying signed numbers. First we recall that the natural numbers 1, 2, 3, 4, . . . are the same as the positive integers +1, +2, +3, +4, ... We will use this fact to explain the rules governing the product of two positive numbers and a positive and negative number. Using a different approach, we will also consider the case when the numbers being multiplied are both negative.

We know from elementary arithmetic that

$$3 \cdot 7 = 21.$$

The numbers 3, 7, and 21 are "the same as" +3, +7, and +21, respectively. It is natural and consistent, therefore, to set the product of +3 and +7 equal to +21. That is,

$$(+3) \cdot (+7) = +21.$$

We extend this idea to the set of all positive real numbers.

The product of two positive real numbers is positive.

Example 1. Evaluate each product.

(a) $(+2)\cdot(+8)$. (b) $(+2)\cdot\left(+\frac{1}{2}\right)$. (c) $(+5)\cdot(+4)\cdot(+3)$.

Solution

(a) $(+2)\cdot(+8) = +16$.

(b) $(+2)\cdot\left(+\frac{1}{2}\right) = +1$.

(c) $(+5)\cdot(+4)\cdot(+3) = (+20)(+3) = +60$.

Multiplication of whole numbers may be defined as an abbreviation for repeated addition. For example, $3\cdot 7$ stands for $7 + 7 + 7$, or $3\cdot 7 = 7 + 7 + 7 = 21$. We use this idea to determine the product of a positive and a negative real number. For example, what is the product of $+3$ and -7? Since $+3$ is the same as 3, we may write

$$(+3)\cdot(-7) = 3\cdot(-7).$$

Viewing multiplication as repeated addition, we write $3\cdot(-7)$ as

$$3\cdot(-7) = -7 + (-7) + (-7).$$

Thus

$$\begin{aligned}(+3)\cdot(-7) &= 3\cdot(-7)\\ &= -7 + (-7) + (-7)\\ &= -21.\end{aligned}$$

Therefore,

$$(+3)\cdot(-7) = -21.$$

Note that since $(+3)\cdot(-7) = (-7)\cdot(+3)$, by the commutative property, it is also true that $(-7)\cdot(+3) = -21$. We extend this result to the product of any two numbers with *opposite* signs, with the following rule:

> To multiply two real numbers with *opposite* signs, multiply the numbers without regard to sign (multiply their absolute values) as we do in elementary arithmetic and **attach to the answer a negative sign.**

Example 2. Evaluate each product.

(a) $(-3)\cdot(+5)$. (b) $(-2)\cdot(+\frac{1}{2})$. (c) $+7\cdot(+2)\cdot(-4)$.

Solution

(a) $(-3)\cdot(+5) = -(3\cdot 5) = -15$.

(b) $(-2)\cdot\left(+\frac{1}{2}\right) = -\left(2\cdot\frac{1}{2}\right) = -1$.

(c) $+7\cdot(+2)\cdot(-4) = +14\cdot(-4) = -56$.

We now examine the product of two negative numbers. This case is a little more difficult to justify, for the following reason. If we think of multiplication of whole numbers as repeated addition, what meaning can we give to the product $(-3) \cdot (-7)$? It does not appear to make sense to say that we add -7 negative 3 times. However, by using the properties of real numbers introduced in the last section we can show that $(-3) \cdot (-7) = +21$. To this end consider the following equalities:

$$(-3)(-7) + (-3) \cdot (+7) = (-3) \cdot [(-7) + (+7)]. \qquad \text{distributive property}$$

Since $(-7) + (+7) = 0$ and $(-3) \cdot 0 = 0$, we see that

$$(-3)(-7) + (-3)(+7) = (-3)[\underbrace{(-7) + (7)}_{0}] = (-3) \cdot 0 = 0$$

Therefore,

$$(-3)(-7) + \underbrace{(-3)(+7)}_{-21} = 0$$

$$(-3)(-7) + (-21) = 0. \qquad \text{the product of a negative number and a positive number is negative}$$

The last equality shows that $(-3)(-7)$ is the additive inverse (or opposite) of -21 because their sum is 0. But we also know that $+21$ is the additive inverse of -21 because

$$(+21) + (-21) = 0.$$

Since a number has *exactly* one additive inverse, it follows that $(-3)(-7)$ and $(+21)$ represent the same number. Thus we have shown that

$$(-3)(-7) = +21.$$

The previous discussion illustrates the following rule.

> To multiply two *negative* real numbers, multiply the numbers without regard to sign (multiply their absolute values) and **attach to the answer a positive sign.**

Example 3. Evaluate each product.
(a) $(-2) \cdot (-7)$. (b) $(-3) \cdot (-4)$. (c) $(-2) \cdot (-3) \cdot (-4)$.

Solution

(a) $(-2) \cdot (-7) = +(|-2| \cdot |-7|) = +(2 \cdot 7) = +14$.
(b) $(-3) \cdot (-4) = +(3 \cdot 4) = 12$.
(c) $(-2) \cdot (-3) \cdot (-4) = (+6) \cdot (-4) = -24$.

In each of the cases discussed for multiplying two signed numbers, we obtained the product by first multiplying the numbers without the signs (the absolute values of the numbers) and then attaching to this product the appropriate sign. We *summarize* the rules given thus far in precise mathematical language.

Rule for Multiplying Real Numbers

The product of two real numbers having

1. *Like* signs (both positive or both negative) is the product of their absolute values, and the sign of this product is *positive*.
2. *Unlike* signs (one positive and the other negative) is the product of their absolute values, and the sign of this product is *negative*.

Example 4. Evaluate each expression.
(a) $(-2)(-3) + (-4)(5)$. (b) $-2(3) + (-4)(-5) + 7(+2)$.

Solution

(a) $(-2)(-3) + (-4)(5) = +6 + (-20)$ rule of signs for multiplication
$= -14.$ addition

(b) $-2(3) + (-4)(-5) + 7(+2) = -6 + (+20) + (14)$
$= +14 + (+14)$
$= 28.$

Example 5. Rewrite the expression $-2 \cdot [x + (-3)]$ without parentheses.

Solution: Using the distributive property, we see that

$$-2 \cdot [x + (-3)] = -2 \cdot x + (-2) \cdot (-3) = -2x + 6.$$

It can be shown that for any number x,

$-1 \cdot x = -x.$ the product of -1 and a number is the opposite of the number

This fact enables us to rewrite algebraic expressions such as $-(2x - y)$ *without parentheses*. Note that

$-(3x + 5) = (-1)(3x + 5)$ the opposite of a quantity is the same as -1 times the quantity
$= (-1)(3x) + (-1)(5)$ distributive property
$= -3x - 5.$ same reason as the first line

Therefore, we see that $-(3x + 5) = -3x - 5$.

In an expression such as $(3x + 5)$, we call $3x$ and 5 the terms within the parentheses. Thus the preceding discussion illustrates the following simple rule for removing parentheses that are preceded by a negative sign.

> Parentheses preceded by a negative sign, may be removed, provided that the sign of each term within the parentheses is changed to the opposite sign.

Example 6. Write the expression

$$-(2xy - 3z - 5w)$$

without parentheses.

Solution: Following the rule stated above, we see that

$$-(2xy - 3z - 5w) = -2xy + 3z + 5w.$$

The following discussion suggests a rule for removing parentheses when they are preceded by a positive sign. The expressions $+(2x - y)$ and $(2x - y)$ are considered the same, just as the numbers $+1$ and 1 are identified with each other. Thus

$$\begin{aligned} (2x - y) &= 1 \cdot (2x - y) && x = 1 \cdot x\text{; identity property for multiplication} \\ &= 1 \cdot (2x) - 1 \cdot (y) && \text{distributive property} \\ &= 2x - y. && \text{identity property for multiplication} \end{aligned}$$

Thus we see that $+(2x - y) = (2x - y) = 2x - y$. Consequently, we have the following rule.

> Parentheses preceded by a positive sign may be removed without changing the sign of any terms within the parentheses.

Example 7. Write the following expression without parentheses.

$$(3xz - 7w + 4r - 9y).$$

Solution: Using the rule stated above, we see that

$$(3xz - 7w + 4r - 9y) = 3xz - 7w + 4r - 9y.$$

EXERCISES 1.8

For Exercises 1–20, find the product. See Examples 1–3.

1. $(-2) \cdot (+7)$ **2.** $(-3) \cdot (+9)$ **3.** $(+5) \cdot (-4)$ **4.** $(+6) \cdot (-3)$

5. $(+2) \cdot (+15)$ 6. $3(+12)$ 7. $(-3) \cdot (-11)$ 8. $(-8) \cdot (-4)$
9. $(-10) \cdot (-2)$ 10. $(-12) \cdot (-3)$ 11. $(-3.2) \cdot (1.5)$ 12. $(-4.3) \cdot (2.2)$
13. $(+5) \cdot (-\pi)$ 14. $(-6) \cdot (\pi)$ 15. $(2) \cdot (\frac{1}{3})$ 16. $(+3) \cdot (\frac{1}{4})$
17. $(-\frac{1}{6}) \cdot (-\frac{1}{2})$ 18. $(-5.1) \cdot (-7.3)$ 19. $(-2.25) \cdot (-4)$ 20. $(-1.2) \cdot (-9.1)$

For Exercises 21–28, simplify the expression.

21. $-1 \cdot (-x)$ 22. $-1 \cdot (-z)$ 23. $-1 \cdot (-6)$ 24. $-1 \cdot (-\pi)$
25. $-1 \cdot (+6)$ 26. $-1 \cdot (+3.1)$ 27. $-1 \cdot r$ 28. $-1 \cdot (y)$

For Exercises 29–46, perform the operations indicated. See Examples 4 and 5.

29. $(-3) \cdot (2 - 7)$ 30. $(-8) \cdot (3 - 6)$
31. $-5 \cdot [3 + (-4)]$ 32. $-4 \cdot [2 - (-3)]$
33. $(-4) \cdot (+5) + (-2) \cdot (-10)$ 34. $(-6) \cdot (+2) + (-5) \cdot (-2)$
35. $(-4) \cdot (-2) \cdot (-1)$ 36. $(-8) \cdot (-1) \cdot (-2)$
37. $(-2)(-2)(-2)(-2)$ 38. $(-1) \cdot (-1) \cdot (-1) \cdot (-1)$
39. $(5 - 2) \cdot (2 - 5)$ 40. $(6 - 4) \cdot (4 - 6)$
41. $(3 + 5) \cdot (3 - 5) + (-2) \cdot (-7)$ 42. $(1 + 4) \cdot (1 - 4) + (-12) \cdot (-1)$
43. $-5(x - 3)$ 44. $-7 \cdot [(-x) + (-y)]$
45. $-1 \cdot [(x + y) - z]$ 46. $-1 \cdot [(r - s) + 3]$

For Exercises 47–56, simplify the expression by removing the grouping symbols. See Examples 6 and 7.

47. $-(x + y)$ 48. $-(r - s)$ 49. $-[(x - w) + s]$ 50. $-[(r + s) - t]$
51. $-(x + y - z)$ 52. $-(r + s - t)$ 53. $+(5x - 2z - 1)$ 54. $+(7y - 3x + 4)$
55. $(8x - y - w)$ 56. $(4z - x + w)$

Solve Exercises 57 and 58.

57. A man owes \$32.20 on his Alpha credit card. Represent this indebtedness by -32.20. He then charges on his credit card four automobile tires each costing \$49.50. Use negative numbers to represent:
(a) His indebtedness for each of the four tires.
(b) The total amount of his indebtedness for the four tires.
(c) The total amount of indebtedness on his credit card after charging the tires.

58. The temperature drops constantly 2°F per hour during a certain 5-hour period of the day. Represent the hourly decrease in temperature by -2°F. Use a negative number to represent the total drop in temperature during the 5-hour period.

1.9 Division of Real Numbers

In Section 1.6 we defined subtraction in terms of addition using the notion of the additive inverse of a number. Similarly, in this section we define division in terms of multiplication, using the concept of the multiplicative inverse (reciprocal) of a number. The following rule tells how to divide real numbers.

Division Rule

> Let x and y be real numbers; then
>
> $$x \div y = x \cdot \frac{1}{y} \qquad \text{provided that } y \neq 0.$$

This rule states that to divide a number x by y, multiply x by the *reciprocal* or *multiplicative inverse* of y, provided that y is not 0. We observe that division by 0 is not permitted because 0, as we have seen in Section 1.7, is the only number that does not have a reciprocal (multiplicative inverse). Therefore,

$$x \div 0$$

is not defined for any *real number* x, including 0.

Recall that the result of dividing x by y is also symbolized by

$$x/y \qquad \text{or} \qquad \frac{x}{y}$$

and is called the quotient of x divided by y.

The notation $\frac{x}{y}$ is also used to represent fractions when x and y are integers. However, in evaluating algebraic expressions, the net effect is the same whether we regard $\frac{x}{y}$ as a fraction or choose to think of $\frac{x}{y}$ as indicating division of x by y. Again, the fraction $\frac{x}{0}$ is not defined.

Since division is defined in terms of multiplication, the rules for the sign of the quotient of two signed numbers are easily obtained from the rules for multiplying signed numbers. Before we state the rule of signs for division, we make some observations on the multiplicative inverse (reciprocal) of a number. Note that the *product of a real number x and its reciprocal $\frac{1}{x}$ is* 1. That is, $x \cdot \frac{1}{x} = 1$, $x \neq 0$.

The multiplicative inverse (reciprocal) of -3 is represented by $\frac{1}{-3}$. It can be shown that

$$\frac{1}{-3} = -\frac{1}{3}.$$

Thus $-\frac{1}{3}$ also represents the multiplicative inverse of -3. Similarly, we could show that $\frac{1}{+3} = +\frac{1}{3}$.

Example 1. Find the multiplicative inverse of each number.

(a) -2. (b) $-\frac{3}{4}$. (c) $\frac{5}{6}$. (d) 7.

Solution: The multiplicative inverse of

(a) -2 is $\frac{1}{-2}$ or $-\frac{1}{2}$, since $(-2) \cdot \left(-\frac{1}{2}\right) = 1$.

(b) $-\frac{3}{4}$ is $-\frac{4}{3}$, since $\left(-\frac{3}{4}\right) \cdot \left(-\frac{4}{3}\right) = +\left(\frac{3}{4} \cdot \frac{4}{3}\right) = 1$.

(c) $\frac{5}{6}$ is $\frac{6}{5}$, since $\frac{5}{6} \cdot \frac{6}{5} = 1$.

(d) 7 is $\frac{1}{7}$, since $7 \cdot \frac{1}{7} = 1$.

Example 2 illustrates computing a quotient in terms of multiplication. Since we already have rules for the sign of the product of two numbers, we can use these to determine the sign of the quotient of two real numbers.

Example 2. Find each quotient using the rule for division.

(a) $\frac{-8}{2}$. (b) $\frac{6}{-3}$. (c) $\frac{-10}{-5}$. (d) $\frac{+9}{+3}$.

Solution

(a) $\frac{-8}{2} = (-8) \cdot \left(\frac{1}{2}\right) = -\left(8 \cdot \frac{1}{2}\right) = -4$.

(b) $\frac{6}{-3} = 6 \cdot \frac{1}{-3} = 6 \cdot \left(-\frac{1}{3}\right) = -\left(6 \cdot \frac{1}{3}\right) = -2$.

(c) $\frac{-10}{-5} = (-10) \cdot \left(\frac{1}{-5}\right) = (-10) \cdot \left(-\frac{1}{5}\right) = +\left(10 \cdot \frac{1}{5}\right) = +2$.

(d) $\frac{+9}{+3} = (+9) \cdot \frac{1}{+3} = 9 \cdot \left(+\frac{1}{3}\right) = +\left(9 \cdot \frac{1}{3}\right) = +3$.

Example 2 illustrates the following rule for determining the sign of a quotient.

Rule for Dividing Real Numbers

> The sign of the quotient of two numbers with
> 1. *Like* signs (both positive or both negative) is positive.
> 2. *Unlike* signs (one positive the other negative) is negative.

When we actually do a division problem, we first disregard the signs and compute the quotient as we would in elementary arithmetic. Then we use the rule for signs given here to attach the correct sign to our answer. The following example illustrates this procedure.

Example 3. Evaluate each quotient.

(a) $\dfrac{+10}{+5}$. (b) $\dfrac{-24}{8}$. (c) $\dfrac{20}{-4}$. (d) $\dfrac{-12}{-3}$.

Solution

(a) $\dfrac{+10}{+5} = +\left(\dfrac{10}{5}\right) = +2.$ like signs, quotient is positive

(b) $\dfrac{-24}{+8} = -\left(\dfrac{24}{8}\right) = -3.$ unlike signs, quotient is negative

(c) $\dfrac{20}{-4} = -\left(\dfrac{20}{4}\right) = -5.$ unlike signs, quotient is negative

(d) $\dfrac{-12}{-3} = +\left(\dfrac{12}{3}\right) = +4.$ like signs, quotient is positive

We note that if x and y are real numbers, it can be shown that the following equalities hold:

$$-\frac{x}{y} = \frac{-x}{y} = \frac{x}{-y}.$$

Example 4. Write the quotient (or fraction)

$$\frac{-2}{3}$$

in two equivalent ways by changing the position of the negative sign.

Solution: Two quotients that are the same as $\dfrac{-2}{3}$ are $\dfrac{2}{-3}$ and $-\dfrac{2}{3}$.

Example 5. Find the multiplicative inverse of $\dfrac{8}{-7}$.

Solution: The multiplicative inverse of $\frac{8}{-7}$ is $\frac{-7}{8}$, since

$$\frac{8}{-7} \cdot \frac{-7}{8} = \left(-\frac{8}{7}\right) \cdot \left(-\frac{7}{8}\right) = +\left(\frac{8}{7} \cdot \frac{7}{8}\right) = 1.$$

Example 6. Evaluate each quotient.

(a) $\frac{1}{3} \div \frac{1}{-6}$. (b) $-\frac{7}{2} \div \frac{7}{-3}$.

Solution

(a) $\frac{1}{3} \div \frac{1}{-6} = \frac{1}{3} \cdot (-6)$ reciprocal of $\frac{1}{-6}$ is -6

$$= -\left(\frac{1}{3} \cdot 6\right)$$

$$= -2.$$

(b) $-\frac{7}{2} \div \frac{7}{-3} = -\frac{7}{2} \cdot \frac{-3}{7}$ reciprocal of $\frac{7}{-3}$ is $\frac{-3}{7}$

$$= \left(-\frac{7}{2}\right) \cdot \left(-\frac{3}{7}\right)$$

$$= +\left(\frac{7}{2} \cdot \frac{3}{7}\right)$$

$$= \frac{3}{2}.$$

Example 7. Evaluate

$$\frac{(-2)(-5) + (-3)(-2)}{(-4)(2)}.$$

Solution: If we view this as indicating a problem in division, we simplify the "numerator" $(-2)(-5) + (-3)(-2)$, and the "denominator" $(-4)(2)$ separately. Thus

$$\frac{(-2)(-5) + (-3)(-2)}{(-4)(2)} = \frac{10 + (+6)}{-8} = \frac{+16}{-8} = -2.$$

EXERCISES 1.9

Find the quotient if possible. See Examples 2 and 3.

1. $\frac{20}{-2}$ **2.** $\frac{10}{-5}$ **3.** $\frac{-40}{8}$ **4.** $\frac{-18}{9}$

5. $\dfrac{-25}{-5}$ 6. $\dfrac{-24}{-12}$ 7. $-12 \div (-3)$ 8. $-15 \div (-5)$

9. $27 \div (-3)$ 10. $18 \div (-6)$ 11. $-50 \div (+10)$ 12. $-42 \div 21$

13. $\dfrac{0}{5}$ 14. $\dfrac{0}{-6}$ 15. $\dfrac{5}{0}$ 16. $\dfrac{-6}{0}$

17. $\dfrac{-1}{-1}$ 18. $\dfrac{-0}{0}$ 19. $\dfrac{0}{0}$ 20. $\dfrac{-1}{1}$

For Exercises 21–32, write the quotient (or fraction) in two equivalent ways by changing the position of the negative sign. See Example 4.

21. $-\dfrac{1}{4}$ 22. $-\dfrac{3}{4}$ 23. $\dfrac{-3}{5}$ 24. $\dfrac{-2}{3}$ 25. $\dfrac{4}{-5}$ 26. $\dfrac{5}{-6}$

27. $-\dfrac{3}{7}$ 28. $-\dfrac{14}{15}$ 29. $\dfrac{7}{-9}$ 30. $\dfrac{5}{-11}$ 31. $\dfrac{-13}{11}$ 32. $\dfrac{-9}{17}$

For Exercises 33–40, find the quotient. See Examples 5 and 6.

33. $\dfrac{1}{2} \div \left(\dfrac{1}{-4}\right)$ 34. $\dfrac{3}{4} \div \left(\dfrac{3}{-2}\right)$ 35. $\dfrac{-1}{2} \div \dfrac{3}{-2}$ 36. $\dfrac{-3}{5} \div \dfrac{4}{-5}$

37. $-\dfrac{7}{9} \div \left(\dfrac{-3}{18}\right)$ 38. $\dfrac{-5}{2} \div \left(-\dfrac{5}{6}\right)$ 39. $\dfrac{3}{-11} \div \dfrac{9}{22}$ 40. $\dfrac{4}{-13} \div \dfrac{2}{26}$

For Exercises 41–46, fill in the blank.

41. $\_\_\_\_ \div (-3) = -8$ 42. $\_\_\_\_ \div (-2) = -3$ 43. $-15 \div \_\_\_\_ = 5$

44. $-20 \div \_\_\_\_ = 10$ 45. $12 \div \_\_\_\_ = +3$ 46. $\_\_\_\_ \div 8 = +2$

For Exercises 47–60, simplify "numerator" and "denominator"; then find the quotient. See Example 7.

47. $[(-3) + 13] \div (2 - 7)$ 48. $[(-4) + (+8)] \div [-3 + (-1)]$

49. $[-10 + (-20)] \div (-7 - 3)$ 50. $(-9 - 6) \div [-3 + (-2)]$

51. $\dfrac{-3[5 + (-3)]}{(-2)(-3)}$ 52. $\dfrac{-5[10 + (-2)]}{2(5)}$

53. $\dfrac{(-4)(-5) - 2(4)}{-4 + (-2)}$ 54. $\dfrac{7(-7) + 4(10)}{-2 + (-1)}$

55. $\dfrac{2^2 + 4^2}{-5}$ 56. $\dfrac{3^2 - 2^2}{1 - 6}$ 57. $\dfrac{4^2 - 5^2}{-3}$

58. $\dfrac{5^2 - 13^2}{-(12)^2}$ 59. $\dfrac{(-3)^2}{-(3^2)}$ 60. $\dfrac{2^2}{(-2)^2}$

Solve Exercises 61 and 62.

61. The temperatures for each day of a cold week in February at 4 A.M. were as follows: Sunday $+4°F$, Monday $+1°F$, Tuesday $+2°F$, Wednesday $-2°F$, Thursday $-6°F$, Friday $-4°F$, and Saturday $-2°F$. What was the average temperature at 4 A.M. for the week?

62. A certain share of stock showed the following gains (+) and losses (−) for a certain business week on the stock exchange: Monday $+2\frac{1}{4}$, Tuesday $+1\frac{1}{4}$, Wednesday $-\frac{1}{4}$, Thursday $-\frac{1}{4}$, and Friday -4. What was the average of the gains and losses for the stock that week?

Key Words

Absolute value
Algebraic expression
Constant
Coordinate
Graph
Inequality
Number line
Operations
- Addition
- Subtraction
- Multiplication
- Division

Properties of real numbers
- Associative
- Commutative
- Distributive
- Identity
  - Additive
  - Multiplicative
- Inverse
  - Additive (opposite)
  - Multiplicative (reciprocal)

Real number
- Integer
- Rational
- Irrational

Signed numbers
- Positive
- Negative

CHAPTER 1 TEST

For Problems 1–3, write the verbal statement in mathematical symbols.

1. The quotient of x by y is 2.

2. Six more than the product of 5 and y is 31.

3. The sum of x and 15 is less than 18.

For Problems 4–8, evaluate the expressions.

4. $6 + 3 \cdot 2$

5. $5 + 8 \cdot 2^2$

6. $8 - 4 \div 2 + 3 \cdot 3$

7. $3 + [(3 \cdot 5 - 2) + 4]$

8. $\dfrac{11 - 2(3)}{5 \div 6 \cdot 6}$

For Problems 9–12, find the numerical value of the algebraic expression when $x = 4$ and $y = 1$.

9. $x(x - y)$ **10.** $2xy + x$ **11.** $(x + y)^2$ **12.** $\dfrac{3xy - 2x + 4y}{x}$

13. Graph the set $\{-5, -2, 0, \frac{1}{2}, +3\}$ on the accompanying number line.

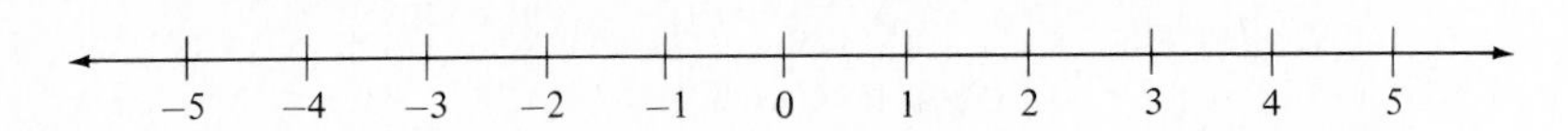

Use set notation to list the members in the set $\{-7, -1, \frac{22}{7}, \sqrt{2}, 2, .3, \pi\}$ that are

14. Integers **15.** Rational numbers **16.** Irrational numbers

For Problems 17 and 18, simplify the expression.

17. $-(-16)$ **18.** $-[+(-2)]$

For Problems 19–21, determine whether the statement is true or false.

19. $-7 < -2$ **20.** $|-6| \leq |-2|$ **21.** $-100 > 2$

For Problems 22–28, evaluate the expression.

22. $15 + (-7) + |-6|$ **23.** $(-5) + (-8)$ **24.** $-7 + (-3) + 10$ **25.** $5 - 12$

26. $-3 - (-5)$ **27.** $-8 - |-4|$ **28.** $-8 - 10 - 2$

Solve Problem 29.

29. The temperature at 1 A.M. is -3°F. Three hours later, the temperature has decreased 4°F. What is the temperature at 4 A.M.?

For each algebraic statement in Problems 30–33 on the left, match the letter on the right that describes the property of real numbers illustrated.

30. $(x + y) + 3 = x + (y + 3)$ A. Commutative property of addition

31. $3 \cdot (2) = 2 \cdot (3)$ B. Commutative property of multiplication

32. $a(x + 2) = a \cdot x + a \cdot 2$ C. Associative property of addition

33. $(3x)y = 3(xy)$ D. Associative property of multiplication

E. Distributive property

For Problems 34 and 35, use the distributive property to rewrite the expression without parentheses.

34. $6(x + 2)$ **35.** $y(x - 1)$

For Problems 36–39, determine whether the statement is true or false. If x represents any real number, then

36. $x + 0 = x$ **37.** $x \cdot 0 = 0$ **38.** $x + 1 = x$ **39.** $x \cdot 1 = x$

For Problems 40–50, find the sum, product, or quotient.

40. $(-5)(+2)$

41. $(2 - 5)(6 - 2) + (-3)(-4)$

42. $\dfrac{-40}{-10}$

43. $\dfrac{0}{-5}$

44. $[10 - (-15)] \div [-2 - 3]$

45. $\dfrac{(-5)(-6) - (-3)(-3)}{-3 - 4}$

46. $(-5)(-6)(-4)$

47. $(-3)(-4)(-2)(-1)$

48. $(-1)^4$

49. $(-1)^5$

50. $\dfrac{6(-6) + (-3)(-2)}{-2(5)}$

CHAPTER 2

First-Degree Equations and Inequalities

CHAPTER 2

First-degree equations can be used to solve practical problems. A woman wishes to display a poster on her house for a political candidate so that the length is twice the width. A town ordinance states that the perimeter of the poster cannot be larger than 36 feet. The equation $2w + w = 18$ will enable her to make her poster with the largest allowable perimeter.

Source: *PAR–NYC. © 1983*

Statements such as

$$2 + 3x = 14$$

and

$$5y - 48 = 2$$

are called *first-degree equations* in one variable. This is because each of these equations contains one *variable* and the variable in each equation has an *exponent* of 1. In this chapter we present methods that enable us to determine what values of the variable are solutions to a first-degree equation in one variable. We will also discuss first-degree inequalities and applications of first-degree equations and inequalities. First-degree equations are fundamental to the study of algebra and it is important to be able to deal with them effectively.

2.1 Simplifying First-Degree Expressions

We begin our study of first-degree equations by learning how to simplify first-degree expressions such as $3x + 7 + 4x - 6$ and $5a - 8 + a - 2a$. The goal of *simplifying* an expression is to obtain another expression that contains *fewer terms* than the original expression. A *term* is either a number or a number multiplied by at least one variable.

For example,

$$7, \quad 3x, \quad -2ab, \quad \text{and} \quad -13xyz$$

are called terms. The expression $3x + 7 + 4x - 6$ contains four terms: $3x$, 7, $4x$, and -6. The numerical part of a term is called the *numerical coefficient* of that term. Thus the numerical coefficients of the terms $3x$, $-2ab$, and $7y$ are 3, -2, and 7, respectively. Observe that coefficients are factors of a term.

To simplify expressions, we shall use the following properties of the real numbers that were discussed in Chapter 1:

1. The associative, commutative, and distributive properties.
2. Multiplicative inverses (reciprocals) to obtain 1 and additive inverses (opposites) to obtain 0.

The following examples illustrate the procedures used for simplifying algebraic expressions.

Example 1. Simplify $2y + 3y$.

Solution:

$$\begin{aligned} 2y + 3y &= (2 + 3)y && \text{distributive property} \\ &= 5y. && \text{addition} \end{aligned}$$

Example 2. Simplify $15z - 6 + 2 + z$.

Solution

$$
\begin{aligned}
15z - 6 + 2 + z &= 15z - 4 + z && \text{addition} \\
&= 15z + z - 4 && \text{commutative property} \\
&= 15z + 1z - 4 && 1z = z;\ \text{multiplicative identity} \\
&= (15 + 1)z - 4 && \text{distributive property} \\
&= 16z - 4. && \text{addition}
\end{aligned}
$$

Notice that the final simplified expression in the last example contains only two *terms,* compared to four *terms* in the original expression. Terms that contain the same variables with the same exponents are called *like terms*. Thus each of the following sets consist of elements that may be called like terms.

$$
\begin{gathered}
\{5a, -2a\} \\
\{6x^2, x^2, -3x^2\} \\
\{-4y^5, 7y^5, y^5\}.
\end{gathered}
$$

In general, the end result of applying the associative, commutative, and distributive properties is to *collect (combine) all like terms*. This is done by adding the numerical coefficients of all like terms in order to get the coefficient of the corresponding term in the simplified form of the expression. We illustrate this approach in the following examples.

Example 3. Simplify $6y + 11 + 3y - 4$ by combining like terms.

Solution: The terms $6y$ and $3y$ are like terms. Adding the coefficients of these terms, we get $6 + 3 = 9$. Thus $6y + 3y = 9y$. Adding the constant terms, we get $11 - 4 = 7$. Therefore, $9y + 7$ is the simplified form of $6y + 11 + 3y - 4$.

Example 4. Simplify $7w + 4 - 2w + 7 + 3w$ by combining like terms.

Solution: First add the terms that contain w as a factor. Next add the constant terms.

$$
\begin{aligned}
7w + 4 - 2w + 7 + 3w &= (7 - 2 + 3)w + (4 + 7) \\
&= 8w + 11.
\end{aligned}
$$

Sometimes we have to use the distributive property to remove grouping symbols before we can combine like terms. The next two examples illustrate this procedure.

Example 5. Simplify $3(2z + 5) - 3z$.

Solution

$$3(2z + 5) - 3z = 6z + 15 - 3z$$ distributive property
$$= 3z + 15.$$ combine like terms; $6z - 3z = 3z$

Example 6. Simplify $4(y - 2) + 3y - 5(-y + 7)$.

Solution

$$4(y - 2) + 3y - 5(-y + 7)$$
$$= 4y - 8 + 3y + 5y - 35$$ distributive property
$$= 12y - 43.$$ combine like terms; $4y + 3y + 5y = 12y$, $-8 - 35 = -43$

Example 7. Simplify $\frac{3}{4}x + \frac{2}{3} - \frac{1}{4}x - \frac{1}{3}$.

Solution

$$\frac{3}{4}x + \frac{2}{3} - \frac{1}{4}x - \frac{1}{3} = \frac{1}{2}x + \frac{1}{3}.$$ combine like terms;
$\frac{3}{4}x - \frac{1}{4}x = \frac{2}{4}x = \frac{1}{2}x$,
$\frac{2}{3} - \frac{1}{3} = \frac{1}{3}$

EXERCISES 2.1

Simplify the following first-degree expressions by combining like terms.

For Exercises 1–30, see Examples 3 and 4.

1. $7x + 4x$
2. $11y + 2y$
3. $14z - 9z$
4. $12w - 5w$
5. $8x - x$
6. $7p - p$
7. $-6q - 3q$
8. $-11z - 12z$
9. $-10x + 3x$
10. $-13z + 5z$
11. $4x + 2x + 5x$
12. $3y + 7y + 6y$
13. $3z - 2z + 5z$
14. $4w + 11w - 2w$
15. $w + 3w - 2w$
16. $-x + 5x + 4x$
17. $5x - x + 3x$
18. $10y + y - 7y$
19. $3x + x - 4x$
20. $5y - 8y + 3y$
21. $13z - 7 + z + 2$
22. $11z + 13 - z - 6$
23. $6q + 11 - 2q - 7$
24. $15p + 8 - 23 - 9p$
25. $-10 + 12v + 4 - 3v$
26. $-3 + 7u - 2u + 5$
27. $5x + 3 - 4x + 5 + x$
28. $3y - 10 + 2y + 11 + y$
29. $12 - 6z + 5 + 8z - 11$
30. $13 + 8w - 7 + w - 2$

For Exercises 31–40, see Examples 5 and 6.

31. $4(x + 1) + 2x$

32. $3(y + 2) + 5y$

33. $5(2z - 3) + 4$

34. $7(3w - 5) + 4w$

35. $-2(3x + 1) + 7x - 5$

36. $-3(4y + 2) + 15y - 1$

37. $3(-2x + 7) - 4(x - 8) + 2x$

38. $-5(3 - 2y) + 8(1 + y) + 2y$

39. $-9(x + 4) + 7x + 3(2x - 5)$

40. $8q + 3(q - 2) - 5(-q + 4)$

For Exercises 41–50, see Example 7.

41. $\frac{2}{5}x + \frac{3}{5}x + 4$

42. $\frac{1}{4}y + \frac{3}{4}y + 11$

43. $0.25x - 0.10x + 0.02x$

44. $0.35w + 0.18w - 0.21w$

45. $0.12y + 0.15y - 0.09y$

46. $0.27z - 0.42z + 0.08z$

47. $(\frac{1}{3})x + (\frac{1}{10}) - (\frac{2}{3})x + (\frac{1}{5})$

48. $(\frac{1}{4})y + (\frac{1}{6}) - (\frac{1}{8})y + (\frac{2}{3})$

49. $(\frac{2}{3})z + (\frac{1}{4})(z - 3) + (\frac{1}{4})$

50. $(\frac{1}{6})(2x + 1) + (\frac{1}{3})x + (\frac{1}{3})(x + 1)$

2.2 Solving First-Degree Equations Using Addition and Subtraction

A statement such as $2 + 3x = 14$ is true or false depending on the value given to the variable x. For example, when we replace x with 5 we get $2 + 3(5) = 17$. Since $17 \neq 14$, the statement $2 + 3x = 14$ is false when x is 5. However, when x is replaced with the value 4, we get $2 + 3(4) = 14$, which is a true statement. When we ask the question "For what values of x will the equation $2 + 3x = 14$ be a true statement?" we are trying to *solve* the equation. The variable x is then called an *unknown*. The set of all values of the unknown that make an equation a true statement is called the *solution set* of the equation. When we have found the solution set, we have *solved* the equation. Any member of the solution set of an equation is called a *solution* of the equation. Thus we say that 4 is a solution to the equation $2 + 3x = 14$. We also say that 4 *satisfies* the equation $2 + 3x = 14$ and that the solution set is $\{4\}$.

In this section we use our ability to simplify algebraic expressions to help us solve first-degree equations. It is usually easier to determine the solution set of a simplified form of an equation rather than to use the original form of the equation.

In each of the following examples we use the following property of equality to obtain simpler equations with the same solution set. Equations with the same solution set are called *equivalent equations*.

> If the same quantity is added to equal quantities, the resulting quantities are equal.

This property is expressed symbolically as follows, where a, b, and c represent real numbers.

Addition Property of Equality

> If $a = b$, then $a + c = b + c$.

Thus we may add the same number to both sides of an equation.

Example 1. Find the solution set of $x + 11 = 20$.

Solution

$x + 11 = 20$	
$x + 11 - 11 = 20 - 11$	add -11 to each side of the equation
$x + 0 = 9$	addition
$x = 9.$	0 is the additive identity

The solution set is $\{9\}$.

Note in Example 1 that adding -11 to each side of the equation is equivalent to subtracting 11 from each side of the equation.

One other property of equality that is sometimes convenient to apply when solving equations is the *symmetric property of equality*. This property is expressed symbolically as follows, where a and b are real numbers.

Symmetric Property of Equality

> If $a = b$, then $b = a$.

Thus, if we want to, the symmetric property allows us to rewrite $10 = x$ as $x = 10$. The following example demonstrates the application of this property of equality.

Example 2. Solve $11 = 2a - 6 - a$.

Solution

$11 = 2a - 6 - a$	
$11 = a - 6$	combine like terms on the right side
$11 + 6 = a - 6 + 6$	add 6 to each side of the equation
$17 = a + 0$	addition
$17 = a$	0 is the additive identity
$a = 17.$	symmetric property of equality

In Example 2 notice that in attempting to solve the original equation we obtained five other equations. All five of these equations are equivalent to the original equation. However, the solution to the last two equations is obviously 17. Since the last equation is equivalent to the original equation, we conclude that 17 is the solution to the original equation. The basic idea in solving equations is to obtain an equivalent equation for which the solution is obvious. In this chapter the techniques that are developed for solving equations produce equivalent equations.

To *check* that we did not make a mistake in solving an equation, we substitute the value for the unknown in the original equation. Thus, in Example 2, we substitute the value 17 for a in the equation

$$11 = 2a - 6 - a.$$

This substitution results in

$$\begin{aligned} 11 &= 2(17) - 6 - 17 \\ &= 34 - 6 - 17 \\ &= 34 - 23 \\ &= 11. \end{aligned}$$

Since the value 17 produced a true statement, it is a solution to the original equation. Thus the solution set is $\{17\}$.

Example 3. Solve $3x - 2 - 2x + 7 = 25$ and check your answer.

Solution

$$\begin{aligned} 3x - 2 - 2x + 7 &= 25 \\ x + 5 &= 25 && \text{combine like terms} \\ x + 5 - 5 &= 25 - 5 && \text{add } -5 \text{ to each side of the equation} \\ x + 0 &= 20 && \text{addition} \\ x &= 20. \end{aligned}$$

We see that 20 is a solution to the last equation.

We now check to see if 20 is a solution to the original equation.

CHECK:

$$\begin{aligned} 3(20) - 2 - 2(20) + 7 &= 25 \\ 60 - 2 - 40 + 7 &= 25 \\ 67 - 42 &= 25 \\ 25 &= 25. \end{aligned}$$

Since 20 produced a true statement, it is a solution of the original equation. Therefore, the solution set is $\{25\}$.

In the final simplified form of each of the three previous examples, notice that the unknown was only on one side of the equality symbol. The general strategy for solving a first-degree equation is to isolate the unknown on one side of the equation and the constant terms on the other. We proceed so that in the end the isolated unknown will have a coefficient of 1.

Example 4. Solve $4 + 2x = -6 + 3x$.

Solution

$4 + 2x = -6 + 3x$	
$4 + 2x - 2x = -6 + 3x - 2x$	add $-2x$ to each side
$4 = -6 + x$	combine like terms
$4 + 6 = -6 + 6 + x$	add 6 to both sides
$10 = x$	addition
$x = 10.$	symmetric property of equality

It is easy to check that 10 is a solution to the original equation. This check will verify that the last equation is equivalent to the original equation.

Example 5. Find the solution set of

$$7\left(w + \frac{1}{2}\right) - 5\left(w - \frac{1}{2}\right) = w + 15.$$

Solution

$7\left(w + \frac{1}{2}\right) - 5\left(w - \frac{1}{2}\right) = w + 15$	
$7w + \frac{7}{2} - 5w + \frac{5}{2} = w + 15$	distributive property
$2w + 6 = w + 15$	combine like terms $\frac{7}{2} + \frac{5}{2} = 6$
$-w + 2w + 6 - 6 = -w + w + 15 - 6$	add $-w$ and -6 to each side
$w = 9.$	combine like terms

Thus the solution set is $\{9\}$. The reader should verify that 9 is the solution to the original equation.

Sometimes a first-degree equation, such as the one in the next example, has no solution.

Example 6. Solve $x + 11 = x + 2$.

Solution

$$\begin{aligned} x + 11 &= x + 2 \\ -x + x + 11 &= -x + x + 2 && \text{add } -x \text{ to each side} \\ 0 \cdot x + 11 &= 0 \cdot x + 2 && \text{combine like terms} \\ 11 &= 2. \quad \text{false} \end{aligned}$$

This procedure shows us that attempting to solve an equation that has no solution will result in a false statement. Since $11 \neq 2$, we conclude that the original equation has no solution.

Any equation that is *true* for all possible values of the variable is called an *identity*. We illustrate this in the next example.

Example 7. Solve $z + 4 = 4 + z$.

Solution

$$\begin{aligned} z + 4 &= 4 + z \\ -z + z + 4 - 4 &= -4 + 4 + z - z && \text{add } -z \text{ and } -4 \text{ to both sides} \\ 0z &= 0z && \text{addition} \end{aligned}$$

or

$$0 = 0. \quad \text{true}$$

Because any value that we assign to the variable z produces a true statement, the original equation has an infinite number of solutions.

Example 8. Solve $4 + \frac{2}{3}y - 3 + \frac{1}{3}y = 4$ and check your answer.

Solution

$$\begin{aligned} 4 + \frac{2}{3}y - 3 + \frac{1}{3}y &= 4 \\ 1 + y &= 4 && \text{combine like terms; } \frac{2}{3} + \frac{1}{3} = 1 \\ 1 - 1 + y &= 4 - 1 && \text{add } -1 \text{ to both sides of the equation} \\ y &= 3. && \text{addition} \end{aligned}$$

It is easy to see that the solution to the last equation is 3. If we did not make any errors, the last equation will be equivalent to the original equation and 3 will be a solution to the original equation. The reader should check this by substituting 3 for y in the original equation.

EXERCISES 2.2

For Exercises 1–10, by substituting the values in the given set for the variable, determine which, if any, of the values make the equation a true statement.

1. $3x = 2 + x$; $\{-1, 0, 1, 2, 3\}$

2. $5z = z - 8$; $\{-2, -1, 0, 1, 2\}$

3. $2y = y + 3$; $\{0, 1, 2, 3, 4, 5\}$

4. $4w = w + 6$; $\{-2, -1, 0, 1, 2\}$

5. $3(x + 4) + 2x = 22$; $\{-2, -1, 0, 1, 2\}$

6. $7(w - 2) + 3w = 6$; $\{-2, -1, 0, 1, 2\}$

7. $y - 8 + 3y = 5y - 9$; $\{-2, -1, 0, 1, 2\}$

8. $2z + 11 + z = 4z + 8$; $\{-6, -3, 0, 3, 6\}$

9. $-2(x - 1) + 3x = -x + 2$; $\{-2, -1, 0, 1, 2\}$

10. $-3(2y + 1) + y = 3y - 3$; $\{-2, -1, 0, 1, 2\}$

For Exercises 11–56, solve the first-degree equation and check your answer. See Examples 1–5 and 8.

11. $x + 7 = 18$

12. $x - 4 = 10$

13. $8 + y = 4$

14. $8 + y = 14$

15. $15 = y + 6$

16. $13 = y - 4$

17. $w + 6 = 13$

18. $w - 7 = 11$

19. $z + 3 = 2$

20. $z + 17 = 9$

21. $4x + 7 - 3x = 11$

22. $7x + 5 - 6x = 13$

23. $5x - 3 - 4x = 7$

24. $11x - 9 - 10x = 1$

25. $12x - 5 - 11x = -4$

26. $15x + 7 - 14x = 6$

27. $5 + 4x = 7 + 5x$

28. $-7 + z = -7 + 2z$

29. $\frac{1}{2}y + 5 = -\frac{1}{2}y + 3$

30. $\frac{3}{4}x - 11 + \frac{1}{4}x = 16$

31. $3x - 3 - 2x - 11 = 20$

32. $5y - 7 - 4y + 3 = 6$

33. $6x + 2 - 5x + 4 = 6$

34. $8w + 5 - 7w - 2 = 1$

35. $4x - 2 + x + 11 = 4x + 15$

36. $3y + 5 + y - 2 = 3y + 5$

37. $-2z + 5 + z - 2 = -2z + 6$

38. $-4w + 1 + w - 3 = -4w + 2$

39. $2x + \frac{3}{4} + x - 2 = 2x + \frac{3}{4}$

40. $3y - \frac{1}{4} + y - \frac{1}{2} = \frac{1}{4} + 3y$

41. $3z + \frac{1}{4} - z - \frac{1}{4} = z + 8$

42. $5z + \frac{2}{3} - 2z + \frac{1}{3} = 2z + 4$

43. $4w - 8 = w + 2 + 2w + 7$

44. $5w + 3 = 3w + 7 + w + 4$

45. $-2(x + 3) + 3(x + 2) = 16$

46. $5(y + 6) - 4(y + 5) + 7 = 0$

47. $7(y - 1) + 11 = 6(y + 2)$

48. $5(z + 2) - 2z = 2(z - 4) - 2$

49. $10x - 5(x - 1) = 4(x + 2) + 10$

50. $17w - 12(w + 4) = 4(w + 1)$

51. $3(x - \frac{1}{2}) + (\frac{5}{2} - x) = 2 + x$

52. $4(y - \frac{1}{3}) + 2(y + \frac{2}{3}) = 6 + 5y$

53. $7(z + \frac{1}{4}) - 3(z + \frac{1}{4}) = 5 + 3z$

54. $9(w + \frac{2}{3}) - 6(w + \frac{1}{3}) = 7 + 2w$

55. $(0.10)(x + 20) + (0.90)(x - 10) = 3$

56. $(0.30)(x - 10) + (0.70)(x + 30) = 8$

For Exercises 57 and 58, show that the equation has no solution. See Example 6.

57. $p + 8 = p + 7$

58. $-x + 3 = -x$

For Exercises 59 and 60, show that the equation has an infinite number of solutions. See Example 7.

59. $p + 7 = 7 + p$

60. $3(x + 1) = x + 3 + 2x$

Solving First-Degree Equations Using Division 2.3

In this section we consider equations such as $3x = 33$. We solve such equations by proceeding as we did in Section 2.2 and by making use of the following property of equality:

> If equal quantities are divided by the same nonzero quantity, the resulting quotients are equal.

This property is expressed symbolically as follows, where a, b, and c represent real numbers with c not equal to zero.

Division Property of Equality

$$\text{If } a = b, \text{ then } \frac{a}{c} = \frac{b}{c} \quad \text{provided that } c \neq 0.$$

The application of this property to an equation produces an equivalent equation.

In order to solve the equation $3x = 33$, we want to get an equivalent equation of the form

$$x = \text{some number.}$$

We can accomplish this by dividing each side of the equation by 3.

$$\begin{aligned} 3x &= 33 \\ \frac{3x}{3} &= \frac{33}{3} && \text{divide both sides of the equation by 3} \\ x &= 11. \end{aligned}$$

Example 1. Solve $-5x + 2 = -33$ and check your answer.

Solution

$$\begin{aligned} -5x + 2 &= -33 \\ -5x + 2 - 2 &= -33 - 2 && \text{add } -2 \text{ to both sides} \\ -5x &= -35 && \text{combine like terms} \\ \frac{-5x}{-5} &= \frac{-35}{-5} && \text{divide both sides by } -5 \\ x &= 7. \end{aligned}$$

The reader should check that 7 is the solution to the original equation.

Example 2. Solve $3p - 11 + 4p - 3 = 0$ and check your answer.

Solution

$$\begin{aligned} 3p - 11 + 4p - 3 &= 0 \\ 7p - 14 &= 0 && \text{combine like terms} \\ 7p - 14 + 14 &= 0 + 14 && \text{add 14 to each side} \\ 7p &= 14 \\ \frac{7p}{7} &= \frac{14}{7} && \text{divide both sides by 7} \\ p &= 2. \end{aligned}$$

We leave it to the reader to verify that 2 is the solution to the original equation.

Example 3. Solve $13 + 7k - 20 = k + 11$.

Solution

$$\begin{aligned} 13 + 7k - 20 &= k + 11 \\ 7k - 7 &= k + 11 && \text{combine like terms} \\ 7k - 7 - k + 7 &= k + 11 - k + 7 && \text{add } -k \text{ and } +7 \text{ to each side} \\ 6k &= 18 && \text{combine like terms} \\ \frac{6k}{6} &= \frac{18}{6} && \text{divide by 6} \\ k &= 3. \end{aligned}$$

It is easy to check that 3 is a solution to the original equation.

EXERCISES 2.3

For Exercises 1–40, solve the equation and check your answer. See Examples 1–3.

1. $7y = 21$

2. $3x = 12$

3. $4x = 20$

4. $6y = 30$

5. $8z = -56$

6. $9w = -54$

7. $-5x = 45$

8. $-8y = 40$

9. $-9z = -72$

10. $-11q = -132$

11. $4x = 2$

12. $9y = 3$

13. $3y = 5$

14. $7y = 11$

15. $8 = 4w$

16. $15 = 3z$

17. $9 = 2x$

18. $1 = 4y$

19. $-7 = 2x$

20. $-10 = 3w$

21. $4 = -3x$

22. $5 = -2z$

23. $3z + 4z = 12$

24. $2x + 3x = 9$

25. $5y + 7y = 17$

26. $w + 3w = 5$

27. $13x + 11 - 4x = 2$

28. $10y + 6 - 2y = -10$

29. $7w + 2 - 5w = -8$

30. $23z + 45 - 3z = 65$

31. $6y + 3 = 2y - 9$

32. $5z - 11 = 2z + 4$

33. $7w + 5 = 2w - 10$

34. $8x - 7 = 5x + 20$

35. $5p + 2 = p - 1$

36. $p - 11 = -8 - p$

37. $4(x - 2) + 5(x + 1) = 2x + 18$

38. $-7(2x + 1) + 3(x - 11) = 2x + 12$

39. $5(x + 2) + 2(x - 2) = 3x - 6$

40. $6(x + 5) - 2(x - 3) = -5x + 18$

2.4 Solving First-Degree Equations Using Multiplication

Closely related to the examples and problems of the preceding section are situations in which it is necessary to multiply both sides of an equation by the same value. This procedure is justified by the following property of equality:

> If two equal quantities are multiplied by the same quantity, the resulting products are equal.

This property is expressed symbolically as follows, where a, b, and c are real numbers:

Multiplication Property of Equality

> If $a = b$, then $ac = bc$.

Thus we may multiply both sides of an equation by the same number. Again, application of this property (with $c \neq 0$) to an equation produces an equivalent equation.

If we want to solve the equation

$$-\frac{1}{4}x = 3,$$

we should find an equivalent equation of the form

$$x = \text{some number}.$$

This can be accomplished by multiplying each side of the equation by -4, which is the reciprocal of the coefficient of the x term. We proceed as follows:

$$-\frac{1}{4}x = 3$$

$$(-4)\left(-\frac{1}{4}x\right) = (-4)3 \qquad \text{multiply both sides of the equation by } -4;\ (-4)\left(-\frac{1}{4}\right) = 1$$

$$x = -12.$$

It is easy to check that -12 is the solution to the original equation.

Example 1. Solve $-\frac{1}{2}x + 3 = 4$.

Solution

$$-\frac{1}{2}x + 3 = 4$$

$$-\frac{1}{2}x + 3 - 3 = 4 - 3 \qquad \text{add } -3 \text{ to both sides}$$

$$-\frac{1}{2}x = 1 \qquad \text{addition}$$

$$(-2)\left(-\frac{1}{2}x\right) = (-2)(1) \qquad \text{multiply both sides by } -2$$

$$x = -2.$$

It is easy to check that -2 satisfies the original equation.

In this book only the solutions to selected examples will be checked. The reader is encouraged to check the solution to those examples that do not include a check.

Example 2. Solve $-\frac{2}{3}x + 7 + \frac{1}{3}x - 3 = 1$.

Solution

$$-\frac{2}{3}x + 7 + \frac{1}{3}x - 3 = 1$$

$$-\frac{1}{3}x + 4 = 1 \qquad \text{combine like terms}$$

$$-\frac{1}{3}x + 4 - 4 = 1 - 4 \qquad \text{add } -4 \text{ to each side}$$

$$-\frac{1}{3}x = -3 \qquad \text{addition}$$

$$(-3)\left(-\frac{1}{3}\right)x = (-3)(-3) \qquad \text{multiply each side by } -3$$

$$x = 9.$$

The reader should check that 9 is a solution to the original equation.

Example 3. Solve $\frac{4}{5}x + \frac{2}{3} - \frac{1}{5}x + \frac{5}{3} = -\frac{2}{3}$.

Solution: To avoid working with fractions, we multiply both sides of the equation by a common denominator of all fractions in the equation. In this case we multiply by 15.

$$\frac{4}{5}x + \frac{7}{3} - \frac{1}{5}x = -\frac{2}{3}$$

$$15\left(\frac{4}{5}x + \frac{7}{3} - \frac{1}{5}x\right) = 15\left(-\frac{2}{3}\right) \qquad \text{multiply both sides by 15}$$

$$12x + 35 - 3x = -10$$

$$9x + 35 = -10 \qquad \text{combine like terms}$$

$$9x + 35 - 35 = -10 - 35 \qquad \text{subtract 35 from each side}$$

$$9x = -45 \qquad \text{addition}$$

$$\frac{1}{9}(9x) = \frac{1}{9}(-45) \qquad \text{multiply each side by } \frac{1}{9}$$

$$x = -5$$

The reader should check that -5 is the desired solution.

EXERCISES 2.4

For Exercises 1–40, solve the equation and check your answer. See Examples 1–3.

1. $\frac{1}{3}x = 5$
2. $\frac{1}{7}y = 6$
3. $\frac{2}{5}w = 4$
4. $\frac{3}{7}z = 6$
5. $\frac{1}{2}y = -7$
6. $-\frac{2}{3}z = 10$
7. $\frac{1}{8}z = -3$
8. $\frac{1}{6}w = -5$
9. $-\frac{1}{3}x = -4$
10. $-\frac{1}{7}y = -7$
11. $\frac{2}{3}q = 4$
12. $\frac{3}{4}x = 6$
13. $\frac{2}{9}w = -4$
14. $\frac{5}{7}x = -10$
15. $-\frac{2}{11}y = 4$
16. $-\frac{3}{8}w = 6$
17. $-\frac{2}{5}x = -4$
18. $-\frac{5}{8}y = -10$
19. $\frac{3}{4} = -\frac{1}{4}q$
20. $\frac{2}{10} = \frac{1}{5}x$
21. $-\frac{1}{3} = \frac{1}{6}p$
22. $-10 = \frac{5}{4}y$
23. $-\frac{1}{9}z = -1$
24. $-\frac{1}{4}x = -10$
25. $-\frac{1}{11}y = -3$
26. $-\frac{1}{6}w = -4$
27. $\frac{1}{3}y + 4 + \frac{1}{3}y = 10$
28. $\frac{1}{4}x - 5 + \frac{1}{4}x = 5$
29. $\frac{1}{5}w + 10 + \frac{2}{5}w = 4$
30. $\frac{2}{5}y + 8 + \frac{1}{5}y = 5$
31. $\frac{3}{7}w + 2 + \frac{1}{7}w = 6$
32. $\frac{4}{7}x + 3 + \frac{1}{7}x = 8$
33. $\frac{4}{9}y - 1 + \frac{1}{9}y = 4$
34. $\frac{2}{9}y + 2 - \frac{1}{9}y = 5$
35. $\frac{2}{5}x + \frac{1}{3} - \frac{1}{5}x = \frac{4}{3}$
36. $\frac{3}{5}y + \frac{1}{4} - \frac{2}{5}y = \frac{5}{4}$
37. $\frac{4}{5}(y + 5) + \frac{1}{5}(y + 5) = 9$
38. $\frac{2}{3}(y + 1) + \frac{1}{3}(y + 1) = 7$
39. $\frac{1}{3}(5y - 2) = \frac{1}{3}(4y + 11)$
40. $\frac{1}{7}(4z - 1) = \frac{1}{7}(3z + 2)$

2.5 Solving First-Degree Equations: The General Case

By using the techniques of the preceding three sections, we are now prepared to solve any first-degree equation that contains just one variable. The following steps are a summary of the techniques presented in Sections 2.2–2.4. The order in which they are listed is arbitrary. The order in which they are applied will depend on the equation that is being solved. It may not be necessary to apply all of them to a particular problem.

Procedures for Solving a First-Degree Equation in One Variable

1. Use the multiplication property of equality to remove fractions.
2. Combine like terms on both sides of the equation.
3. Use the addition property of equality to isolate the unknown terms on one side of the equation and the constant terms on the other side.
4. Use the division property of equality to obtain a coefficient of 1 for the variable term.

Example 1. Solve $\dfrac{3(x + 5)}{7} = 2x + 1 + x.$

Solution

$$\frac{3(x + 5)}{7} = 2x + 1 + x$$

$$\frac{3(x + 5)}{7} = 3x + 1$$ combine like terms on the right side of the equation

$$3(x + 5) = 7(3x + 1)$$ multiply both sides by 7 to eliminate fractions

$$3x + 15 = 21x + 7$$ distributive property

$$3x + 15 - 21x - 15 = 21x + 7 - 21x - 15$$ add $-21x$ and -15 to both sides

$$-18x = -8$$ combine like terms

$$\frac{-18x}{-18} = \frac{-8}{-18}$$ divide both sides by -18

$$x = \frac{4}{9}.$$

Example 2. Solve $14y - 3 + 2y = 4 + 6y + 11$ and check your answer.

Solution

$$14y - 3 + 2y = 4 + 6y + 11$$

$$16y - 3 = 6y + 15$$ combine like terms

$$16y - 3 - 6y + 3 = 6y + 15 - 6y + 3$$ add $-6y$ and 3 to both sides

$$10y = 18$$ combine like terms

$$\frac{10y}{10} = \frac{18}{10}$$ divide each side by 10

$$y = 1.8.$$

Notice that the solution to Example 1 was left as the fraction $\frac{4}{9}$, whereas the solution to Example 2 was written in decimal form as 1.8. Since the decimal representation of $\frac{4}{9}$ is 0.444 . . . , the fractional form is easier to work with than the decimal form. If a fraction has an exact, one- or two-decimal place representation, the decimal representation may be preferable to the fractional representation.

Example 3. Solve $\frac{2x - 7}{3} = \frac{x - 11}{2} + \frac{1}{4}$.

Solution

$$\frac{2x - 7}{3} = \frac{x - 11}{2} + \frac{1}{4}$$

$12\left(\frac{2x - 7}{3}\right) = 12\left(\frac{x - 11}{2} + \frac{1}{4}\right)$	multiply both sides by 12 to remove the fractions
$4(2x - 7) = 6(x - 11) + 3$	result of multiplication
$8x - 28 = 6x - 66 + 3$	distributive property
$8x - 28 = 6x - 63$	combine like terms
$8x - 28 - 6x + 28 = 6x - 63 - 6x + 28$	add $-6x + 28$ to both sides
$2x = -35$	combine like terms
$\frac{2x}{2} = \frac{-35}{2}$	divide both sides by 2
$x = \frac{-35}{2} = -17.5.$	

EXERCISES 2.5

For Exercises 1–30, solve the equation and check your answer. See Examples 1–3.

1. $3x + 7 = 1$

2. $3y + 5 = 17$

3. $5 + 2x = 17$

4. $11 - 4w = 35$

5. $7x + 15 = x - 3$

6. $5z - 13 = 2z + 2$

7. $p + 4 = 3p + 18$

8. $4q + 3 = 7q + 21$

9. $3(y + 2) = 2(y + 5)$

10. $7(w - 3) = 11(w + 1)$

11. $-4(x - 3) = 3(2x + 1) + 4x - 5$

12. $3(z + 2) = 4z + 3 - 2(3z + 1)$

13. $5x - 2 = \frac{x + 30}{4}$

14. $13(y + 1) = \frac{y + 51}{2}$

15. $\frac{3x - 1}{7} = x + 5$

16. $\frac{7w - 1}{3} = 2w + 7$

17. $\frac{3q + 5}{2} = \frac{q + 6}{5}$

18. $\frac{y - 5}{11} = \frac{2y + 5}{7}$

19. $\frac{z - 3}{7} = \frac{z + 1}{3}$

20. $\frac{2x + 5}{9} = \frac{x + 6}{4}$

21. $\frac{4(y - 3)}{5} = 3y + 2 + \frac{2(y + 2)}{3}$

22. $\frac{7(w - 2)}{3} = \frac{4(w + 2)}{5} + 2w - 10$

23. $\frac{x + 5}{2} + \frac{x - 3}{3} = \frac{x - 1}{2}$

24. $\frac{w + 7}{3} - \frac{w + 1}{5} = \frac{w - 2}{3}$

25. $\frac{p - 2}{7} = \frac{p + 7}{2} - \frac{p + 10}{7}$

26. $\frac{q + 3}{4} = \frac{q + 2}{3} - \frac{q + 1}{4}$

27. $0.4(10x + 8) = 3x + 7.2$

28. $0.4(y + 5) = 4.4y + 10$

29. $0.60(4x + 10) = 0.4(x + 10)$

30. $(0.2)(3x - 20) = (0.1)(16x + 80)$

Applications I: Writing and Solving First-Degree Equations 2.6

It is a fact of life that the mathematical problems we encounter outside the classroom are seldom in equation form. Most often, the equation or equations must be extracted from written or verbal descriptions associated with a given situation. It is important to be able to translate a written or verbal situation to a mathematical statement. Table 2.1 contains some common English phrases and words with their usual mathematical interpretation. In Table 2.1, x and y represent unknown numbers.

Table 2.1

English Phrase or Word	Mathematical Expression
The difference of x and y	$x - y$
The product of x and y	xy
The quotient of x by y	$\frac{x}{y}$
The sum of x and y	$x + y$
is	$=$
of	Multiply
Two-thirds of a number x	$\frac{2}{3}x$
7 less than a number x	$x - 7$
7 more than a number x	$x + 7$
7 times a number x	$7x$

In this section we learn how to write an equation that describes a problem involving a relationship between numbers. Even though there is no fixed method for solving applied problems, the following steps are often suggested:

1. Read the problem carefully in order to understand the situation that is being described.
2. Identify the unknown quantity and represent it by x or some other appropriate symbol.
3. If there is more than one unknown, try to express them all in terms of the unknown in step 2.
4. Write an equation that represents the information given in the problem.
5. Solve for the unknown and check your answer.

Example 1. The sum of an unknown number and 18 is 43. Find the number.

Solution: Let x denote the unknown number; then $x + 18$ represents the sum of the unknown number and 18. Thus

$$x + 18 = 43.$$

Once we have translated our verbal problem to a mathematical equation, we proceed to solve the equation.

$$\begin{aligned} x + 18 - 18 &= 43 - 18 && \text{add } -18 \text{ to both sides} \\ x &= 25. \end{aligned}$$

CHECK: It is easy to see that 25 is the correct solution because

$$25 + 18 = 43.$$

Example 2. If 11 is subtracted from five times an unknown number, the result is 24. Find the unknown number.

Solution: Let n denote the unknown number; then

$$\begin{aligned} 5n - 11 &= 24 \\ 5n &= 24 + 11 && \text{add 11 to both sides} \\ 5n &= 35 && \text{divide each side by 5} \\ n &= 7. \end{aligned}$$

Thus the unknown number is 7.

CHECK: $5(7) - 11 = 35 - 11 = 24.$

Example 3. One-third of a number is added to 11 and the result is 15. Find the number.

Solution: Let y denote the number; then

$$\begin{aligned} \frac{1}{3}y + 11 &= 15 \\ y + 33 &= 45 && \text{multiply each side by 3} \\ y &= 45 - 33 && \text{subtract 33 from each side} \\ y &= 12. \end{aligned}$$

Thus the required number is 12.

CHECK: $\left(\frac{1}{3}\right) 12 + 11 = 4 + 11 = 15.$

Example 4. Find three consecutive integers with the sum of 132.

Solution: Let x denote the first number. Consecutive integers such as 1, 2 or 23, 24 differ by 1. Let

x represent the first number,
$x + 1$ represent the second number,
$x + 2$ represent the third number.

We know that

$$\begin{aligned} x + (x + 1) + (x + 2) &= 132 \\ 3x + 3 &= 132 && \text{combine like terms} \\ 3x &= 129 && \text{subtract 3 from each side} \\ x &= 43. && \text{divide each side by 3} \end{aligned}$$

Thus the numbers are 43, 44, and 45.

CHECK: $43 + 44 + 45 = 132$.

EXERCISES 2.6

Solve Exercises 1–20. See Examples 1–4.

1. The sum of a number and 13 is 31. Find the number.
2. The sum of a number and 14 is 86. Find the number.
3. When a number is multiplied by 11, the result is 187. Find the number.
4. When a number is multiplied by 13, the result is 91. Find the number.
5. When a number is multiplied by $\frac{5}{3}$, the result is 20. Find the number.
6. When a number is multiplied by $\frac{4}{7}$, the result is 44. Find the number.
7. When 6 is added to seven times a number, the result is 62. Find the number.
8. When 13 is added to five times a number, the result is 58. Find the number.
9. When 9 is subtracted from $\frac{1}{5}$ of a number, the result is 3. Find the number.
10. When 11 is subtracted from $\frac{1}{8}$ of a number, the result is 100. Find the number.
11. When a number is added to four times itself, the result is 80. Find the number.
12. When a number is added to seven times itself, the result is 120. Find the number.
13. The sum of three consecutive integers is 54. Find the integers.
14. The sum of three consecutive even integers is 78. Find the integers.
15. After subtracting 9 from a number, the difference is multiplied by 11. If the resulting product is 165, find the number.
16. After subtracting 3 from a number, the difference is multiplied by 7. If the resulting product is 2 less than 100, find the number.
17. One number is 19 more than another number. If the sum of the two numbers is 47, find the numbers.

18. One number is 6 less than another number. When their sum is divided by 3, the result is 16. Find the two numbers.

19. When four times a number is decreased by 6 and this difference is divided by 2, the result is 9. Find the number.

20. When seven times a number is decreased by 5 and this difference is divided by 9, the result is 8. Find the number.

2.7 Formulas and Word Problems

A formula is an equation that contains more than one variable. Perhaps you are already familiar with formulas such as those used to compute areas, perimeters, temperature conversions, and simple interest. Formulas are used in many fields, such as business, economics, chemistry, and physics. A list of common formulas with an explanation of their variables is included inside the front cover of this book.

Example 1. \$1000 is invested for 2 years in an account that pays simple interest at the rate of 14% per year. How much interest is earned?

Solution: Using the formula $i = Prt$, where r is the rate of interest, P is the principal, and t is the time in years, we get

$$i = P \cdot r \cdot t$$
$$i = (1000)(.14)(2) = 280.$$

Therefore, the interest earned is \$280.

Example 2. Jim wants to earn \$70 per year in interest from the same type of account that was described in Example 1. How much principal will he need to deposit to reach this goal?

Solution: Using $i = Prt$ as explained in Example 1 and substituting, we get

$$i = Prt$$
$$70 = P(.14)(1)$$
$$70 = (.14)P.$$

Using the symmetric property of equality and solving for P results in

$$.14P = 70$$
$$\frac{.14P}{.14} = \frac{70}{.14} \qquad \text{divide each side by .14}$$
$$P = 500.$$

Thus the principal needed is \$500.

Example 3. Find the area of a circular region that has a diameter of 9 feet.

Solution: The formula for the area of a circle is $A = \pi r^2$, where r is the radius. Since the radius of a circle is half the diameter, the radius is 4.5 feet. Thus $A = \pi(4.5)^2 = 20.25\pi$ square feet. If we use 3.14 as an approximation for π, then the area of the circular region is approximately 63.59 square feet.

Example 4. Jip-Jack's Record and Tape Store is planning to have a big sales promotion in June. They expect to sell everything at a 40% discount. In preparing for this sale, the store manager first increases the price of all items by 25%. What is the true percentage discount?

Solution: Let P denote the price of an item before it is marked up and let S denote the sale price. Observe that after the increase the new price is

$$\begin{array}{ccccc} P & + & (.25)P & = & (1.25)P \\ \begin{pmatrix}\text{initial}\\\text{price}\end{pmatrix} & + & \begin{pmatrix}\text{increase in}\\\text{price before}\\\text{the sale}\end{pmatrix} & = & \begin{pmatrix}\text{price after}\\\text{the initial}\\\text{markup}\end{pmatrix} \end{array}$$

When an item is discounted 40%, it is the same as selling it for 60% of the previous price. Thus the sale price is

$$\begin{aligned} S &= (.60)[(1.25)P] \\ &= .75P. \end{aligned}$$

Therefore, the discount is actually 25%.

Example 5. You decide to take your young neighbor who weighs 48 pounds to the playground. When you get to the teeter-totter (seesaw), there are no other youngsters available and you agree to be his partner. If he sits 9 feet from the pivot and you weigh 120 pounds, how far from the pivot should you sit in order to balance the teeter-totter?

Solution: The teeter-totter will be balanced when the weights and distances shown in Figure 2.1 satisfy the equation $w_1 d_1 = w_2 d_2$. Thus we want $48 \cdot 9 = 120 \cdot d_2$. Solving for d_2 results in 3.6 feet.

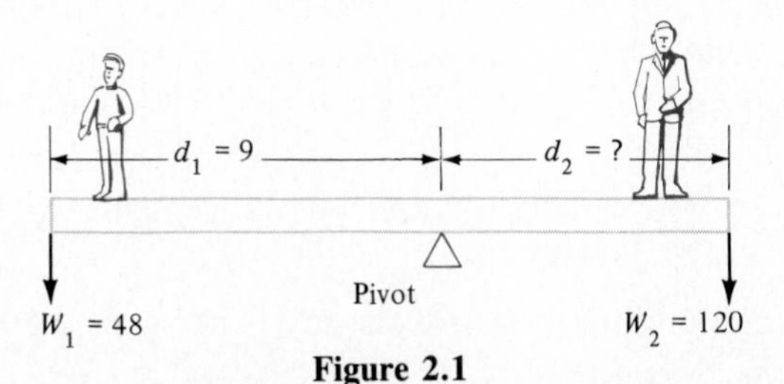

Figure 2.1

EXERCISES 2.7

Solve Exercises 1–25.

1. If \$500 is invested for 3 years in an account that pays simple yearly interest at the rate of 11%, what is the total interest earned? See Example 1.

2. If \$4000 is invested for 2 years in an account that pays simple annual interest of 9.5%, what is the total interest earned? See Example 1.

3. Dawn wants to earn \$140 per year in interest from an account that pays simple interest at the rate of 14% per year. How much principal must be invested in order to earn \$140? See Example 2.

4. Loretta wants to earn \$150 per year in interest from an account that pays simple interest at the rate of 12% per year. How much principal must be invested in order to reach this goal? See Example 2.

5. Find the area and circumference of a circular region with radius 8 feet. The circumference of a circle is given by $C = \pi d$, where d is the diameter of the circle and π is approximately 3.14. See Example 3.

6. Find the circumference and the area of a circular region with a diameter of 6 feet. The circumference of a circle is given by $C = \pi d$, where d is the diameter of the circle and π is approximately 3.14. See Example 3.

7. A suit that is priced at \$200 is marked up by 20%. After 2 months at the new price the suit has not sold, so it has a "half-price" tag placed on it. What is the true percentage discount from the original price of \$200? See Example 4.

8. A home entertainment unit was originally priced at \$1500. It is first marked up 20% and then has a "$\frac{1}{3}$-off" tag placed on it. What is the true percentage discount from the original price of \$1500? See Example 4.

9. A furniture store computes the price at which it sells an item at 250% of the cost of the item to the store.
(a) If the store buys a chair from a warehouse for \$400, at what price will the store sell the chair?
(b) The store later has an inventory sale and reduces every item by 40%. Express the inventory sale price of the chair as a percentage of the price in part (a). See Example 4.

10. A jewelry store sells a particular line of rings at 275% of the cost of the ring to the store.
(a) If the store purchases a ring for \$400, at what price will it sell the ring?
(b) If the store reduces the price of the ring by \$600, what will be the percentage profit? See Example 4.

11. Brad weighs 180 pounds and Jane weighs 120 pounds. If Jane sits 6 feet from the pivot on a teeter-totter, how far from the pivot should Brad sit if they want it to be balanced? (HINT: See Example 5.)

12. Ricky weighs 42 pounds and Eddie weighs 48 pounds. If Ricky sits 4 feet from the pivot of a teeter-totter, how far from the pivot should Eddie sit if they want to balance the teeter-totter? See Example 5.

13. Use the formulas $A = LW$ and $P = 2(L + W)$ to find the area A and perimeter P of a rectangle with length $L = 16$ feet and width $W = 9$ feet.

14. Find the perimeter P of a rectangular room that is 12 feet wide and 26 feet long. What is the area of the room?

15. Find the area and perimeter of each triangle in the accompanying figures. The area of a triangle is given by $A = (\frac{1}{2}) \cdot (\text{base}) \cdot (\text{height})$.

(a)

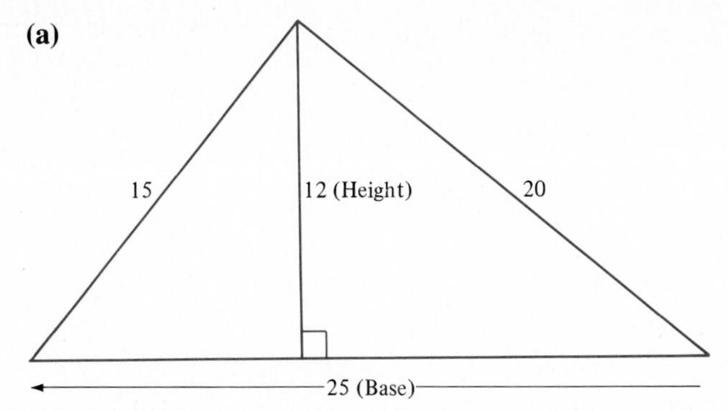

(b)

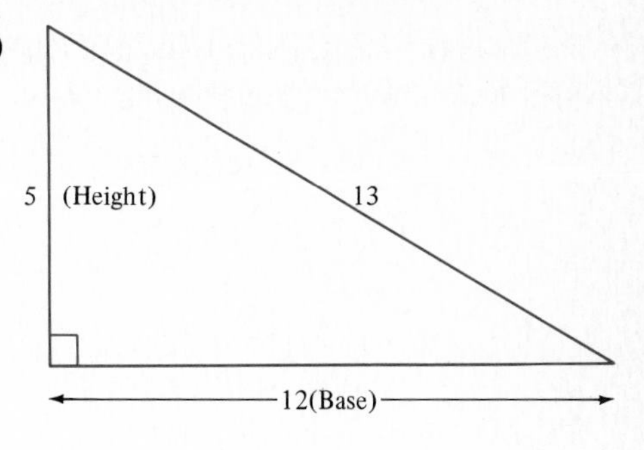

16. Find the area and perimeter of each triangle in the accompanying figures. The area of a triangle is given by $A = (\frac{1}{2}) \cdot (\text{base}) \cdot (\text{height})$.

(a)

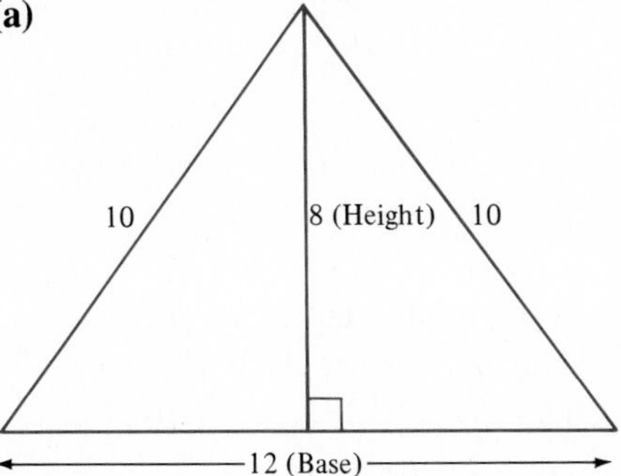

(b)

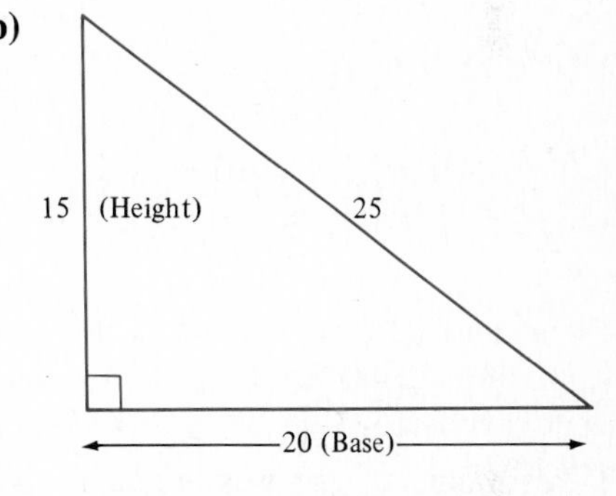

17. Find the area of the trapezoid shown. The area of a trapezoid is given by $A = (\frac{1}{2}) \cdot (\text{height}) \cdot (\text{sum of the bases})$.

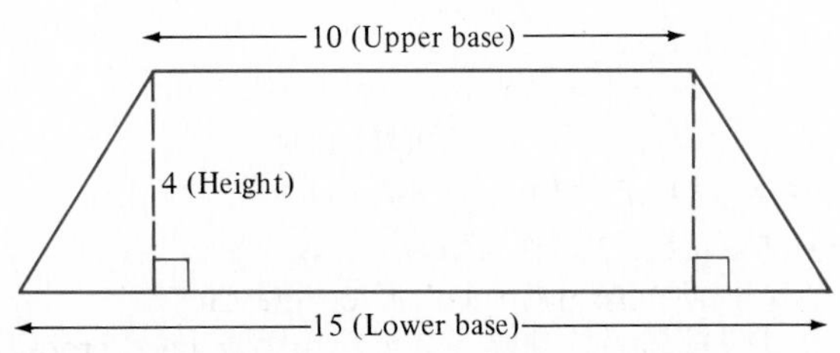

18. Find the area of the trapezoid shown. The area of a trapezoid is given by $A = (\frac{1}{2}) \cdot (\text{height}) \cdot (\text{sum of the bases})$.

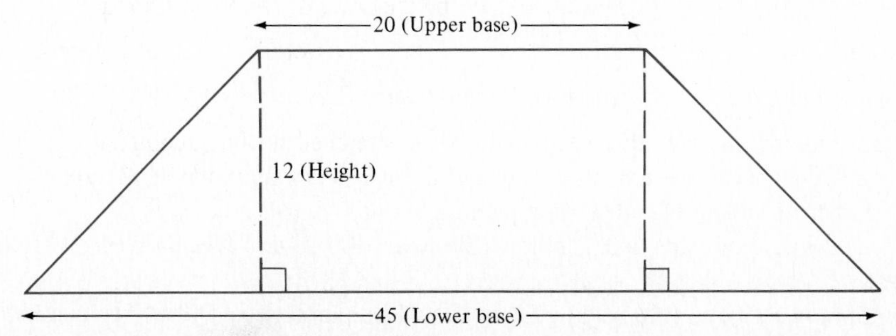

19. Find the total surface area and volume of a box that has length 18 inches, width 10 inches, and height 8 inches as shown in the accompanying figure.

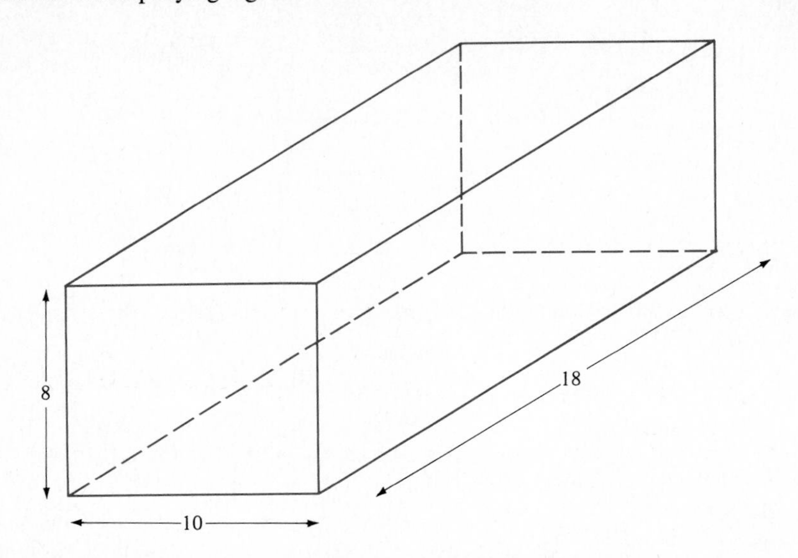

20. The formula for the lateral surface area of a cylinder is $A = 2\pi rh$, where r is the radius and h is the height. A can of roofing tar has the shape of a cylinder with radius 10 inches and height 30 inches. Find the area of a paper label that is wrapped around the can shown in the accompanying figure. Assume that there is no overlap.

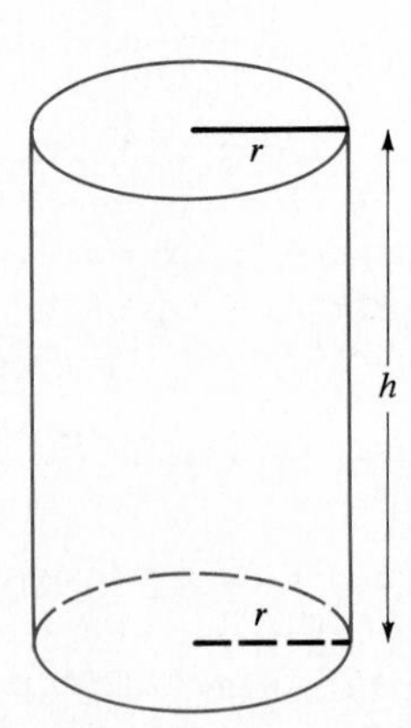

21. What is the total surface area of the can described in Exercise 20?

22. The area on the front and side of a garage that needs painting is sketched in the accompanying figure. Assume that the front and back have the same area and that the two sides have the same area.

(a) What is the total number of square feet that need painting?

(b) If the paint is only sold in gallon cans and 1 gallon will cover 400 square feet, how many gallons should be purchased?

(c) If two coats of paint are necessary, how many gallons should be purchased? Discuss your answers.

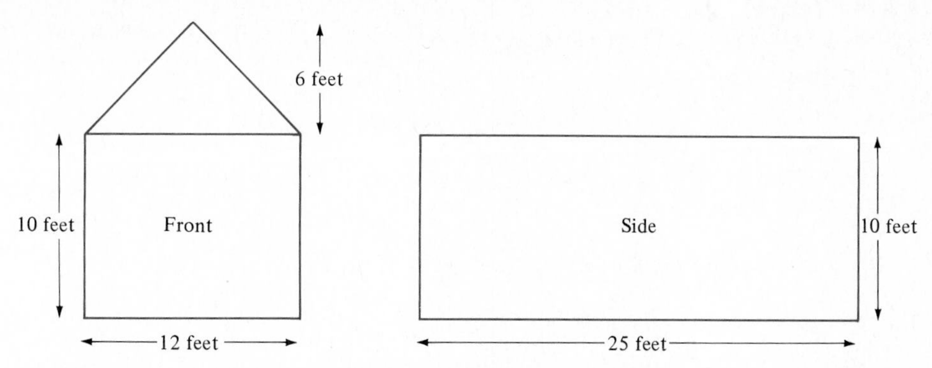

23. Use the formula $d = rt$, where d is distance, r is rate, and t is time, to find how long it takes to drive a distance of 330 miles at a constant rate of 55 miles per hour.

24. Use the formula $d = rt$, where d is distance, r is rate, and t is time, to find out how long it takes to drive a distance of 275 miles at a constant rate of 50 miles per hour.

25. A local delivery service will handle any package for a flat rate of $2, provided that it weighs less than 20 pounds and has a combined linear measurement of length plus girth of at most 84 inches. (Girth is the distance around the box; i.e., girth $= 2W + 2H$.) If a box that weighs 15 pounds is 36 inches long and has equal width and height, what is the largest the width and height may be if the box is to be eligible for the flat rate?

Applications II: Advanced Word Problems (Optional) 2.8

In this section we study applications that are often encountered in homes, on the job, or in a recreational setting. Some of the examples and exercises require the straightforward application of one or more formulas. Other examples and exercises require both the use of formulas and the solution of a first-degree equation. Most of these problems are more complex than the previous applications. However, the suggestions given in Section 2.6 still apply. Review those suggestions before reading the following examples.

Example 1. During the spring semester break, the occupants of room 1313 in Jinx Dormitory plan to drive 2420 miles to a popular college student rendezvous. The car they plan to take gets 220 miles to each full tank of gas. How many times will the tank be filled during the round trip?

Solution: Let x denote the number of times the tank will need to be filled. Thus

$$\underbrace{220x}_{\text{miles per full tank of gas}} = \underbrace{4840}_{\text{round-trip mileage}}$$

To solve this equation, we divide each side of the equation by 220. Thus

$$\frac{220x}{220} = \frac{4840}{220}$$

or

$$x = 22.$$

Therefore, we need 22 full tanks of gas.

Example 2. Each day, Jessica parks her car in the designated freshman parking area, which is located 1.5 miles from the building in which she has her first class. At what minimum rate must she walk if she arrives in the parking lot 20 minutes before class and wants to arrive in class on time?

Solution: Let r denote the rate at which she walks. Since distance is given in miles, we compute her rate of walking in miles per hour. To do this, we change 20 minutes to $\frac{20}{60} = \frac{1}{3}$ of an hour and use the formula distance equals rate times time, or $d = r \cdot t$. Thus we may write

$$1.5 = r \cdot \frac{20}{60} \qquad \text{or} \qquad 1.5 = \frac{1}{3}r.$$

Multiplying both sides by 3 results in $r = 4.5$ miles per hour.

CHECK:

$$1.5 = 4.5 \cdot \frac{1}{3}$$

$$1.5 = 1.5.$$

Example 3. A pharmacologist wants a 40% (by volume) alcohol solution. She has available 8 liters of a 30% solution and 10 liters of an 80% solution. How much of the 80% solution must be added to the 30% solution to obtain the desired 40% solution?

Solution: Consider the following diagram, in which x denotes the amount of the 80% solution that is necessary.

8 liters	+	x liters	=	$(x + 8)$ liters
30% alcohol		80% alcohol		40% alcohol

Thus

$$\underbrace{(.30)8}_{\substack{\text{volume of}\\ \text{alcohol in}\\ \text{8 liters of}\\ \text{30\% solution}}} + \underbrace{(.80)x}_{\substack{\text{volume of}\\ \text{alcohol in}\\ x\text{ liters of}\\ \text{80\% solution}}} = \underbrace{.40(x + 8)}_{\substack{\text{volume of}\\ \text{alcohol in}\\ (x+8)\text{ liters}\\ \text{of the 40\% solution}}}$$

The equation relates the amount of pure alcohol involved in the problem. It may be rewritten as

$$2.4 + .8x = .4x + 3.2$$

or

$$.4x = .8$$
$$x = 2 \text{ liters of the 80\% solution.}$$

CHECK: $(.30)(8) + (.80)(2) = 4$ liters of alcohol in a total of $8 + 2 = 10$ liters of solution. Observe that 4 is 40% of 10.

EXERCISES 2.8

Solve Exercises 1–20.

1. If the car in Example 1 gets 198 miles to each tankful, how many times will the tank be filled?

2. If the car in Example 1 gets 180 miles to each tankful, how many times will the tank be filled?

3. If the student in Example 2 has 30 minutes to walk to class, at what rate should she walk?

4. A student parks his car 1.6 miles from the building in which he has his first class. If he walks at a rate of 4 miles per hour, how much time should he allow to walk to class? See Example 2.

5. If the pharmacologist in Example 3 has 8 liters of a 20% solution and 10 liters of an 80% solution, how much of the 80% solution must she add to the 20% solution to obtain a 40% solution?

6. A chemistry student wants a 10% solution of sodium chloride. He has available 100 liters of a 5% solution and 300 liters of a 15% solution. How many liters of the 15% solution must be added to the 5% solution to obtain the desired 10% solution? See Example 3.

7. A 6-gallon radiator is half full with a solution of 20% antifreeze. How much pure antifreeze must be added to obtain a 50% solution? See Example 3.

8. A 26-quart radiator contains 16 quarts of a 40% antifreeze solution. How much pure antifreeze must be added to obtain a 60% solution? See Example 3.

9. After the output of a factory was increased by $\frac{1}{5}$, the factory was producing 3000 items per day. What was the daily output prior to the increase in production?

10. The number of daily orders that a company received has decreased by $\frac{1}{8}$. The company is now receiving 630 orders each day. How many orders was it receiving daily prior to the decrease?

11. A poster is designed so that the width is $\frac{2}{3}$ of the length. If the perimeter is restricted to a measurement of 120 inches, what should be the size of the poster?

12. A poster is designed so that the width is six-tenths (.6) of the length. If the perimeter is restricted to a measurement of 120 inches, what are the dimensions of the poster?

13. When gasoline cost $1.25 per gallon, it was proposed that a $.50 per gallon surcharge would encourage conservation. What percentage increase would the surcharge represent?

14. If the cost of unleaded premium gasoline drops from $1.40 per gallon to $1.19 per gallon, what percentage decrease is this from the original price per gallon?

15. A passenger train going 60 miles per hour passes a freight train going 45 miles in the same direction. In how many hours will the trains be 30 miles apart if they both maintain their given speeds?

16. A passenger train going 70 miles per hour passes a freight train going 45 miles per hour in the same direction. In how many hours will the trains be 40 miles apart if they both maintain their speeds?

17. A steamboat travels upstream for 5 hours. The speed of the river's current is 6 miles per hour and the speed of the boat in still water is 14 miles per hour. How long does the return trip take?

18. A steamboat travels upstream for 3 hours. The speed of the boat in still water is 16 miles per hour. If the speed of the river's current is 4 miles per hour, how long does the return trip take?

2.9 First-Degree Inequalities

We are often interested in solving an inequality rather than an equality. An example of a first-degree inequality is

$$5x - 2 > 13.$$

This is read "$5x - 2$ is greater than 13." The symbols $\neq$, $>$, $<$, $\geq$, and $\leq$ are all used to denote inequalities and are read as indicated in Section 1.1 and discussed in Section 1.4.

A solution to a first-degree inequality is any value of the variable that makes the inequality a true statement. Consider the inequality $3x + 2 > 10$. The number 4 is a solution of $3x + 2 > 10$ because $3(4) + 2 = 14$ and $14 > 10$. The number 1 is *not* a solution of $3x + 2 > 10$ because $3(1) + 2 = 5$, but 5 is not greater than 10.

The set of all values of a variable that make an inequality a true statement is called the *solution set* of the inequality. In this section we learn how to determine the solution set of a first-degree inequality. The rules for solving first-degree inequalities are similar to the rules for solving first-degree equations. The only differences occur when we multiply or divide each side of an inequality by a negative number. We shall demonstrate how to solve inequalities by stating each rule and then giving an example to clarify the rule. Each rule will be stated in terms of the $\leq$ symbol, but the rule is also true if $\leq$ is replaced by either $>$, $<$, or $\geq$.

As with equations, it is permissible to add the same quantity to each side of an inequality. When this is done, the inequality symbol remains the same. This is expressed symbolically as follows, where a, b, and c are real numbers.

Addition Property of Inequality

> If $a \leq b$, then $a + c \leq b + c$.

Example 1. Solve $x - 2 \leq 13$.

Solution

$$
\begin{aligned}
x - 2 &\leq 13 \\
x - 2 + 2 &\leq 13 + 2 && \text{add 2 to each side} \\
x &\leq 15.
\end{aligned}
$$

Thus any number less than, or equal to, 15 is in the solution set of $x - 2 \leq 13$.

The graph of the solution set of an inequality is often a helpful and convenient visual representation of the solution set. The solution set of the inequality of Example 1 is $\{x : x \leq 15\}$, that is, all real numbers less than or equal to 15. The graph of this set is indicated in Figure 2.2. The closed circle at 15 indicates that 15 is included in the solution set. If the original inequality had been $<$, the solution would have been $x < 15$. The graph would have had an open circle at 15 to indicate that 15 was not part of the solution set.

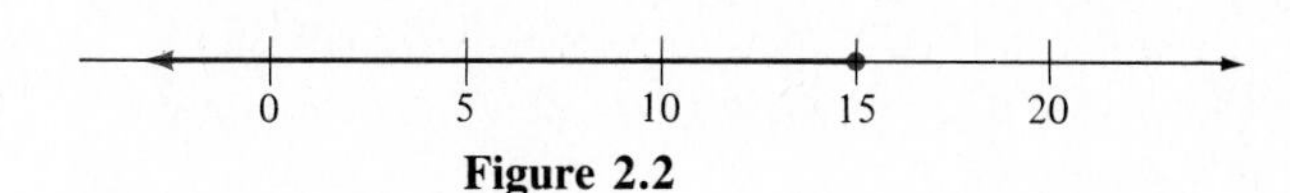

Figure 2.2

The following rule states that if the same quantity is subtracted from each side of an inequality, the inequality symbol remains the same. This is expressed symbolically as follows, where a, b, and c are real numbers.

Subtraction Property of Inequality

> If $a \leq b$, then $a - c \leq b - c$.

We apply this property in the next example.

Example 2. Solve $2x - 3 < x + 7$ and graph the solution set.

Solution

$$
\begin{aligned}
2x - 3 &< x + 7 \\
2x - 3 + 3 &< x + 7 + 3 && \text{add 3 to each side} \\
2x &< x + 10 \\
2x - x &< x + 10 - x && \text{subtract } x \text{ from each side} \\
x &< 10.
\end{aligned}
$$

The graph of $x < 10$ is shown in Figure 2.3. The open circle at 10 indicates that 10 is not part of the solution set.

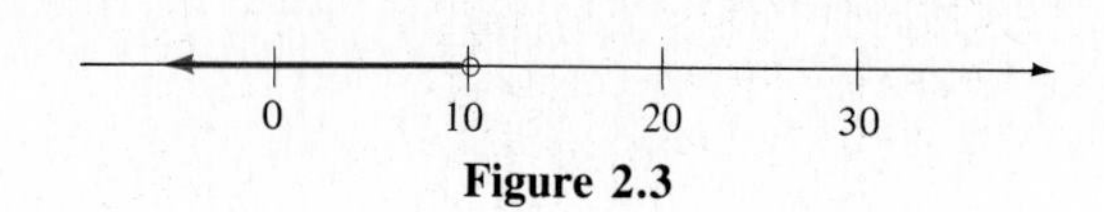

Figure 2.3

When we multiply or divide each side of an inequality by a positive number, the inequality symbol remains the same. This is expressed symbolically as follows, where a and b are real numbers and c is a positive real number.

Multiplication Property of Inequality for Positive Multipliers

If $a \le b$ and $c > 0$, then $\quad ac \le bc \quad$ and $\quad \dfrac{a}{c} \le \dfrac{b}{c}$.

Example 3. Solve $6x > 24$ and graph the solution set.

Solution

$$6x > 24$$

$$\frac{6x}{6} > \frac{24}{6} \qquad \text{divide each side by 6}$$

$$x > 4.$$

The graph of $x > 4$ is shown in Figure 2.4.

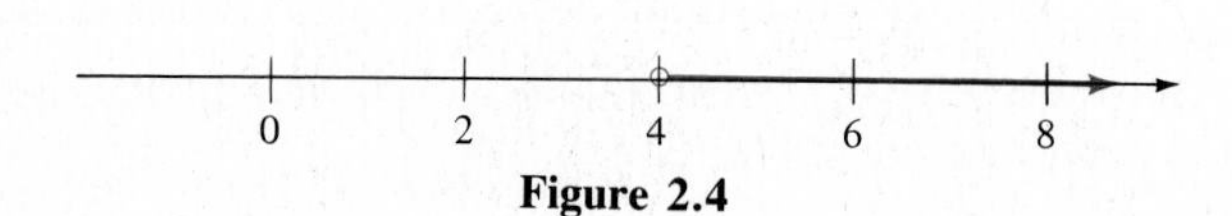

Figure 2.4

Example 4. Solve $\dfrac{x}{5} \leqslant \dfrac{1}{3}$.

Solution

$$\frac{x}{5} \leqslant \frac{1}{3}$$

$$5\left(\frac{x}{5}\right) \leqslant 5\left(\frac{1}{3}\right) \qquad \text{multiply each side by 5}$$

$$x < \frac{5}{3}.$$

The graph of $x \leqslant \dfrac{5}{3}$ is shown in Figure 2.5.

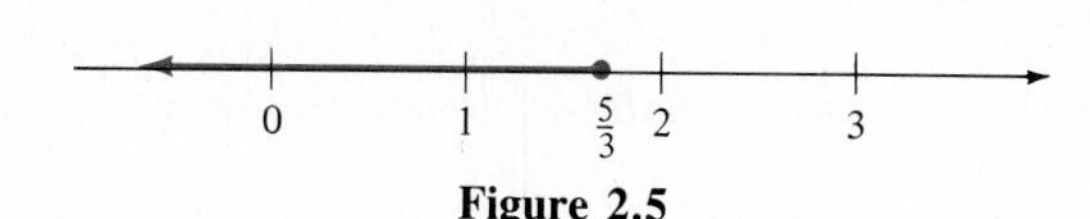

Figure 2.5

Example 5. Solve $4x + 1 \leq x - 3$ and graph the solution set.

Solution

$$4x + 1 \leq x - 3$$
$$4x + 1 - 1 \leq x - 3 - 1 \qquad \text{subtract 1 from each side}$$
$$4x \leq x - 4$$
$$4x - x \leq x - 4 - x \qquad \text{subtract } x \text{ from each side}$$
$$\frac{1}{3}(3x) \leq \frac{1}{3}(-4) \qquad \text{multiply each side by } \frac{1}{3}$$
$$x \leq \frac{-4}{3}.$$

The graph of $x \leq \frac{-4}{3}$ is shown in Figure 2.6.

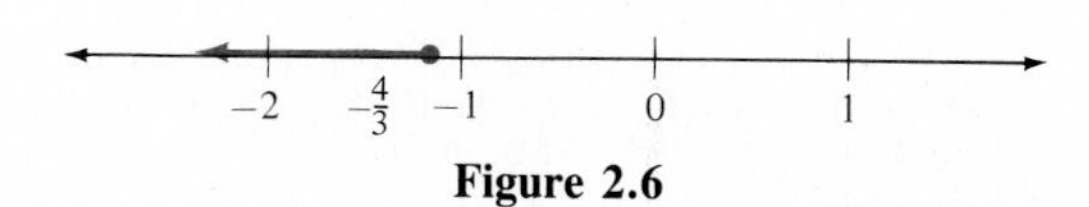

Figure 2.6

We now consider what happens when we multiply each side of an inequality by a negative number. We begin our discussion by looking at a particular example. We know that $3 < 5$. Suppose that we multiply 3 by -2 and 5 by -2. The results are -6 and -10. But -10 is to the left of -6 on the number line and therefore $-6 > -10$. Notice that the direction of the inequality sign was reversed when we multiplied by -2.

The preceding discussion motivates the following rule.

> When we multiply each side of an inequality by a negative number, the sense of the inequality symbol is reversed.

This is expressed symbolically as follows, where a and b are real numbers and c is a negative number.

Multiplication Property of Inequality for Negative Multipliers

> If $a \leq b$ and $c < 0$, then $ac \geq bc$.

Now we know that if c is a negative number, then its reciprocal, $\frac{1}{c}$, is also a negative number. Furthermore, since dividing both sides of an inequality by c is the same as multiplying both sides by $\frac{1}{c}$, we may state the following rule.

Division Property of Inequality for Negative Multipliers

If $a \le b$ and $c < 0$, then $\frac{a}{c} \ge \frac{b}{c}$.

Example 6. Solve $-4x + 2 > 10$.

Solution

$-4x + 2 > 10$

$-4x > 8$ result of subtracting 2 from each side

$\frac{-4x}{-4} < \frac{8}{-4}$ divide each side by -4; reverse inequality

$x < -2.$

The following example demonstrates the use of inequalities in a practical situation.

Example 7. A local school signed up 380 students to attend a soccer game being played at another school. To transport the students, it was decided to hire buses. If each bus has a maximum capacity of 40 students, what is the minimum number of buses that are necessary?

Solution: Let x denote the required number of buses; then $40x$ is the number of students these buses can carry and we want $40x \ge 380$. Dividing both sides of this equation by 40 results in

$$x \ge \frac{380}{40} \quad \text{or} \quad x \ge 9\frac{1}{2}.$$

Thus $x = 10$, since it is difficult to hire half-buses.

CHECK: $(40)(10) = 400 \ge 380$.

EXERCISES 2.9

Solve the following inequalities and graph each solution set.

For Exercises 1–12, see Example 1.

1. $x - 2 \ge 5$ **2.** $x - 4 \ge 11$ **3.** $x + 3 \le 8$ **4.** $x + 6 \le 7$

5. $x + 2 < 1$ 6. $x + 6 < 4$ 7. $x + 13 > 15$ 8. $x + 9 > 5$

9. $x - 2 \leq -6$ 10. $x - 7 \leq -5$ 11. $x + 4 > -7$ 12. $x + 8 > -2$

For Exercises 13–18, see Examples 3 and 4.

13. $3x \geq 12$ 14. $5x \geq 15$ 15. $5x < -10$

16. $4x < -20$ 17. $\frac{2}{5}x \geq -5$ 18. $\frac{3}{7}x \geq -6$

For Exercises 19–24, see Example 6.

19. $-3x > 12$ 20. $-5x > 15$ 21. $-6x \leq -9$

22. $-5x > -8$ 23. $-\frac{1}{4}x < 2$ 24. $-\frac{2}{3}x < 4$

For Exercises 25–38, see Examples 2, 5, and 6.

25. $3x + 8 \geq -4$
26. $5x + 1 > -4$
27. $-2x + 3 < 7$
28. $-5x - 2 \leq 8$
29. $-x + 5 < x + 7$
30. $7x + 2 > 2x - 13$
31. $\frac{2}{3}x + 1 < -1$
32. $-\frac{1}{5}x + \frac{1}{4} > \frac{1}{2}$
33. $3(x - 2) \leq 21$
34. $14x + 5 \geq 3(x + 20)$
35. $4(x - 2) < 5(3x + 5)$
36. $(x - 1) + 4 \geq 2x + 9$
37. $2(x + 1) + 4 < 3(2x + 1) - 1$
38. $3(x - 1) + x > 2(x + 3) + 1$

For Exercises 39 and 40, see Example 7.

39. The math club of a local college sponsored a day trip to Atlantic City. Three hundred and ninety students signed up for the trip. How many buses should be ordered if the maximum capacity of each bus is 45 and each bus is to be used to its capacity when possible?

40. A math class decides to celebrate and have a pizza party. It is estimated that each student (and the instructor) will eat three slices of pizza. The pizzas are prepared and cut into eight slices each. If there are 28 hungry people to feed, how many pizzas should be ordered?

Key Words

Coefficient
Combining like terms
First-degree equation
First-degree inequality
Identity
Solution
Solution set
Term
Unknown
Variable

CHAPTER 2 TEST

For Problems 1–6, simplify the expression.

1. $-4x + 18x$ 2. $\frac{1}{2}z - \frac{1}{4}z$ 3. $5k - 10 + k + 2$

4. $\frac{3}{4}y + 2 - \frac{1}{4}y - 1$ 5. $.74 - .21x + .16 - .06x$ 6. $8(5x + 4) + 4x - 3(4x + 1)$

For Problems 7–14, use the addition and subtraction properties of equality to solve the equation.

7. $x + 12 = 20$ **8.** $22 = z + 5$ **9.** $5 + 3x = 10 + 2x$

10. $\frac{2}{3}x + 4 + \frac{1}{3}x = 11$ **11.** $1.2x - 5 - .2x = 1$ **12.** $4(z - \frac{1}{2}) + 5 = 3(z + 2)$

13. $-5(x + 2) + 6(x + 1) = 5$ **14.** $\frac{1}{2}(y - 6) + \frac{1}{2}(y + 2) = 12$

For Problems 15–22, solve the equation.

15. $8x = 40$ **16.** $4w + 2 = 14$ **17.** $5z - 8 = 3z + 2$

18. $\frac{1}{7}x = 2$ **19.** $\frac{3}{4}y = 9$ **20.** $5(x + 3) + x = 21$

21. $\frac{2}{3}(x + 6) - \frac{8}{3}(x - 9) = 50$ **22.** $.4(x + 5) + .6(x - 8) = 2.8$

For Problems 23–26, solve the inequality and graph the solution set.

23. $2x + 3 \geq x + 5$ **24.** $\frac{1}{3}y > 2$ **25.** $-\frac{2}{5}x \leq -3$ **26.** $(z + 4) + 2z < 7(z + 8)$

Solve Problems 27–30.

27. A toy manufacturer wants to market a set of blocks consisting of red and green blocks. It is decided that the total number of blocks in each set will be 60 and that there will be three times as many red blocks as green blocks. How many blocks of each color will be in a set?

28. Jean ran a 4-mile race in 40 minutes. What is her average rate of running in miles per hour?

29. The volume of a cone is given by $V = \frac{1}{3}\pi r^2 h$, where r is the radius of the base and h is the height. Find the volume of a cone with height 8 inches and radius $1\frac{1}{2}$ inches.

30. A floor is to be covered with tiles that measure 1 foot by 1 foot. The dimensions of the floor are 9 feet by 11 feet. What is the least number of boxes of tiles that are necessary if the tiles are shipped 12 to a box?

CHAPTER 3

Exponents and Polynomials

Exponents and polynomials can be used in the quality control of a production process. An inspector takes a sample of two fuses from a box coming off the production line. The polynomial expansion, $(p + q)^2 = p^2 + 2pq + q^2$, can be used to tell her the probability of getting 2, 1, or 0 bad fuses.

Source: *International Telephone & Telegraph Corporation*

We have already encountered formulas such as $A = \pi r^2$ and $A = s^2$ in which a variable has an exponent greater than 1. In this chapter we develop rules for working with integer exponents. We then show how to use these rules when adding, subtracting, and multiplying polynomials. We also learn how to divide a polynomial by a monomial or a binomial. Because the ideas presented in this chapter are used throughout the remainder of this text, it is important to understand them.

3.1 Positive Integer Exponents

As noted in Chapter 1, when we write $7^2 = 49$ we are using the base 7 with the exponent 2 to denote the product $7 \cdot 7$. Similarly, we may write

$$2^5 = 2 \cdot 2 \cdot 2 \cdot 2 \cdot 2 = 32$$

by using the base 2 and the exponent 5. We say that 32 is 2 raised to the fifth power. When we write $2^5 = 2 \cdot 2 \cdot 2 \cdot 2 \cdot 2 = 32$, we say we have *evaluated* 2^5. In general, we have the following definition.

nth Power of b

For any real number b and any positive integer n,

$$b^n = \underbrace{b \cdot b \cdot \cdots \cdot b}_{n \text{ factors of } b}$$

The expression b^n is called the nth *power of b; b is the base;* and *n is the exponent.* Expressions such as 7^2 and 2^5 are said to be in *exponent form.*

Example 1. For each expression, identify the base and the exponent, and evaluate. (a) 3^4. (b) $(-5)^4$. (c) -5^4.

Solution

(a) The base is 3, the exponent is 4, and

$$3^4 = 3 \cdot 3 \cdot 3 \cdot 3 = 81.$$

(b) The base is -5, the exponent is 4, and

$$(-5)^4 = (-5)(-5)(-5)(-5) = 625.$$

(c) The base is 5, the exponent is 4, and

$$-5^4 = -(5 \cdot 5 \cdot 5 \cdot 5) = -625.$$

In Example 1, notice the difference between the values of $(-5)^4 = 625$ and $-5^4 = -625$. In the former, the base is -5 and the exponent is 4. In the latter, the base is 5, the exponent is 4, and the negative sign is applied after performing the multiplication. Thus $-5^4 = -(5^4)$. In general, an exponent is applied to the *one* number or symbol that *immediately* precedes it. Thus, in the expression $5x^2$, the exponent 2 is applied only to x.

We now consider the product of two powers each with the same base. For example, we know that $3^2 \cdot 3^4$ is the same as

$$\begin{aligned}(3 \cdot 3) \cdot (3 \cdot 3 \cdot 3 \cdot 3) &= 3 \cdot 3 \cdot 3 \cdot 3 \cdot 3 \cdot 3 \\ &= 3^6.\end{aligned}$$

Thus $3^2 \cdot 3^4 = 3^{2+4} = 3^6$.

More generally, if a is any real number, we may write

$$a^3 \cdot a^6 = a^{3+6} = a^9.$$

These observations motivate the fundamental rule of exponents, which is stated as follows.

Rule 1: Multiplication of Powers with a Common Base

If a is any real number and m and n are positive integers, then

$$a^m \cdot a^n = a^{m+n}.$$

Notice that the exponents are added and the common base does not change.

Example 2. Use the preceding rule to rewrite each expression as one power. (a) $3^4 \cdot 3^2$. (b) $(-2)^5(-2)^2$. (c) $x^6 \cdot x^9$.

Solution

(a) $3^4 \cdot 3^2 = 3^{4+2} = 3^6$.
(b) $(-2)^5(-2)^2 = (-2)^{5+2} = (-2)^7$.
(c) $x^6 \cdot x^9 = x^{6+9} = x^{15}$.

We now consider powers with a base that is the product of two numbers. For example, consider $(3 \cdot 2)^4$. If we first perform the multiplication in the parentheses, we get $(3 \cdot 2)^4 = 6^4$. However, we could also rewrite $(3 \cdot 2)^4$ as

$$\begin{aligned}(3 \cdot 2)^4 &= (3 \cdot 2)(3 \cdot 2)(3 \cdot 2)(3 \cdot 2) \\ &= (3 \cdot 3 \cdot 3 \cdot 3)(2 \cdot 2 \cdot 2 \cdot 2) = 3^4 \cdot 2^4.\end{aligned}$$

Thus we see that $(3 \cdot 2)^4 = 3^4 \cdot 2^4$.

Notice that in $(3 \cdot 2)^4$ the base is $3 \cdot 2$ and the exponent is 4, whereas in the equivalent form of $3^4 \cdot 2^4$, there are two bases, 3 and 2, both with the exponent 4. This example motivates the following rule.

Rule 2: Multiplication of Powers with a Common Exponent

For any two real numbers a and b, and any positive integer n,

$$(a \cdot b)^n = a^n \cdot b^n.$$

If we use symmetric property of equality, Rule 2 may also be written as $a^n \cdot b^n = (a \cdot b)^n$.

Rule 2 gives us two methods for evaluating the power of a product. For example, $(3 \cdot 5)^2$ may be evaluated as

$$(3 \cdot 5)^2 = (15)^2 = 225$$

or as

$$(3 \cdot 5)^2 = 3^2 \cdot 5^2 = 9 \cdot 25 = 225.$$

To be flexible, you should be able to apply the rule using either method.

Example 3. Use Rule 2 to evaluate

(a) $5^3 \cdot 2^3$. (b) $10^4\left(\frac{1}{5}\right)^4$.

Solution

(a) $5^3 \cdot 2^3 = (5 \cdot 2)^3$
$= 10^3$
$= 1000.$

(b) $10^4\left(\frac{1}{5}\right)^4 = \left(10 \cdot \frac{1}{5}\right)^4$
$= 2^4$
$= 16.$

Although Rule 2 is stated in terms of just two factors a and b, it may be extended to expressions with more than two factors. For example,

$$(4 \cdot 5 \cdot x)^2 = 4^2 \cdot 5^2 \cdot x^2$$

and

$$\begin{aligned} 7^3 \cdot 3^3 \cdot a^3 \cdot b^3 &= (7 \cdot 3 \cdot a \cdot b)^3 \\ &= (21ab)^3. \end{aligned}$$

As the following example demonstrates, there are instances in which we use both Rule 1 and Rule 2 to rewrite an expression.

Example 4. Rewrite $(3x)^2(5x)^3$ as ax^n, where a and n are positive integers.

Solution

$(3x)^2(5x)^3 = (3^2 \cdot x^2)(5^3 \cdot x^3)$	Rule 2
$= 3^2 \cdot 5^3 \cdot x^2 \cdot x^3$	commutative property
$= 3^2 \cdot 5^3 \cdot x^{2+3}$	Rule 1
$= 9 \cdot 125 \cdot x^5$	simplify
$= 1125x^5.$	

We now turn our attention to the quotient of two powers with a common base. The quotient $\frac{5^7}{5^3}$ may be simplified by writing

$$\frac{5^7}{5^3} = \frac{5 \cdot 5 \cdot 5 \cdot 5 \cdot 5 \cdot 5 \cdot 5}{5 \cdot 5 \cdot 5}$$

$$= \frac{(5 \cdot 5 \cdot 5 \cdot 5)(5 \cdot 5 \cdot 5)}{5 \cdot 5 \cdot 5} \qquad \text{associate factors}$$

$$= (5 \cdot 5 \cdot 5 \cdot 5) \cdot \frac{5 \cdot 5 \cdot 5}{5 \cdot 5 \cdot 5} \qquad \frac{ab}{c} = \frac{a}{1} \cdot \frac{b}{c} = a\frac{b}{c}$$

$$= (5 \cdot 5 \cdot 5 \cdot 5) \cdot 1 \qquad \frac{5 \cdot 5 \cdot 5}{5 \cdot 5 \cdot 5} = 1$$

$$= 5^4.$$

Note that the new exponent is 4 and that $7 - 3 = 4$. Similarly, the quotient $\frac{13^5}{13^4}$ may be rewritten as

$$\frac{13^5}{13^4} = \frac{13(13 \cdot 13 \cdot 13 \cdot 13)}{13 \cdot 13 \cdot 13 \cdot 13} = 13 \cdot \frac{13 \cdot 13 \cdot 13 \cdot 13}{13 \cdot 13 \cdot 13 \cdot 13} = 13 \cdot 1 = 13.$$

In each of these examples, the simplified form has the common base of the original form. Also, the exponent appearing in the simplified form is obtained by subtracting the exponent in the denominator from the exponent in the numerator. We generalize these observations with Rule 3 and then illustrate the rule with an example.

Rule 3: Division of Powers with a Common Base

For any real number $a \neq 0$ and any two positive integers m and n with $m > n$,

$$\frac{a^m}{a^n} = a^{m-n}.$$

We know that $0^2 = 0 \cdot 0 = 0$. Similarly, if n is any positive integer, $0^n = 0$. Since division by zero is not permitted, we do not allow a to equal zero in Rule 3.

Example 5. Rewrite each quotient as one power and evaluate if possible.
(a) $\frac{7^8}{7^6}$. (b) $\frac{(9.1)^4}{(9.1)^3}$. (c) $\frac{x^{23}}{x^{17}}$.

Solution

(a) $\dfrac{7^8}{7^6} = 7^{8-6} = 7^2 = 49.$

(b) $\dfrac{(9.1)^4}{(9.1)^3} = (9.1)^{4-3} = (9.1)^1 = 9.1.$

(c) $\dfrac{x^{23}}{x^{17}} = x^{23-17} = x^6.$

EXERCISES 3.1

For Exercises 1–12, rewrite the expression using exponent notation.

1. $5 \cdot 5 \cdot 5 \cdot 5$ **2.** $3 \cdot 3 \cdot 3 \cdot 3$ **3.** $x \cdot x \cdot x \cdot x \cdot x \cdot x$ **4.** $y \cdot y \cdot y \cdot y$

5. $(-2)(-2)(-2)$ **6.** $(-4)(-4)(-4)$ **7.** $-4 \cdot 4$ **8.** $-9 \cdot 9$

9. $-7 \cdot 7 \cdot 7 \cdot 7 \cdot 7$ **10.** $-5 \cdot 5 \cdot 5 \cdot 5$ **11.** $(-x)(-x)$ **12.** $(-z)(-z)(-z)$

For Exercises 13–24, (a) identify the base, (b) identify the exponent, and (c) evaluate the expression. See Example 1.

13. 4^3 **14.** 11^2 **15.** 2^4 **16.** 6^3 **17.** $(-4)^2$ **18.** $(-5)^2$

19. -2^5 **20.** -3^3 **21.** $-(-2)^2$ **22.** $-(-3)^2$ **23.** $-(-2)^3$ **24.** $-(-3)^3$

For Exercises 25–42, use Rule 1 to rewrite the expression as one power. See Example 2.

25. $5^2 \cdot 5$ **26.** $7 \cdot 7^2$ **27.** $2^3 \cdot 2^4$ **28.** $2^5 \cdot 2^3$

29. $3^2 \cdot 3^8$ **30.** $4^3 \cdot 4^7$ **31.** $(\frac{1}{4})^2(\frac{1}{4})^3$ **32.** $(\frac{2}{3})^5(\frac{2}{3})^6$

33. $3^2 \cdot 3^4 \cdot 3^6$ **34.** $7 \cdot 7^2 \cdot 7^3$ **35.** $6^2 \cdot 6^4 \cdot 6 \cdot 6^3$ **36.** $5 \cdot 5^8 \cdot 5^3 \cdot 5^2$

37. $x^3 \cdot x^5$ **38.** $a^4 \cdot a^7$ **39.** $z^2 \cdot z \cdot z^5$ **40.** $x^3 \cdot x \cdot x^2$

41. $x^2 \cdot x^3 \cdot x^4$ **42.** $y^3 \cdot y^4 \cdot y^5$

For Exercises 43–60, use Rule 2 to rewrite the expression as one power. See Example 3.

43. $2^2 \cdot 5^2$ **44.** $2^2 \cdot 3^2$ **45.** $7^2 \cdot 2^2$ **46.** $9^2 \cdot 2^2$

47. $5^4 \cdot 2^4$ **48.** $25^3 \cdot 4^3$ **49.** $7^3(\frac{1}{7})^3$ **50.** $8^2(\frac{1}{8})^2$

51. $(-\frac{1}{2})^4 2^4$ **52.** $(-3)^4(\frac{1}{3})^4$ **53.** $(-\frac{1}{2})^3(2)^3$ **54.** $-11^3(\frac{1}{11})^3$

55. $9^4(\frac{1}{3})^4$ **56.** $(12)^7(\frac{1}{6})^7$ **57.** $x^5 \cdot y^5$ **58.** $a^3 \cdot b^3$

59. $p^5 \cdot q^5 \cdot r^5$ **60.** $a^7 \cdot b^7 \cdot c^7$

For Exercises 61–72, use Rule 2 to rewrite each expression. See Example 4.

61. $(5xy)^2$ **62.** $(3ab)^2$ **63.** $(4uv)^2$ **64.** $(5yz)^2$ **65.** $(2pq)^3$ **66.** $(4yz)^3$

67. $(5uv)^3$ **68.** $(10ab)^3$ **69.** $(3abc)^4$ **70.** $(5wxy)^4$ **71.** $(10xy)^5$ **72.** $(10rs)^4$

For Exercises 73–90, use Rule 3 to rewrite the expression as one power. See Example 5.

73. $\frac{3^8}{3^7}$ **74.** $\frac{5^5}{5^3}$ **75.** $\frac{2^5}{2^2}$ **76.** $\frac{4^5}{4^3}$ **77.** $\frac{14^3}{14^2}$ **78.** $\frac{13^4}{13^3}$

79. $\frac{10^8}{10^6}$ **80.** $\frac{10^7}{10^5}$ **81.** $-\frac{4^7}{4^4}$ **82.** $-\frac{6^4}{6^2}$ **83.** $\frac{y^5}{y^3}$ **84.** $\frac{z^8}{z^5}$

85. $\frac{b^{15}}{b^{10}}$ **86.** $\frac{x^{14}}{x^{11}}$ **87.** $\frac{a^{12}}{a^5}$ **88.** $\frac{x^{14}}{x^6}$ **89.** $\frac{b^{11}}{b^8}$ **90.** $\frac{w^9}{w^5}$

For Exercises 91–100, use the rules developed in this section to rewrite the expression as one power.

91. $\frac{4^2 \cdot 4^3}{4^4}$ **92.** $\frac{3^5 \cdot 3^6}{3^8}$ **93.** $\frac{7^2 \cdot 7^3 \cdot 7^5}{7 \cdot 7^4}$ **94.** $\frac{11 \cdot 11^6 \cdot 11^3}{11^2 \cdot 11^5}$

95. $\frac{5^3 \cdot 5^2 \cdot 5^4}{5^2 \cdot 5^2 \cdot 5}$ **96.** $\frac{9^5 \cdot 9^4 \cdot 9^3}{9 \cdot 9 \cdot 9^2}$ **97.** $\frac{x^2 \cdot x^3}{x^4}$ **98.** $\frac{a^7 \cdot a^8}{a^{10}}$

99. $\frac{y^2 \cdot y^4 \cdot y^5}{y^3 \cdot y}$ **100.** $\frac{z^6 \cdot z^4 \cdot z^3}{z^2 \cdot z}$

Integer Exponents 3.2

Thus far we have considered only expressions with positive-integer exponents. In this section we show how to interpret an exponent form with an exponent of zero and exponent forms with negative integer exponents.

We know that $\frac{5^2}{5^2} = \frac{25}{25} = 1$. However, if we use Rule 3 to evaluate $\frac{5^2}{5^2}$, we get $5^{2-2} = 5^0$. Since

$$\frac{5^2}{5^2} = 1 \quad \text{and} \quad \frac{5^2}{5^2} = 5^0,$$

we see that $5^0 = 1$. This example motivates the Zero-Exponent rule.

Rule 4: Zero Exponent

> If a is any nonzero real number, then
>
> $$a^0 = 1.$$

This rule is very easy to apply. Whenever we see a nonzero number with an exponent of 0, the answer is 1. For example, $117^0 = 1$.

Example 1. Evaluate each expression.
(a) 3^0. (b) $(-3)^0$. (c) -3^0.

Solution
(a) $3^0 = 1$.
(b) $(-3)^0 = 1$.
(c) $-3^0 = -1$.

It is also possible to give meaning to a negative-integer exponent. For example, if we apply rule 3 to $3^2/3^4$, the result is $3^{2-4} = 3^{-2}$. But

$$\begin{aligned}\frac{3^2}{3^4} = \frac{3 \cdot 3}{3 \cdot 3 \cdot 3 \cdot 3} &= \frac{3}{3} \cdot \frac{3}{3} \cdot \frac{1}{3} \cdot \frac{1}{3} \\ &= 1 \cdot 1 \cdot \frac{1}{3} \cdot \frac{1}{3} \\ &= \frac{1}{3^2}.\end{aligned}$$

Since $\frac{3^2}{3^4} = 3^{-2}$ and $\frac{3^2}{3^4} = \frac{1}{3^2}$, we conclude that $3^{-2} = \frac{1}{3^2}$. This example motivates the following rule for negative exponents.

Rule 5: Negative-Integer Exponents

If a is any nonzero real number and n is a positive integer, then

$$a^{-n} = \frac{1}{a^n}.$$

Thus a negative-integer exponent tells us to raise the base to the power without the negative sign, and then to take the reciprocal.

Example 2. Evaluate each expression.
(a) 2^{-3}. (b) 10^{-2}. (c) -5^{-3}.

Solution

(a) $2^{-3} = \frac{1}{2^3} = \frac{1}{8}$.

(b) $10^{-2} = \frac{1}{10^2} = \frac{1}{100}$.

(c) $-5^{-3} = -\frac{1}{5^3} = -\frac{1}{125}$.

It is important to notice that in order to evaluate an expression that contains a negative exponent such as 3^{-2}, we must first obtain an equivalent expression that contains positive exponents. Thus 3^{-2} must first be written as $\frac{1}{3^2}$, which can now be evaluated as $\frac{1}{3 \cdot 3} = \frac{1}{9}$.

Recall that $\frac{a}{b}$ is defined for all real numbers a and b, with $b \neq 0$. Also, for any positive integer m,

$$\begin{aligned} \left(\frac{a}{b}\right)^m &= \frac{a}{b} \cdot \frac{a}{b} \cdot \cdots \cdot \frac{a}{b} \qquad m \text{ factors of } \frac{a}{b} \\ &= \frac{a \cdot a \cdot \cdots \cdot a}{b \cdot b \cdot \cdots \cdot b} \\ &= \frac{a^m}{b^m}. \end{aligned}$$

This result motivates the following rule.

Rule 6: Division of Powers with a Common Exponent

If a and b are nonzero real numbers and m is any integer, then

$$\left(\frac{a}{b}\right)^m = \frac{a^m}{b^m}.$$

Note that Rule 6 may also be written as

$$\frac{a^m}{b^m} = \left(\frac{a}{b}\right)^m.$$

Example 3. Evaluate each power using Rule 6.

(a) $\left(\frac{3}{5}\right)^3$. (b) $\left(\frac{8}{7}\right)^2$.

Solution: Using Rule 6, we may rewrite and then evaluate each expression as follows:

(a) $\left(\frac{3}{5}\right)^3 = \frac{3^3}{5^3} = \frac{27}{125}$.

(b) $\left(\frac{8}{7}\right)^2 = \frac{8^2}{7^2} = \frac{64}{49}$.

The following discussion demonstrates how to proceed when the exponent on a quotient is a negative integer. By using the rules of exponents, we can show that

$$\left(\frac{2}{5}\right)^{-4} = \left(\frac{5}{2}\right)^4.$$

More generally, if a and b are nonzero real numbers and m is a positive integer, we have

$$\left(\frac{a}{b}\right)^{-m} = \left(\frac{b}{a}\right)^m.$$

Example 4. Evaluate $\left(\frac{3}{7}\right)^{-2}$.

Solution

$$\left(\frac{3}{7}\right)^{-2} = \left(\frac{7}{3}\right)^2 = \frac{7^2}{3^2} = \frac{49}{9}.$$

The last rule we develop in this section is concerned with simplifying expressions such as $(5^3)^2$. Using the rules for exponents that have already been stated, we may write

$$(5^3)^2 = 5^3 \cdot 5^3 = 5^{3+3} = 5^{2\cdot 3} = 5^6.$$

Similarly,

$$(3^2)^4 = 3^2 \cdot 3^2 \cdot 3^2 \cdot 3^2 = 3^{2+2+2+2} = 3^{4\cdot 2} = 3^8.$$

Also, $(2^3)^{-1} = \frac{1}{2^3}$. But $\frac{1}{2^3} = 2^{-3}$ and therefore $(2^3)^{-1} = 2^{-3}$. Summarizing these three examples, we note that

$$(5^3)^2 = 5^6, \qquad (3^2)^4 = 3^8, \qquad \text{and} \qquad (2^3)^{-1} = 2^{-3}.$$

In each of these examples, the exponent on the right is the product of the two exponents on the left. We generalize these observations by stating the following rule.

Rule 7: Power Rule

If a is any nonzero real number and m and n are integers, then

$$(a^m)^n = a^{m\cdot n}.$$

Example 5. Apply Rule 7 and then evaluate each exponential.
(a) $(2^3)^2$. (b) $-(3^2)^2$. (c) $(4^2)^{-4}$.

Solution

(a) $(2^3)^2 = 2^{3\cdot 2} = 2^6 = 64$.
(b) $-(3^2)^2 = -3^{2\cdot 2} = -3^4 = -81$.
(c) $(4^2)^{-4} = 4^{2(-4)} = 4^{-8} = \frac{1}{4^8} = \frac{1}{65{,}536}$.

It may be shown that all the rules developed in this section and Section 3.1 are valid for all integer exponents, provided that the base is not zero. The rules of these two sections are summarized in Table 3.1.

Table 3.1

Rules for Working with Exponents	
a and b are nonzero real numbers. m and n are integers.	
Rule 1	$a^m a^n = a^{m+n}$
Rule 2	$(ab)^n = a^n \cdot b^n$
Rule 3	$\dfrac{a^m}{a^n} = a^{m-n}$
Rule 4	$a^0 = 1$
Rule 5	$a^{-m} = \dfrac{1}{a^m}$
Rule 6	$\left(\dfrac{a}{b}\right)^m = \dfrac{a^m}{b^m}$
Rule 7	$(a^m)^n = a^{m \cdot n}$

The next example contains expressions that involve variables. You should be able to apply Rules 1–7 and simplify them. That is, you should be able to rewrite them in the form x^n or ax^n.

Example 6. Use Rules 1–7 to simplify by rewriting the following expressions in the form x^n or ax^n, where a is a real number and n is a positive integer.

(a) $\dfrac{x^7 \cdot x^4 \cdot x^3}{(x^4)^3}$. (b) $\dfrac{(z^4)^2 z^3}{z^2 \cdot z^5}$. (c) $\dfrac{(3w)^3 w^4}{9w^2}$.

Solution

(a) $\dfrac{x^7 \cdot x^4 \cdot x^3}{(x^4)^3} = \dfrac{x^{7+4+3}}{x^{4 \cdot 3}}$ Rule 1 and Rule 7

$= \dfrac{x^{14}}{x^{12}}$

$= x^{14-12}$ Rule 3

$= x^2.$

Thus

$$\frac{x^7 \cdot x^4 \cdot x^3}{(x^4)^3} \quad \text{simplifies to} \quad x^2.$$

(b)
$$\begin{aligned}
\frac{(z^4)^2 \cdot z^3}{z^2 \cdot z^5} &= \frac{z^{4 \cdot 2} \cdot z^3}{z^2 \cdot z^5} && \text{Rule 7} \\
&= \frac{z^8 \cdot z^3}{z^2 \cdot z^5} \\
&= \frac{z^{8+3}}{z^{2+5}} && \text{Rule 1} \\
&= \frac{z^{11}}{z^7} \\
&= z^{11-7} && \text{Rule 3} \\
&= z^4.
\end{aligned}$$

Thus

$$\frac{(z^4)^2 \cdot z^3}{z^2 \cdot z^5} = z^4.$$

(c)
$$\begin{aligned}
\frac{(3w)^3 \cdot w^4}{9w^2} &= \frac{3^3 \cdot w^3 \cdot w^4}{9w^2} && \text{Rule 7} \\
&= \frac{3^3 \cdot w^{3+4}}{9 \cdot w^2} && \text{Rule 1} \\
&= \frac{27}{9} \cdot \frac{w^7}{w^2} \\
&= 3 \cdot w^{7-2} && \text{Rule 3} \\
&= 3w^5.
\end{aligned}$$

EXERCISES 3.2

For Exercises 1–12, use Rules 4 and 5 to evaluate the expression. See Examples 1 and 2.

1. 5^0 **2.** 7^0 **3.** -8^0 **4.** -5^0 **5.** $(-5)^0$ **6.** $(-8)^0$

7. 3^{-2} **8.** 7^{-2} **9.** 9^{-2} **10.** 2^{-3} **11.** 5^{-3} **12.** 2^{-4}

For Exercises 13–24, use Rule 6 to evaluate the expression. See Example 3.

13. $(\frac{2}{3})^2$ **14.** $(\frac{3}{4})^2$ **15.** $(\frac{5}{7})^2$ **16.** $(\frac{8}{9})^2$ **17.** $(\frac{4}{5})^3$ **18.** $(\frac{3}{4})^3$

19. $(\frac{2}{5})^3$ **20.** $(\frac{5}{7})^3$ **21.** $(\frac{1}{2})^4$ **22.** $(\frac{1}{3})^4$ **23.** $(\frac{2}{5})^4$ **24.** $(\frac{3}{5})^4$

For Exercises 25–36, evaluate the expression. See Example 4.

25. $(\frac{2}{3})^{-2}$ **26.** $(\frac{3}{5})^{-2}$ **27.** $(\frac{3}{4})^{-2}$ **28.** $(\frac{5}{2})^{-2}$ **29.** $(\frac{4}{5})^{-3}$ **30.** $(\frac{3}{2})^{-3}$

31. $(\frac{5}{2})^{-3}$ **32.** $(\frac{3}{5})^{-3}$ **33.** $(\frac{1}{2})^{-1}$ **34.** $(\frac{1}{5})^{-1}$ **35.** $\left(\frac{1}{2^2}\right)^{-1}$ **36.** $\left(\frac{1}{3^2}\right)^{-1}$

For Exercises 37–50, use Rule 7 to evaluate the expression. See Example 5.

37. $(2^2)^3$ **38.** $(3^2)^2$ **39.** $(5^2)^3$ **40.** $(4^2)^3$ **41.** $(5^0)^4$ **42.** $(3^0)^7$

43. $(5^2)^0$ **44.** $(6^4)^0$ **45.** $(3^2)^{-1}$ **46.** $(5^2)^{-1}$ **47.** $(5^{-1})^2$ **48.** $(8^{-1})^2$

49. $(11^2)^{-1}$ **50.** $(12^2)^{-1}$

For Exercises 51–86, simplify the expression. Write your answer with positive exponents.

51. $x^7 \cdot x^{-11}$ **52.** $b^4 \cdot b^{-15}$ **53.** $a^0 \cdot a^{-4}$ **54.** $z^0 \cdot z^{-3}$

55. $\frac{12^{-11}}{12^3}$ **56.** $\frac{9^{-7}}{9^2}$ **57.** $\frac{7^{15}}{7^{-4}}$ **58.** $\frac{5^{10}}{5^{-2}}$

59. $\frac{8^{-20}}{8^{-18}}$ **60.** $\frac{7^{-15}}{7^{-13}}$ **61.** $6^{-4} \cdot 6^3 \cdot 6^{-2}$ **62.** $11^{-5} \cdot 11^3 \cdot 11^{-1}$

63. $5^{-8} \cdot 5^5 \cdot 5^0$ **64.** $7^{-13} \cdot 7^0 \cdot 7^8$ **65.** $(2^7)^5$ **66.** $(6^4)^3$

67. $(5^4)^4$ **68.** $(3^2)^4$ **69.** $(x^2)^4$ **70.** $(z^3)^4$

71. $(y^7)^2$ **72.** $(a^8)^5$ **73.** $\frac{z^{10}}{z^{-2}}$ **74.** $\frac{w^8}{w^{-4}}$

75. $\frac{b^{-5}}{b^3}$ **76.** $\frac{a^{-6}}{a^4}$ **77.** $\frac{c^3}{c^{-5}}$ **78.** $\frac{z^4}{z^{-6}}$

79. $(p^2)^3$ **80.** $(q^4)^2$ **81.** $(a^5)^4$ **82.** $(x^3)^4$

83. $(y^2)^2$ **84.** $(z^3)^3$ **85.** $(w^{-2})^3$ **86.** $(x^{-3})^4$

For Exercises 87–104, evaluate the expression.

87. $2^3 + 3^2$ **88.** $2^4 + 3^2$ **89.** $2^5 - 5^2$ **90.** $3^4 - 4^3$

91. $3^0 - 7^0$ **92.** $5^0 - 4^0$ **93.** $\frac{4^{-1}}{2^{-1}}$ **94.** $\frac{2^{-3}}{2^{-2}}$

95. $13^{-4} \cdot 13^2 \cdot 13^3$ **96.** $5^{-5} \cdot 5^{-2} \cdot 5^9$ **97.** $\frac{2^2 \cdot 2 \cdot 2^3}{(2^2)^3}$ **98.** $\frac{3 \cdot 3^2 \cdot 3^3}{(3^2)^3}$

99. $\frac{(3^3)^2}{3 \cdot 3^2 \cdot 3^3}$ **100.** $\frac{(2^2)^3}{2^3 \cdot 2 \cdot 2^2}$ **101.** $\left(\frac{3}{5}\right)^{-4}$ **102.** $\left(\frac{2}{3}\right)^{-6}$

103. $\left(\frac{7}{8}\right)^{-1}$ **104.** $\left(\frac{7}{8}\right)^{-2}$

For Exercises 105–116, simplify the expression. See Example 6.

105. $\dfrac{w^2 \cdot w^3 \cdot w^6}{(w^4)^2}$ **106.** $\dfrac{y^5 \cdot y^3 \cdot y^8}{(y^3)^4}$ **107.** $\dfrac{a^4 \cdot a^6 \cdot a^2}{(a^2)^4}$ **108.** $\dfrac{b^7 \cdot b^2 \cdot b^5}{(b^3)^2}$

109. $\dfrac{(x^5)^2 x^4}{x^3 \cdot x^7}$ **110.** $\dfrac{(z^4)^2 z^5}{z^3 \cdot z^2}$ **111.** $\dfrac{(q^3)^5 q^4}{q^7 \cdot q^8}$ **112.** $\dfrac{(p^4)^3 p^5}{p^9 \cdot p^4}$

113. $\dfrac{(2w)^4 w^2}{8w^3}$ **114.** $\dfrac{(2w)^2 w^4}{4w^2}$ **115.** $\dfrac{(5x)^3 x^5}{25x^4}$ **116.** $\dfrac{(5x)^4 x^3}{25x^5}$

3.3 Polynomials

In this section we learn how to identify some special polynomials and how to add or subtract two polynomials. We also learn how to evaluate polynomials for particular values of the variable or variables.

A *monomial* is a single term such as x, $5y$, 21, $2x^2y^5$, or $58x^3yz^2$. Thus a monomial is a constant or a constant times a variable or variables. In a monomial, the exponents on all variables must be nonnegative integers. A *binomial* is the sum or difference of two monomials. For example, $x + y$, $a - b$, $2x + 3y$, $4x^2 - 9$, and $21x^2y - 10xy^4z^3$ are all binomials. A *trinomial* is the sum of three monomials. Thus $a^3 + b^2 + 2ab$, $x^2 + 6x - 5$ and $7x^5 - 2xy + z^8$ are all trinomials. In general, a *polynomial* is a *monomial* or a *finite sum of monomials*. Thus $x^7 + 3x^2 - 4xy + y^6$ is a polynomial with four terms.

Example 1. Identify each polynomial as a monomial, binomial, or trinomial. (a) $2x^4 - 50$. (b) 3. (c) $7x^2 - 2x + 11$.

Solution

(a) Binomial. (b) Monomial. (c) Trinomial.

For most of this chapter we shall be concerned with polynomials that contain a single variable. For example, $15y^4 - 13y^3 + 2y^2 - 7y + 12$ is called a polynomial in y. Note also that we have written this polynomial in *decreasing powers* of y. That is, as we read the terms from left to right, the exponents are decreasing. When a polynomial in one variable is written in decreasing powers of the variable, we say that it is in *standard form*.

The constant factor of a monomial is called the *numerical coefficient* or *coefficient* of the monomial. Thus the coefficient of $13x^3$ is 13. When a polynomial is in standard form, the first coefficient is called the *leading coefficient*. Thus the leading coefficient is the coefficient of the term with the largest exponent on the variable.

Example 2. Write each polynomial in standard form and identify the leading coefficient.
(a) $1 - x + 3x^2 + 4x^3 + 7x^5$.
(b) $11 - x^2 + 6x + x^5 - 2x^3$.
(c) $3x - x^5 + 4 + x^2$.

Solution
(a) $7x^5 + 4x^3 + 3x^2 - x + 1$; the leading coefficient is 7.
(b) $x^5 - 2x^3 - x^2 + 6x + 11$; the leading coefficient is 1.
(c) $-x^5 + x^2 + 3x + 4$; the leading coefficient is -1.

Monomials that contain the same variables with the same exponents are called *like terms*. Thus $13x^2y^3$ and $-2x^2y^3$ are like terms. Similarly, $7a^4$ and $-11a^4$ are like terms. To add like terms, we make use of the distributive property, as explained in Chapter 2. For example, to add $13x^2y^3$ and $2x^2y^3$, we proceed as follows:

$$\begin{aligned} &13x^2y^3 + 2x^2y^3 \\ &= (13 + 2)x^2y^3 && \text{distributive property} \\ &= 15x^2y^3. && \text{addition} \end{aligned}$$

Notice that *to add two like terms, we simply add the corresponding coefficients to get the new coefficient*. The variable terms remain the same. Thus, to add $5y^2z^4$ and $3y^2z^4$, we simply add 5 and 3 to get 8 and our answer is $8y^2z^4$. Similarly, to subtract two like terms, we subtract the coefficients to get the coefficient of the difference of the two monomials. For example, to subtract $7a^4$ from $11a^4$, we subtract 7 from 11 to get 4. Thus $11a^4 - 7a^4 = 4a^4$.

Since a polynomial is the finite sum of monomials, to add or subtract two polynomials, we simply add or subtract the corresponding like terms. The following example demonstrates how to add two polynomials.

Example 3. Add $3x^3 + 7x^2 - 2x + 11$ and $5x^3 - 2x^2 + 5x + 8$.

Solution: First write the two polynomials in the vertical form by lining up the like terms as follows:

$$\begin{array}{rr} & 3x^3 + 7x^2 - 2x + 11 \\ \text{(add)} & \underline{5x^3 - 2x^2 + 5x + 8}. \end{array}$$

Next add the like terms and obtain the sum

$$8x^3 + 5x^2 + 3x + 19.$$

The method that was just demonstrated will allow us to find the sum of any two polynomials. However, since we only add the corresponding like terms, many people prefer the horizontal form of addition. For example, to add $7x^5 - 4x^3 + 5x^2 - 2x + 1$ and $2x^5 + x^3 + 3x^2 + 5x - 7$ we identify and group like terms as indicated.

$$(7x^5 + 2x^5) + (-4x^3 + x^3) + (5x^2 + 3x^2) + (-2x + 5x) + (1 - 7).$$

Adding these like terms, we obtain the answer of $9x^5 - 3x^3 + 8x^2 + 3x - 6$. Of course, the same answer would have been obtained using the vertical form of addition.

Example 4. Add $5x^4 + x + 1$ and $x^3 + 2x + 4$.

Solution: Associating like terms and adding, we get

$$\begin{aligned} &5x^4 + x^3 + (x + 2x) + (1 + 4) \\ &= 5x^4 + x^3 + 3x + 5. \end{aligned}$$

Example 5. Subtract $3x^2 - 2x + 5$ from $7x^2 + x - 4$.

Solution: Recall that to subtract one number from another, we change the sign of the number being subtracted and add. Thus the problem may be written as

$$\begin{aligned} (7x^2 + x - 4) - (3x^2 - 2x + 5) &= (7x^2 + x - 4) + (-3x^2 + 2x - 5) \\ &= 4x^2 + 3x - 9. \end{aligned}$$

We could obtain the same answer by eliminating the parentheses after changing the sign of each term of the polynomial that is being subtracted. Thus

$$\begin{aligned} (7x^2 + x - 4) - (3x^2 - 2x + 5) &= 7x^2 + x - 4 - 3x^2 + 2x - 5 \\ &= 4x^2 + 3x - 9. \end{aligned}$$

The *degree* of a polynomial in *one variable* is the largest exponent on the variable that appears on the nonzero terms of the polynomial. Therefore, the degree of $13x^5 - 6x^3 + 2x - 7$ is 5 because it is the largest exponent appearing in the polynomial. Similarly, the degree of $7x - 4$ is 1. We say that a nonzero constant, such as 5, has degree 0. This is because 5 may be written as $5x^0 = 5(1) = 5$. We say that 0 has no degree.

Example 6. Identify (a) the degree of $7x^5 - 4x^3 + 5x^2 - 2x + 1$, (b) the degree of $2x^3 - 3x^2 + x + 7$.

Solution

(a) The degree of $7x^5 - 4x^3 + 5x^2 - 2x + 1$ is 5.
(b) The degree of $2x^3 - 3x^2 + x + 7$ is 3.

The degree of a polynomial is often used to classify the polynomial. In particular, if the degree is 2 or 3, the polynomial is called a *quadratic* or *cubic,* respectively.

Example 7. Classify each polynomial as quadratic or cubic.
(a) $2x^3 - 7x^2 + 5$. (b) $x^2 - 2x + 11$. (c) $-5x^3 + 4$.

Solution

(a) Cubic. (b) Quadratic. (c) Cubic.

The degree of a nonzero monomial in more than one variable is the sum of the exponents on all the variables. For example, the degree of $15x^2y^3z$ is

$$2 + 3 + 1 = 6.$$

The *degree* of a polynomial *in more than one variable* is the largest degree of all the nonzero terms of the polynomial.

Example 8. Determine the degree of each polynomial.
(a) $3x^2y + 5xy^3 + 7x^2y^3 + y^4$.
(b) $4x^6 + 12x^2y^2 + 7y^3$.

Solution

(a) The degree of $3x^2y$ is $2 + 1 = 3$, the degree of $5xy^3$ is $1 + 3 = 4$, the degree of $7x^2y^3$ is $2 + 3 = 5$, and the degree of y^4 is 4. Thus the degree of the polynomial is 5.
(b) The degree of $4x^6$ is 6, the degree of $12x^2y^2$ is $2 + 2 = 4$, and the degree of $7y^3$ is 3. Thus the degree of the polynomial is 6.

Polynomials are sometimes denoted as $p(x)$ or $q(x)$. These expressions are read as "p of x" and "q of x," respectively. For example, we write

$$p(x) = 4x^2 - 2x + 1$$

or

$$q(x) = -3x^2 + 7x + 11.$$

This notation is very convenient for evaluating polynomials. For example, if we wish to evaluate $p(x) = 4x^2 - 2x + 1$ when $x = 1$, we write $p(1)$. This tells us to replace x with 1 in the polynomial and thus

$$\begin{aligned} p(1) &= 4(1)^2 - 2(1) + 1 \\ &= 4 - 2 + 1 \\ &= 3. \end{aligned}$$

Similarly, if $q(x) = -3x^2 + 7x + 11$, then

$$\begin{aligned} q(-2) &= -3(-2)^2 + 7(-2) + 11 \\ &= -12 - 14 + 11 \\ &= -15. \end{aligned}$$

Example 9. Evaluate each polynomial as indicated.
(a) $p(2)$ if $p(x) = x^3 - 3x^2 + x + 5$.
(b) $q(-1)$ if $q(x) = x^4 + 2x^3 - x^2 + x + 7$.

Solution

(a) $$\begin{aligned} p(2) &= (2)^3 - 3(2)^2 + 2 + 5 \\ &= 8 - 12 + 2 + 5 \\ &= 3. \end{aligned}$$

(b) $q(-1) = (-1)^4 + 2(-1)^3 - (-1)^2 + (-1) + 7$
$= 1 - 2 - 1 - 1 + 7$
$= 4.$

EXERCISES 3.3

For Exercises 1–12, identify the polynomial as a monomial, binomial, or trinomial. See Example 1.

1. $4x^3 - 1$ **2.** $2x^2 + 3x + 7$ **3.** $3xy$ **4.** 13
5. $7 - xy$ **6.** $x^3 + 11x - 1$ **7.** $-17x^5$ **8.** $y^3 + 11y - y^2$
9. $y^2 - 16$ **10.** $-5x^2y$ **11.** $x^3 + 2x^2 + 1$ **12.** $y^2 - 7y$

For Exercises 13–30, combine like terms where possible and then write the polynomial in standard form and identify the leading coefficient. See Example 2.

13. $3x + 7 - x + x^2$ **14.** $x^3 - 4 + x + 1$ **15.** $7 + x + 3x^2 + x^3$
16. $x^5 - x + 7 + x^4$ **17.** $-3x + x^2 - 2x + x^2 + 1$ **18.** $y - y^4 + 2y + y^4$
19. $3 + 3w + w^2 - 3w$ **20.** $7x - 6x^2 + 5x$ **21.** $4y + y^2 + 2y$
22. $3x^2 - 4x + x^2 + x^3$ **23.** $x^3 - 2x^2 + 1 + x^2$ **24.** $5 - x^2 + 4x^3 + x^2$
25. $7 + x^3 - x^2 - x^3$ **26.** $12 + 2x^2 + x - 2x^2$ **27.** $y^4 - 5 + y^4 + 1$
28. $w^3 + 7 + w^3 + 3$ **29.** $2w^5 + w + w^5 + 4$ **30.** $3w^6 - w + w^6 - 8$

For Exercises 31–50, perform the indicated operation. See Examples 3–5.

31. $(x^3 + 2x - 1) + (2x^3 - 3x + 1)$ **32.** $(x^4 - 2x^2 + 7) + (x^4 + x^2 + 2)$
33. $(x^3 + 2x - 1) - (2x^3 - 3x + 1)$ **34.** $(2x^2 + 3x + 7) - (x^2 + 2x + 7)$
35. $(3y^3 + 2y) + (y^2 + y + 5)$ **36.** $(4y^4 + 2y^3 + y) + (y^4 - y^3 - 2y)$
37. $(3y^3 + 2y) - (y^2 + y + 5)$ **38.** $(5w^3 + 3w^2 + w) - (3w^3 + 2w^2 - w)$
39. $(4x^2y - 3xy) + (2x^2y + 2xy)$ **40.** $(6x^2y + 4xy) + (4x^2y - 3xy)$
41. $(4x^2y - 3xy) - (2x^2y + 2xy)$ **42.** $(6x^2y + 4xy) - (4x^2y - 3xy)$
43. $(y^2 + 5) + (y^3 + 2y)$ **44.** $(y^3 + 4) + (y^3 + 3y)$
45. $(y^2 + 5y) - (y^3 + 2y)$ **46.** $(x^3 + 10) - (x^3 + 3x)$
47. $(y^2 - 2y + 1) + (y^2 + 2y + 1)$ **48.** $(3y^2 + 2y + 1) + (y^2 + 6y + 1)$
49. $(5y^2 - 3y + 5) - (3y^2 - 2y + 4)$ **50.** $(3x^2 + 4x - 2) - (2x^2 - 3x + 1)$

For Exercises 51–60, identify the polynomial as a quadratic or cubic and then evaluate it as indicated. See Examples 7 and 9.

51. $p(x) = 2x^2 - x + 3,$ $p(1) =$
52. $p(x) = 5x^2 - 2x - 3,$ $p(-2) =$

53. $q(x) = 4x^2 - 9x + 1,\quad q(2) =$

54. $p(y) = y^2 - 2y + 1,\quad p(1) =$

55. $h(x) = 3x^3 + 2x^2 + x + 1,\quad h(-1) =$

56. $p(y) = y^3 + 2y^2 + 7y - 5,\quad p(3) =$

57. $h(y) = 3y^3 - 2y + 4,\quad h(-1) =$

58. $h(x) = 5x^3 - 3x + 1,\quad h(2) =$

59. $p(z) = 5z^2 - 2z + 1,\quad p(-1) =$

60. $p(z) = 4z^2 + 3z - 5,\quad p(2) =$

For Exercises 61–70, identify the degree of the polynomial. See Example 8.

61. $7x^4 - 3x^3 + 2$

62. $\frac{1}{2}y - y^5 + 3y^8$

63. $11xy^2z$

64. $2x^2y^3z^4$

65. $5x^2y + 11xy^3 + 7x^2y^2 + 9x^3y^4$

66. $7xy^2 + 5xy + 6x^2y^2 + 2$

67. $7 - 2x + x^3 - 5x^5$

68. $3 + 5x + x^4 - 2x^5$

69. $100 + 50x + \frac{1}{2}x^2$

70. $75 - 25x + \frac{1}{4}x^2$

Multiplication of Polynomials 3.4

In this section we learn how to multiply one polynomial by another. We begin by considering the product of two monomials. For example, consider the product of $3x^2y$ and $2x^3y^2$. We can obtain this product as follows:

$$\begin{aligned}
&(3x^2y)(2x^3y^2) \\
&= 3 \cdot 2 \cdot x^2 \cdot x^3 \cdot y \cdot y^2 && \text{commutative property of multiplication} \\
&= 6x^{2+3}y^{1+2} && \text{multiplication and Rule 1 for exponents} \\
&= 6x^5y^3.
\end{aligned}$$

Example 1. Multiply $7a^3b^4$ and $6ab^3$.

Solution

$$\begin{aligned}
(7a^3b^4)(6ab^3) &= 7 \cdot 6 \cdot a^3 \cdot a \cdot b^4 \cdot b^3 && \text{commutativity} \\
&= 42a^{3+1}b^{4+3} && \text{multiplication and Rule 1 for exponents} \\
&= 42a^4b^7.
\end{aligned}$$

To multiply a monomial by a polynomial, we use the distributive law. For example, to multiply $2x$ and $x + y$, we write

$$\begin{aligned}
2x(x + y) &= 2x(x) + 2x(y) && \text{distributive property} \\
&= 2x^2 + 2xy.
\end{aligned}$$

Similarly, the product of $3ax^2$ and $2a - 5x^3$ is

$$\begin{aligned} 3ax^2(2a - 5x^3) &= (3ax^2)(2a) - (3ax^2)(5x^3) \\ &= 6a^2x^2 - 15ax^5. \end{aligned}$$

If the polynomial has more than two terms, the preceding approach is still appropriate.

Example 2. Find the product of $2x^2y$ and $9x - 2y + 3xy^2 - 7$.

Solution

$$\begin{aligned} &2x^2y(9x - 2y + 3xy^2 - 7) \\ &= (2x^2y)(9x) - (2x^2y)(2y) + (2x^2y)(3xy^2) - (2x^2y)(7) \\ &= 2 \cdot 9 \cdot x^2 \cdot x \cdot y - (2)(2)x^2 \cdot y \cdot y + 2 \cdot 3 \cdot x^2 \cdot x \cdot y \cdot y^2 - (2)(7)x^2 \cdot y \\ &= 18x^3y - 4x^2y^2 + 6x^3y^3 - 14x^2y. \end{aligned}$$

When we multiply two polynomials, we treat one polynomial as a single term and make repeated use of the distributive property. This procedure is illustrated in the following examples.

Example 3. Find the product of $x - y$ and $x^2 + xy + y^2$.

Solution: Treat the binomial $x - y$ as a single term $(x - y)$.

$$\begin{aligned} &(x - y)(x^2 + xy + y^2) \\ &= (x - y)x^2 + (x - y)xy + (x - y)y^2 \\ &= x^3 - x^2y + x^2y - xy^2 + xy^2 - y^3. \end{aligned}$$

After we combine like terms in this last expression, we get $x^3 - y^3$ as the desired product.

Example 4. Multiply $a + b$ and $a^2 + 2ab + b^2$.

Solution: Treat the binomial $a + b$ as a single term $(a + b)$.

$$\begin{aligned} &(a + b)(a^2 + 2ab + b^2) \\ &= (a + b)a^2 + (a + b)2ab + (a + b)b^2 \\ &= a^3 + a^2b + 2a^2b + 2ab^2 + ab^2 + b^3 \\ &= a^3 + 3a^2b + 3ab^2 + b^3. \end{aligned}$$

The products in Examples 3 and 4 can be obtained by multiplying each term of the first polynomial by each term of the second polynomial. For example, in Example 4,

$$\begin{aligned} &(a + b)(a^2 + 2ab + b^2) \\ &= a(a^2) + a(2ab) + a(b^2) + b(a^2) + b(2ab) + b(b^2) \\ &= a^3 + 2a^2b + ab^2 + a^2b + 2ab^2 + b^3 \\ &= a^3 + 3a^2b + 3ab^2 + b^3. \end{aligned}$$

The preceding observation motivates the following rule for finding the product of two polynomials.

> To multiply two polynomials, multiply each term of the second polynomial (the one on the right) by each term of the first polynomial and then combine like terms.

Example 5. Use the preceding rule to multiply $2y + 5$ and $y^2 - 3y + 7$.

Solution: First multiply each term of $y^2 - 3y + 7$ by $2y$. The result is $2y^3 - 6y^2 + 14y$. Next multiply each term of $y^2 - 3y + 7$ by 5. The result is $5y^2 - 15y + 35$. Now add and combine like terms to get the product

$$\begin{aligned}&(2y^3 - 6y^2 + 14y) + (5y^2 - 15y + 35)\\ &= 2y^3 - y^2 - y + 35.\end{aligned}$$

Thus far, when we have multiplied two polynomials, the polynomials were written horizontally, one next to the other. For this reason, the method is often called the *horizontal form* of multiplication. Another way of applying the rule that immediately precedes Example 5 for multiplying two polynomials is to write one polynomial below the other. This is the *vertical form* of multiplication and it is similar to the way two large numbers are multiplied. For example, to multiply $3x + 7$ and $4x^2 + 2x + 3$, we write

$$\begin{array}{lrl}
 & 4x^2 + 2x + 3 & \\
\text{(mult)} & \underline{\qquad 3x + 7} & \\
 & 28x^2 + 14x + 21 & 7 \text{ times } 4x^2 + 2x + 3\\
 & \underline{12x^3 + 6x^2 + 9x \qquad} & 3x \text{ times } 4x^2 + 2x + 3\\
 & 12x^3 + 34x^2 + 23x + 21. & \text{addition of like terms}
\end{array}$$

Example 6. Use the vertical form of multiplication to get the product of $5x^2 - 3$ and $11x^3 - 2x + 4$.

Solution

$$\begin{array}{r}
11x^3 - 2x + 4\\
\underline{5x^2 - 3}\\
-33x^3 \qquad + 6x - 12\\
\underline{55x^5 - 10x^3 + 20x^2 \qquad\qquad}\\
55x^5 - 43x^3 + 20x^2 + 6x - 12.
\end{array}$$

To be successful at algebra, it is helpful to be able to quickly, and correctly, compute the product of two polynomials. Although the horizontal form of multiplication is preferred, at this time it may not be the best for you. Select a method and practice it until you have mastered it.

EXERCISES 3.4

For Exercises 1–20, multiply the polynomial by the monomial. See Examples 1 and 2.

1. $x(y + z)$

2. $2y(3y - z)$

3. $rs(2r + 3s)$

4. $7xy(x - 8y)$

5. $xy(y^2 - 3y)$

6. $-7(m - 3n)$

7. $-2x(x^2 - y^2)$

8. $3a^2b(ax - by)$

9. $21mn(7m - n + mn)$

10. $6x(2x^2 - xy - y^2)$

11. $8x^2y^3(6x + 17y)$

12. $-3mn(-2m + 3n - m^2n)$

13. $7xy(3x^2 - 2xy + y^2)$

14. $-6x^2y^3(5x^3y^2 - 3x^2y^3 + 9xy)$

15. $-6m^2n\left(\frac{m}{3} + \frac{n^2}{2} - \frac{mn}{6}\right)$

16. $\frac{2}{3}y\left(\frac{3}{5}y - \frac{2}{3}z\right)$

17. $\frac{x}{5}(20x - 15x^2y - y)$

18. $0.4z(0.25 + 0.08z + 1.3z^2)$

19. $0.2y(12y^2 - 20y + 10)$

20. $\frac{3x}{4}(.08x + 1.2)$

For Exercises 21–40, use the horizontal form of multiplication to find the product of the polynomials. See Examples 3–5.

21. $(x + 3)(x + 2)$

22. $(y + 1)(y + 5)$

23. $(z - 2)(z - 5)$

24. $(a - 1)(a - 7)$

25. $(x + 1)(x + 2)$

26. $(x + 1)(x + 3)$

27. $(y - 2)(y^2 + 2y - 1)$

28. $(y + 2)(y^2 - y + 1)$

29. $(w^2 + 6w)(w^2 - 7w + 1)$

30. $(w^2 - 4w)(w^2 + 2w + 2)$

31. $(p^2 - 3p)(p^2 + 3p - 2)$

32. $(p^2 + 4p)(p^2 + 2p - 1)$

33. $(x + 1)(x^2 + x + 1)$

34. $(y^2 - 1)(y^2 + 2y + 1)$

35. $(w - 1)(w^2 - w + 1)$

36. $(w + 1)(w^2 - 2w + 2)$

37. $(x + 4)(x^3 + x^2 + 3)$

38. $(y^2 + 2)(y^6 + 3y^4 + y^2)$

39. $(y - 2)(y^2 + y - 5)$

40. $(z - 3)(z^4 - z^2 - 5)$

In Exercises 41–50, use the vertical form of multiplication to find the product of the polynomials. See Example 6.

41. $(x + 6)(x^2 + 2x + 1)$

42. $(x - 7)(x^2 + 3x + 5)$

43. $(y + 3)(y^2 + 2y - 5)$

44. $(y - 2)(y^2 + 3y + 4)$

45. $(w - 2)(w^2 - 3w + 2)$

46. $(w + 5)(w^2 - w + 6)$

47. $(x - 8)(2x^3 + x^2 + 5x)$ **48.** $(y - 2)(y^3 + 2y^2 + y)$ **49.** $(x^3 + x^2 + x)(x^2 - x)$
50. $(x^3 - x^2 - x)(x^3 + x^2)$

For Exercises 51–60, find the product and write your answer in standard form.

51. $(x^2 - x + 1)(x^2 + x - 4)$ **52.** $(y^2 + y + 2)(y^2 - y - 1)$
53. $(2a^2 + 2a + 1)(a^2 - a + 1)$ **54.** $(b^3 + 2b - 1)(b^2 - 2b + 5)$
55. $(z^2 + z + 2)(z^2 - z + 1)$ **56.** $(w^2 + 2w + 1)(-w^2 + w - 1)$
57. $(x^2 - x + 3)(x^2 + x - 2)$ **58.** $(2x^2 + x + 2)(x^2 + x - 2)$
59. $(3x^2 + x + 1)(3x^2 - x - 1)$ **60.** $(y^2 + y + 1)(2y^2 - y + 1)$

Multiplication of Two Binomials 3.5

The multiplication of two binomials is a special case of the multiplication of two polynomials. In this section we develop a technique, known as the *FOIL method,* that will help us quickly calculate the product of two binomials. Two special cases that involve the product of two binomials will also be presented.

Consider the product of $(x + 3)$ and $(x + 5)$. Using the horizontal method of multiplication, we get

$$\begin{aligned}(x + 3)(x + 5) &= x^2 + 5x + 3x + 15\\ &= x^2 + 8x + 15.\end{aligned}$$

Notice that the first term, x^2 of the polynomial $x^2 + 5x + 3x + 15$ is the product of the *First* terms of $(x + 3)$ and $(x + 5)$. That is,

$$\overbrace{(x + 3)(x}^{x^2} + 5).$$

Similarly, the term $5x$ is the product of the *Outermost* terms of the two binomials. That is,

$$\overbrace{(x + 3)(x + 5}^{5x}).$$

Also, the term $3x$ is the product of the *Innermost* terms of the two binomials. Finally, the term 15 is the product of the *Last* terms of the two binomials. In summary, we have

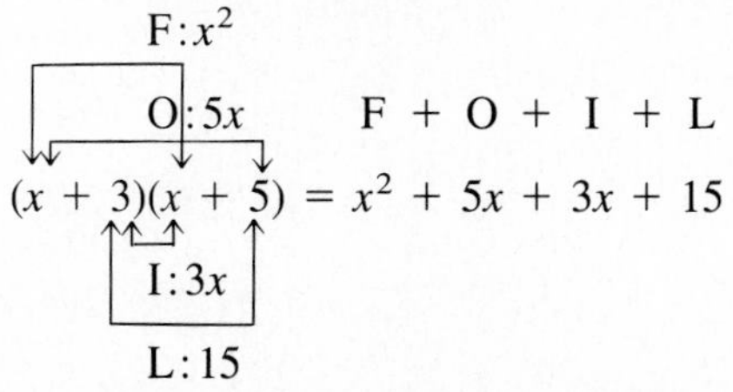

F: x^2

O: $5x$ F + O + I + L

$$(x + 3)(x + 5) = x^2 + 5x + 3x + 15$$

I: $3x$

L: 15

$$= x^2 + 8x + 15.$$

As in this example, it is often the case that the terms associated with O and I are like terms and may be combined. When this is possible, the product of the two binomials is usually a trinomial. When a trinomial is written in standard form, the second term is called the *middle term*. Thus $8x$ is the middle term of $x^2 + 8x + 15$.

Example 1. Compute the product $(x + 2)(x + 7)$.

Solution

$$(x + 2)(x + 7) = \overset{F}{x^2} + \overset{O}{7x} + \overset{I}{2x} + \overset{L}{14}$$

$$= x^2 + 9x + 14.$$

Example 2. Compute the product $(2x - 3)(x + 5)$.

Solution

$$(2x - 3)(x + 5) = \overset{F}{2x^2} + \overset{O}{10x} - \overset{I}{3x} - \overset{L}{15}$$

$$= 2x^2 + 7x - 15.$$

Example 3. Compute the product $(3x - 4)(2x - 1)$.

Solution

$$(3x - 4)(2x - 1) = \overset{F}{6x^2} - \overset{O}{3x} - \overset{I}{8x} + \overset{L}{4}$$

$$= 6x^2 - 11x + 4.$$

Example 4. Compute the product $(3x + 2a)(2y - 5b)$.

Solution

$$\overbrace{(3x + 2a)(2y - 5b)}^{F \quad O}_{I \quad L} = \overset{F}{6xy} - \overset{O}{15bx} + \overset{I}{4ay} - \overset{L}{10ab}.$$

Since there are no like terms to combine, the product is $6xy - 15bx + 4ay - 10ab$.

When the two binomials being multiplied are the sum and difference of the same two terms, their product has a special form. To see this, consider the product $(x + y)(x - y)$.

$$\overbrace{(x + y)(x - y)}^{F \quad O}_{I \quad L} = \overset{F}{x^2} - \overset{O}{xy} + \overset{I}{xy} - \overset{L}{y^2}$$

$$= x^2 - y^2.$$

Notice that the product of $(x + y)$ and $(x - y)$ is the square of x minus the square of y. Binomials of this type, $x^2 - y^2$, are called the difference of two squares. We summarize this observation with the following rule.

Difference of Two Squares

$$(a + b)(a - b) = a^2 - b^2.$$

Example 5. Multiply $(m + 3)$ and $(m - 3)$.

Solution

$$(m + 3)(m - 3) = m^2 - 3^2$$
$$= m^2 - 9.$$

Example 6. Multiply $(2a + 3b)$ and $(2ab - 3b)$.

Solution

$$\begin{aligned}(2a + 3b)(2a - 3b) &= (2a)^2 - (3b)^2 \\ &= 4a^2 - 9b^2.\end{aligned}$$

When the two binomials being multiplied are the same, the product also has a special form. For example,

$$\begin{aligned}(x + y)^2 &= (x + y)(x + y) \\ &= x^2 + xy + yx + y^2 \\ &= x^2 + 2xy + y^2\end{aligned}$$

and

$$\begin{aligned}(x - y)^2 &= (x - y)(x - y) \\ &= x^2 + x(-y) + (-y)x + (-y)^2 \\ &= x^2 - 2xy + y^2\end{aligned}$$

Notice that the square of binomials such as $(x + y)^2$ or $(x - y)^2$ can be obtained by adding the results of the following three steps.

1. Square the first term of the binomial.
2. Take twice the product of the two terms of the binomial.
3. Square the last term of the binomial.

Example 7. Find $(x + 7)^2$.

Solution: Applying the steps outlined above, we get

x^2	+	$2(7x)$	+	7^2.
the square of the first term of the binomial	+	twice the product of the two terms of the binomial	+	the square of the last term of the binomial

Thus $(x + 7)^2 = x^2 + 14x + 49$.

Example 8. Find $(3x - 5)^2$.

Solution

$(3x)^2$	+	$2(3x)(-5)$	+	$(-5)^2$.
the square of the first term of the binomial	+	twice the product of the two terms of the binomial	+	the square of the last term of the binomial

Thus $(3x - 5)^2 = 9x^2 - 30x + 25$.

EXERCISES 3.5

Multiply the following binomials and combine all like terms.

For Exercises 1–42, refer to Examples 1–4.

1. $(x + 3)(x + 2)$ **2.** $(a + 1)(a + 5)$ **3.** $(y + 4)(y - 2)$
4. $(m + 6)(m - 1)$ **5.** $(p - 3)(p + 2)$ **6.** $(z - 7)(z + 5)$
7. $(x + 4)(x + 2)$ **8.** $(y + 3)(y + 5)$ **9.** $(w + 5)(w - 2)$
10. $(z + 4)(z - 1)$ **11.** $(v - 4)(v + 2)$ **12.** $(u - 3)(u + 4)$
13. $(x - 3)(x - 2)$ **14.** $(y - 2)(y - 4)$ **15.** $(w - 2)(w - 1)$
16. $(z - 4)(z - 5)$ **17.** $(v - 3)(v - 6)$ **18.** $(u - 4)(u - 6)$
19. $(x + 7)(x + 3)$ **20.** $(y + 6)(y + 7)$ **21.** $(y - 9)(y - 8)$
22. $(x - 10)(x - 12)$ **23.** $(2a + 3)(a - 5)$ **24.** $(x - 4)(3x + 4)$
25. $(y - 2)(11y - 4)$ **26.** $(3z + 6)(z + 5)$ **27.** $(7x + 2)(5x + 7)$
28. $(3y + 5)(8y + 3)$ **29.** $(5x + 6)(8x - 7)$ **30.** $(9z - 4)(7z - 3)$
31. $(-2x + 5)(x - 3)$ **32.** $(-8y + 7)(-5y + 6)$ **33.** $(3x - 5)(2x + 3)$
34. $(2y + 4)(3y - 4)$ **35.** $(2x - 1)(5x + 1)$ **36.** $(3y - 1)(5y + 1)$
37. $(6x + 2)(3x + 4)$ **38.** $(7y + 3)(2y + 5)$ **39.** $(x + 2y)(x - y)$
40. $(m + n)(2m - n)$ **41.** $(y + \frac{2}{3}z)(y + \frac{1}{3}z)$ **42.** $\left(\frac{a}{5} - 3\right)\left(\frac{a}{3} + 5\right)$

For Exercises 43–66, refer to Examples 5 and 6.

43. $(x + 4)(x - 4)$ **44.** $(y + 1)(y - 1)$ **45.** $(z - 1)(z + 1)$
46. $(x - 5)(x + 5)$ **47.** $(x - 3)(x + 3)$ **48.** $(y + 6)(y - 6)$
49. $(z + 10)(z - 10)$ **50.** $(w - 9)(w + 9)$ **51.** $(u - 8)(u + 8)$
52. $(v - 12)(v + 12)$ **53.** $(2x + 3)(2x - 3)$ **54.** $(5a - 2)(5a + 2)$
55. $(1 + 8m)(1 - 8m)$ **56.** $(7 - 2m)(7 + 2m)$ **57.** $(7x + 3y)(7x - 3y)$
58. $(11y - 10z)(11y + 10z)$ **59.** $(x + \frac{1}{2})(x - \frac{1}{2})$ **60.** $(y - \frac{1}{3})(y + \frac{1}{3})$
61. $(\frac{1}{3}x + \frac{2}{5})(\frac{1}{3}x - \frac{2}{5})$ **62.** $(\frac{2}{7}y - \frac{1}{4})(\frac{2}{7}y + \frac{1}{4})$ **63.** $(3mn - 7)(3mn + 7)$
64. $(ab + x)(ab - x)$ **65.** $(.2 + x)(.2 - x)$ **66.** $(.4y - .3)(.4y + .3)$

For Exercises 67–90, refer to Examples 7 and 8.

67. $(x + 4)^2$ **68.** $(x + 5)^2$ **69.** $(y + 7)^2$ **70.** $(y + 10)^2$
71. $(w + 12)^2$ **72.** $(w + 9)^2$ **73.** $(y - 2)^2$ **74.** $(y - 5)^2$
75. $(x - 8)^2$ **76.** $(y - 9)^2$ **77.** $(w - 6)^2$ **78.** $(w - 4)^2$

79. $(x - 10)^2$ **80.** $(x - 12)^2$ **81.** $(2a + 3)^2$ **82.** $(3x + 2)^2$

83. $(2y - 5)^2$ **84.** $(3w - 4)^2$ **85.** $(\frac{1}{2}x + 1)^2$ **86.** $(\frac{1}{2}y + 3)^2$

87. $(\frac{1}{3}y - 1)^2$ **88.** $(\frac{1}{2}w - 3)^2$ **89.** $(.5a + .2)^2$ **90.** $(.3a - .7)^2$

3.6 Dividing a Polynomial by a Monomial or a Binomial

In previous mathematics courses we learned how to divide one number by another number: for example, $24 \div 6 = 4$ and $84 \div 4 = 21$. We are also familiar with the idea of a remainder. Thus, when 25 is divided by 7, the result is 3 plus a remainder of 4. In this example, 25 is the dividend, 7 is the divisor, 3 is the quotient, and 4 is the remainder. To check our division, we multiply the quotient by the divisor and then add the remainder. If our division was correct, the result of the check will be the dividend. Since

$$7 \cdot 3 + 4 = 21 + 4 = 25,$$

our answer is correct. Recall that the quotient is the largest multiple of the divisor that is less than or equal to the dividend.

When a division problem involves larger numbers, we use long division. For example, to divide 739 by 9 we proceed as follows:

$$\begin{array}{r} 8 \\ 9\overline{)739} \\ \underline{72} \end{array}$$

9 divides 73 eight times because $9 \cdot 8 = 72$ (actually 9 divides 739 eighty times because $9 \cdot 80 = 720$)

$$\begin{array}{r} 8(0) \\ 9\overline{)739} \\ \underline{72(0)} \\ 19 \end{array}$$

subtract 72 from 73 and bring down the next digit in the divisor (actually, we are subtracting 720 from 739)

$$\begin{array}{r} 82 \\ 9\overline{)739} \\ \underline{72} \\ 19 \\ \underline{18} \end{array}$$

9 divides 19 two times because $9 \cdot 2 = 18$

$$\begin{array}{r} 82 \\ 9\overline{)739} \\ \underline{72} \\ 19 \\ \underline{18} \\ 1 \end{array}$$

subtract 18 from 19; Since the result of this subtraction is less than the divisor, this is the remainder

Thus, when 739 is divided by 9, the result is 82 plus a remainder of 1.

CHECK: $9 \cdot 82 + 1 = 738 + 1 = 739$.

In each of these division problems we worked only with integers. We now extend this technique to divide a polynomial by a monomial. The method will only use polynomials. For example, to divide $32x^4 + 24x^2$ by $8x$, we proceed as follows:

$$\begin{array}{r l} & 4x^3 \\ 8x \big) & 32x^4 + 24x^2 \\ & \underline{32x^4 \qquad\quad} \end{array}$$

$8x$ divides $32x^4$ exactly $4x^3$ times, because $(8x)(4x^3) = 32x^4$

$$\begin{array}{r l} & 4x^3 \\ 8x \big) & 32x^4 + 24x^2 \\ & \underline{32x^4 \qquad\quad} \\ & \qquad\quad 24x^2 \end{array}$$

subtract $32x^4$ from $32x^4$ and bring down the next term

$$\begin{array}{r l} & 4x^3 + \ \ 3x \\ 8x \big) & 32x^4 + 24x^2 \\ & \underline{32x^4 \qquad\quad} \\ & \qquad\quad 24x^2 \\ & \qquad\quad \underline{24x^2} \end{array}$$

$8x$ divides $24x^2$ exactly $3x$ times, because $(8x)(3x) = 24x^2$

since the result of the subtraction is 0, there is no remainder

Thus, when $(32x^4 + 24x^3)$ is divided by $8x$, the result is $4x^3 + 3x$.

CHECK: $8x(4x^3 + 3x) = 32x^4 + 24x^2$.

Since division by zero is not defined, the preceding problem is meaningless when $8x = 0$ or when $x = 0$. For the rest of this chapter we will assume that the divisor is not zero.

When dividing a polynomial by a monomial, we continue the process until the result of the subtraction is a polynomial *with degree less than the degree of the divisor*. The next example involves division with a remainder.

Example 1. Divide $10x^3 - 5x^2 + 15x$ by $5x^2$.

Solution

$$\begin{array}{r l} & 2x - 1 \\ 5x^2 \big) & 10x^3 - 5x^2 + 15x \\ & \underline{10x^3 \qquad\qquad} \\ & \qquad -5x^2 \\ & \qquad \underline{-5x^2 \qquad\quad} \\ & \qquad\qquad\qquad 15x \end{array}$$

The result of the last subtraction is a polynomial ($15x$) of lesser degree than the divisor ($5x^2$). Thus, using polynomials, we cannot divide $15x$ by $5x^2$. We conclude that when $(10x^3 - 5x^2 + 15x)$ is divided by $5x^2$, the result is $2x - 1$ plus a remainder of $15x$.

CHECK: $5x^2(2x - 1) + 15x = 10x^3 - 5x^2 + 15x$.

When using long division to divide a polynomial by a monomial, it is important to write the polynomial in standard form.

Example 2. Divide $-48y^5 + 30y^3 + 78y^8$ by $6y^3$.

Solution: First write the polynomial in standard form as $78y^8 - 48y^5 + 30y^3$. Then use long division.

$$\begin{array}{r|l} & 13y^5 - 8y^2 + 5 \\ \hline 6y^3 & 78y^8 - 48y^5 + 30y^3 \\ & \underline{78y^8} \\ & \qquad -48y^5 \\ & \qquad \underline{-48y^5} \\ & \qquad\qquad 30y^3 \\ & \qquad\qquad \underline{30y^3} \end{array}$$

CHECK: $6y^3(13y^5 - 8y^2 + 5) = 78y^8 - 48y^5 + 30y^3$.

When a polynomial is divided by a monomial and there is no remainder, we say that the monomial divides the polynomial. Thus $6y^3$ divides $78y^8 - 48y^5 + 30y^3$.

To divide a polynomial by a binomial, we simply extend the preceding ideas. Thus, to divide $8x^3 + 16x^2 + 8x + 1$ by $2x + 1$, we proceed as follows.

$$\begin{array}{r|l} & 4x^2 \\ \hline 2x + 1 & 8x^3 + 16x^2 + 8x + 1 \\ & \underline{8x^3 + \ \ 4x^2} \end{array}$$

divide $8x^3$ by $2x$ to obtain $4x^2$. Next multiply $(2x + 1)$ by $4x^2$ to obtain $8x^3 + 4x^2$

$$\begin{array}{r|l} & 4x^2 \\ \hline 2x + 1 & 8x^3 + 16x^2 + 8x + 1 \\ & \underline{8x^3 + \ \ 4x^2} \\ & \qquad\ 12x^2 + 8x \end{array}$$

subtract $8x^3 + 4x^2$ from $8x^3 + 16x^2$ and bring down the next term

$$\begin{array}{r} 4x^2 + 6x \\ 2x + 1 \overline{)8x^3 + 16x^2 + 8x + 1} \\ \underline{8x^3 + 4x^2} \\ 12x^2 + 8x \\ \underline{12x^2 + 6x} \\ 2x + 1 \end{array}$$

$2x$ divides $12x^2$ exactly $6x$ times; $6x$ times $(2x + 1)$ is $12x^2 + x$; subtract and bring down the next term

$$\begin{array}{r} 4x^2 + 6x + 1 \\ 2x + 1 \overline{)8x^3 + 16x^2 + 8x + 1} \\ \underline{8x^3 + 4x^2} \\ 12x^2 + 8x \\ \underline{12x^2 + 6x} \\ 2x + 1 \\ \underline{2x + 1} \end{array}$$

$2x$ divides $2x$ once; 1 times $(2x + 1)$ is $2x + 1$; note that there is no remainder

The following example involves division when there is a remainder.

Example 3. Divide $14x^3 + 9x^2 + 2x + 3$ by $7x^2 + x$.

Solution

$$\begin{array}{r} 2x + 1 \\ 7x^2 + x \overline{)14x^3 + 9x^2 + 2x + 3} \\ \underline{14x^3 + 2x^2} \\ 7x^2 + 2x \\ \underline{7x^2 + x } \\ x + 3 \end{array}$$

Since the result of the last subtraction, $x + 3$, is a polynomial with degree less than the degree of the divisor, $7x^2 + x$, the remainder is $x + 3$.

CHECK:

$$\begin{aligned} &(2x + 1)(7x^2 + x) + (x + 3) \\ &= 14x^3 + 2x^2 + 7x^2 + x + x + 3 \\ &= 14x^3 + 9x^2 + 2x + 3. \end{aligned}$$

When using long division to divide a polynomial by a binomial, we write both the divisor and the dividend in standard form. If the dividend does not include a power of the variable that is less than the degree of the leading term, it is helpful to write the missing power with a coefficient of zero for the division process. The next example illustrates these ideas.

Example 4. Perform the division:

$$(-41x + 7 + 18x^3) \div (3x + 5).$$

Solution: First rewrite the dividend as $18x^3 + 0x^2 - 41x + 7$ and then proceed as before.

$$\begin{array}{r}
6x^2 - 10x + 3 \\
3x + 5\overline{)18x^3 + 0x^2 - 41x + 7} \\
\underline{18x^3 + 30x^2} \\
-30x^2 - 41x \\
\underline{-30x^2 - 50x} \\
9x + 7 \\
\underline{9x + 15} \\
-8
\end{array}$$

Thus the quotient is $6x^2 - 10x + 3$ and the remainder is -8.

EXERCISES 3.6

Perform the following divisions and check your answers.

For Exercises 1–22, see Example 1.

1. $(18x^2 + 6x) \div 3x$

2. $(25y^2 + 10y) \div 5y$

3. $(40w^2 + 16w) \div 8w$

4. $(45z^2 + 30z) \div 15z$

5. $(14x^3 + 7x^2) \div 7x$

6. $(18y^3 + 9y^2) \div 3y$

7. $(49x^3 + 35x^2) \div 7x$

8. $(50w^3 + 20w^2) \div 10w$

9. $(100c^5 + 50c^3) \div 25c^2$

10. $(60a^5 + 50a^3) \div 10a^2$

11. $(20x^5 + 10x^3) \div 5x^2$

12. $(45a^4 + 30a^3) \div 15a^2$

13. $(15y^4 - 10y^3) \div 5y^2$

14. $(36b^4 - 48b^2) \div 12b^2$

15. $(8y^6 - 4y^4) \div 2y^5$

16. $(24b^3 - 18b^2) \div 6b^3$

17. $(12x^3 + 8x^2 + 4x) \div 4x$

18. $(15c^4 + 10c^3 + 5c^2) \div 5c$

19. $(39y^4 + 26y^3 - 13y^2) \div 13y^2$

20. $(60b^5 + 48b^3 - 24b^2) \div 12b^2$

21. $(100z^5 - 50z^3 + 25z^2) \div 25z^3$

22. $(80a^4 - 40a^3 + 20a^2) \div 20a^3$

For Exercises 23–50, see Examples 1–4.

23. $(100x^3 + 50x^2 + 5x) \div 10x^2$

24. $(30b^3 + 15b^2 + 5b) \div 15b^2$

25. $(50y^3 + 30y^2 + 15y) \div 10y^2$

26. $(48w^3 + 24w^2 + 18w) \div 12w^2$

27. $(72y^4 - 36y^3 + 24y) \div 12y^2$

28. $(56c^5 + 28c^4 - 14c) \div 7c^2$

29. $(144z^5 - 96z^4 + 72z^3 + 48z^2) \div 12z^3$

30. $(500b^6 + 350b^5 - 200b^4 + 150b^2) \div 50b^3$

31. $(12x^5 + 6x^4 - 6x + 18) \div 3x^2$

32. $(75y^9 - 100y^5 + 25y^2 + 50y) \div 25y^3$

33. $(24a^7 + 16a^3 + 12a^{10} - 4a^5) \div 4a^6$

34. $(-14a + 7 + 21a^4 - 42a^3) \div 7a^2$

35. $(6x^2 + 25x + 24) \div (2x + 3)$

36. $(8x^2 + 14x + 5) \div (4x + 5)$

37. $(6x^2 + 19x + 10) \div (3x + 2)$

38. $(12x^2 + 15x + 6) \div (4x + 3)$

39. $(8x^2 + 8x - 6) \div (2x - 1)$

40. $(15x^2 - x - 6) \div (3x - 2)$

41. $(8x^2 - 2x - 3) \div (4x - 3)$

42. $(15x^2 + 4x - 4) \div (5x - 2)$

43. $(6x^3 + 7x^2 + 5x + 2) \div (3x + 2)$

44. $(8x^3 + 16x^2 + 8x + 3) \div (2x + 3)$

45. $(6x^3 - 17x^2 + 18x - 6) \div (3x - 4)$

46. $(10x^3 - x^2 - 17x - 4) \div (5x + 2)$

47. $(3x^4 + 13x^2 + 5) \div (3x^2 + 1)$

48. $(2x^3 - 2x^2 + 3x + 6) \div (x^2 - 1)$

49. $(x - 1 + x^3) \div (x - 1)$

50. $(-2x^2 + 4x^3 + 5) \div (2x + 1)$

Key Words

Base
Binomial
Coefficient
Cubic
Degree
Difference of two squares
Exponent
FOIL method
Horizontal form
Leading coefficient
Like terms
Middle term
Monomial
Polynomial
Power
Quadratic
Quotient
Remainder
Standard form
Trinomial
Vertical form

CHAPTER 3 TEST

For Problems 1–6, identify the base, the exponent, and then evaluate or rewrite the expression by performing the indicated multiplication using the rules of exponents.

1. 5^3 2. 3^4 3. -2^5 4. -3^2 5. $(4xy)^3$ 6. $(3ab)^2$

For Problems 7–21, evaluate the expression.

7. $(3 \cdot 2)^3$

8. 11^0

9. 0^{11}

10. $(-2)^2$

11. -2^2

12. $\left(\frac{1}{4}\right)^{-2}$

13. $3^{-2} + 2^{-3}$

14. $(7 \cdot 2)^3$

15. $\frac{17^4}{17^3}$

16. $\left(\frac{2}{5}\right)^{-1}$

17. 3^{-2}

18. $(2^3)^{-1}$

19. $5^4 \cdot 5^{-3} \cdot 5^2$

20. $\frac{9^3}{9^4}$

21. $(-2)^3(2)^2(-2)^{-2}$

For Problems 22–31, perform the operation indicated and write your answer in standard form. Identify the degree of your answer.

22. $(x^3 - 3x^2 + x - 2) + (2x^5 + 3x^2 + 2x + 1)$

23. $(y^4 + 2y^2 - y + 7) - (y^4 + y^3 + y^2 + 6)$

24. $(a^2 + 2a + 1)(3a)$

25. $(b^3 - b^2 + 3)(-2b^2)$

26. $(a + 7)^2$

27. $(x - 4)^2$

28. $(y + 3)(y - 3)$

29. $(y - 11)(y + 11)$

30. $(x + 5)(x + 6)$

31. $(3w - 2)(w + 5)$

For Problems 32–34, let $p(x) = x^3 - 3x^2 + x + 5$ *and evaluate.*

32. $p(1)$ **33.** $p(-1)$ **34.** $p(3)$

For Problems 35–40, perform the division indicated.

35. $(121x^3 + 77x^2 - 33) \div 11x$

36. $(100y^2 + 50y + 25) \div 25y$

37. $(a^3 - a^5 + a + a^2) \div a$

38. $(25b - 10b^2 + 5b^3 + 10) \div 5b$

39. $(3x^3 + x^2 + x - 2) \div (3x - 2)$

40. $(27x^3 - 15x^2 - 7x + 4) \div (9x - 2)$

CHAPTER 4

Factoring

Factoring proves useful in solving certain types of equations that arise in a practical situation. For example, a paving contractor has to pave a 1500-square-foot area so that the length is 20 feet more than the width. Factoring the equation $w^2 + 20w - 1500 = 0$ provides the solution to his problem.

Source: *Asphalt Institute*

CHAPTER 4
CHAPTER 4
CHAPTER 4
CHAPTER 4
CHAPTER 4
CHAPTER 4
CHAPTER 4
CHAPTER 4
CHAPTER 4
CHAPTER 4
CHAPTER 4
CHAPTER 4
CHAPTER 4
CHAPTER 4
CHAPTER 4
CHAPTER 4
CHAPTER 4
CHAPTER 4
CHAPTER 4
CHAPTER 4
CHAPTER 4
CHAPTER 4
CHAPTER 4
CHAPTER 4
CHAPTER 4
CHAPTER 4
CHAPTER 4
CHAPTER 4
CHAPTER 4
CHAPTER 4
CHAPTER 4
CHAPTER 4
CHAPTER 4

In order to understand the process of factoring polynomials, we will first factor integers and learn how to remove the greatest common factor of a set of monomials. We then examine some special products and identify patterns that are useful in factoring these types of products. Next, we show that factoring can be a useful tool for solving equations. Finally, we learn how to deal with applied (word) problems whose solution involves solving equations by factoring.

4.1 Factoring Integers

When the numbers 5 and 7 are multiplied, they form the product 35. We call 5 and 7 factors of 35. In general, we say that an integer n is a *factor* of an integer m if m can be written as a product $n \cdot q$, where q is also an integer. Thus we have $m = n \cdot q$. The integer q is also called a factor of m. For example, the number 3 is a factor of 51 because $51 = 3 \cdot 17$.

Example 1. Find all the positive integer factors of 12.

Solution: Since $12 = 1 \cdot 12$, $12 = 2 \cdot 6$, and $12 = 3 \cdot 4$, we see that 1, 2, 3, 4, 6, and 12 are the positive integer factors of 12.

Example 2. Find all the integer factors of 9.

Solution: Since $9 = 1 \cdot 9$, $9 = (-1)(-9)$, $9 = 3 \cdot 3$, and so on, the integer factors of 9 are 1, -1, 3, -3, 9, and -9. A convenient way of listing these factors is ± 1, ± 3, ± 9, where, for example, ± 3 is read as "plus or minus three."

> A positive integer greater than 1 that has only itself and 1 as positive integer factors is called a *prime number*.

Example 3. Find all the positive integer factors of 17.

Solution: The only positive integer factors of 17 are 1 and 17. Thus we see that 17 is a prime number.

Example 3 establishes 17 as a prime number. The first eight prime numbers are 2, 3, 5, 7, 11, 13, 17, and 19.

Positive integers, other than 1, that are not prime, are called *composite*. Some composite numbers are 4, 6, 15, 91, 1000, 2132, and 5811.

When a positive integer is written as the product of powers of primes, it is said to be *factored completely*. When an integer is factored completely, we say that it cannot be factored any further. Note that an integer has only one set of prime factors.

Example 4. Factor 30 completely.

Solution

$$\begin{aligned} 30 &= 6 \cdot 5 \\ &= \overbrace{2 \cdot 3} \cdot 5. \end{aligned}$$

Since 2, 3, and 5 are all prime numbers, they cannot be factored any further. Therefore, 30 has been factored completely as the product of 2, 3, and 5.

Example 5. Factor 300 completely.

Solution

$$\begin{aligned} 300 &= 10 \cdot 30 \\ &= \overbrace{2 \cdot 5} \cdot \overbrace{5 \cdot 6} \\ &= 2 \cdot 5 \cdot 5 \cdot \overbrace{2 \cdot 3} \\ &= 2^2 \cdot 3 \cdot 5^2. \end{aligned}$$

The *greatest common factor (GCF)* of a set of integers is the largest integer that is a factor of each of the integers. The next example demonstrates that when each integer of a set is factored completely, it is easy to find the GCF of those integers.

Example 6. Find the GCF of 30 and 42.

Solution: First we factor 30 and 42 completely into prime factors as follows:

$$\begin{aligned} 30 &= 5 \cdot 6 & \qquad 42 &= 2 \cdot 21 \\ &= 5 \cdot 2 \cdot 3 & &= 2 \cdot 3 \cdot 7 \\ &= (2 \cdot 3) \cdot 5. & &= (2 \cdot 3) \cdot 7. \end{aligned}$$

Since $2 \cdot 3 = 6$ is the largest integer that is a factor of 30 and 42, we conclude that 6 is the GCF of 30 and 42.

To obtain the GCF of a set of larger integers, we proceed as follows:

1. Factor each integer completely.
2. Select the different prime factors that are *common to all* the given integers.
3. Write the highest power of each prime factor that is common to each of the given integers.
4. The product of these prime powers is the GCF of the numbers.

We illustrate the above procedure for finding the GCF of a set of numbers by applying it to

$$60, 72, \text{ and } 84.$$

That is, we find the largest positive integer that is a factor of all three numbers.
First, we factor the numbers completely.

$$\begin{aligned} 60 &= 2 \cdot 30 = 2 \cdot 5 \cdot 6 = 2 \cdot 5 \cdot 2 \cdot 3 = 2^2 \cdot 3 \cdot 5 \\ 72 &= 2 \cdot 36 = 2 \cdot 2 \cdot 18 = 2 \cdot 2 \cdot 2 \cdot 3 \cdot 3 = 2^3 \cdot 3^2 \\ 84 &= 2 \cdot 42 = 2 \cdot 2 \cdot 21 = 2 \cdot 2 \cdot 3 \cdot 7 = 2^2 \cdot 3 \cdot 7. \end{aligned}$$

The prime factors that are common to the numbers are 2 and 3. Next, we determine the largest power of 2 that is a factor of each of the given numbers. Notice that $2^2 = 4$ is the largest power of 2 that is a factor of each of the given integers. Continuing with this type of reasoning, we take each prime factor raised to the largest power that is a factor of each of the three numbers. The product of those prime powers is the GCF of the original numbers. Thus the GCF of 60, 72, and 84 is $2^2 \cdot 3 = 4 \cdot 3 = 12$. Since 12 is the GCF of the three numbers, they may be factored as

$$\begin{aligned} 60 &= 12 \cdot 5 \\ 72 &= 12 \cdot 6 \\ 84 &= 12 \cdot 7. \end{aligned}$$

EXERCISES 4.1

For Exercises 1–10, find all the positive integer factors of the number. See Example 1.

1. 8 **2.** 10 **3.** 43 **4.** 37 **5.** 35

6. 77 **7.** 28 **8.** 63 **9.** 30 **10.** 105

For Exercises 11–20, determine whether or not the number is a prime. See Example 3.

11. 23 **12.** 4 **13.** 26 **14.** 29 **15.** 27

16. 17 **17.** 13 **18.** 39 **19.** 37 **20.** 42

For Exercises 21–30, factor the number completely. See Examples 4 and 5.

21. 18 **22.** 44 **23.** 50 **24.** 28 **25.** 76

26. 144 **27.** 200 **28.** 216 **29.** 300 **30.** 325

For Exercises 31–40, find the greatest common factor of the numbers. See Example 6.

31. 40, 70, 100 **32.** 36, 56, 60 **33.** 7, 13, 19 **34.** 11, 13, 16

35. 7, 287, 1036 **36.** 12, 48, 60 **37.** 45, 60, 75 **38.** 30, 54, 84

39. 21, 210, 2100 **40.** 11, 110, 1100

Common Monomial Factors 4.2

As with numbers, when the algebraic expressions x and $2y + z$ are multiplied, they form the product $x(2y + z) = 2xy + xz$. We call x and $(2y + z)$ *factors* of $2xy + xz$. The process of expressing a number or a polynomial in terms of its factors is called *factoring*. Note that factoring is the reverse of multiplying. In this section we extend the ideas in Section 4.1 to finding the greatest common factor of a set of monomials.

> The greatest common factor (GCF) of a set of monomials is a factor of each of the monomials. It is the monomial of highest degree and with the largest numerical coefficient that is a factor of each of the monomials.

The easiest case occurs when the monomials are all different powers of the same variable.

Example 1. Find the GCF of x^2, x^5, and x^9.

Solution: We select the largest power of the variable that is a factor of *all* three monomials. Therefore, the GCF of x^2, x^5, and x^9 is x^2.

Example 2. Find the GCF of x^2y^6, x^5y^5, and x^9y^3.

Solution: The largest power of x that is a common factor of all the monomials is x^2. The largest power of y that is a common factor of all the monomials is y^3. Therefore, the GCF is their product, x^2y^3.

Example 3. Find the GCF of $18x^2y^6$, $24x^3y^4$, and $180x^7y^3$.

Solution: First factor each numerical coefficient completely. Next group together the largest power of each factor that is common to all three monomials:

$$
\begin{aligned}
18x^2y^6 &= 2 \cdot 3^2 \cdot x^2 \cdot y^6 = (2 \cdot 3 \cdot x^2 \cdot y^3)(3 \cdot y^3) \\
24x^3y^4 &= 2^3 \cdot 3x^3 \cdot y^4 = (2 \cdot 3 \cdot x^2 \cdot y^3)(2^2 \cdot x \cdot y) \\
180x^7y^3 &= 2^2 \cdot 3^2 \cdot 5x^7 \cdot y^3 = (2 \cdot 3 \cdot x^2 \cdot y^3)(2 \cdot 3 \cdot 5 \cdot x^5).
\end{aligned}
$$

The GCF is the product $2 \cdot 3 \cdot x^2 \cdot y^3 = 6x^2y^3$. Note that the remaining factors contain no common factor.

We now show how finding the greatest common factor of a set of monomials can be used to find the greatest common factor of a polynomial. This involves factoring *each term* of the polynomial and identifying the greatest common factor of the terms. We illustrate the technique with the polynomial

$$9x^2y^2 + 18x^3y.$$

First we completely factor the numerical coefficients of each term and then group together the GCF of the terms:

$$\begin{aligned} 9x^2y^2 + 18x^3y &= 3^2x^2y^2 + 2 \cdot 3^2x^3y \\ &= (3^2x^2y)(y) + (3^2x^2y)(2x). \end{aligned}$$

Thus the greatest common monomial factor of these two terms is $3^2x^2y = 9x^2y$.

Finally, using the distributive property, note that we may rewrite the polynomial as

$$\begin{aligned} 9x^2y^2 + 18x^3y &= (3^2x^2y)(y + 2x) \\ &= (9x^2y)(y + 2x). \end{aligned}$$

We say that $9x^2y(y + 2x)$ is the factored form of $9x^2y^2 + 18x^3y$ and that we have *factored out* $9x^2y$ from $9x^2y^2 + 18x^3y$. We also say that we have *removed* $9x^2y$ from $9x^2y^2 + 18x^3y$.

Example 4. Remove the GCF from each expression.
(a) $ax + a^2y$. (b) $4x^3y + 10x^2y^2$.

Solution

(a) $ax + a^2y = ax + a \cdot ay = a(x + ay)$.

(b)
$$\begin{aligned} 4x^3y + 10x^2y^2 &= 2^2x^3y + 2 \cdot 5x^2y^2 \\ &= (2x^2y) \cdot 2x + (2x^2y) \cdot 5y \qquad \text{GCF} = 2x^2y \\ &= 2x^2y(2x + 5y). \end{aligned}$$

A close examination of Example 4(a) reveals the importance of the distributive law relative to multiplication and factoring. For instance, if the expression

$$ax + a^2y = a(x + ay)$$

is read from left to right, we are factoring. If it is read from right to left we are multiplying.

Example 5. Remove the GCF from $8a^4b^2 - 12a^3b^3 + 18a^2b^4$.

Solution: First we completely factor the numerical coefficients of each term. Thus

$$\begin{aligned} &8a^4b^2 - 12a^3b^3 + 18a^2b^4 \\ &= 2^3a^4b^2 - 2^2 \cdot 3a^3b^3 + 2 \cdot 3^2a^2b^4. \end{aligned}$$

Next observe that the greatest common monomial factor of each term is $2a^2b^2$. Thus we may write the above as

$$(2a^2b^2)2^2a^2 - (2a^2b^2)2 \cdot 3ab + (2a^2b^2)3^2b^2.$$

Using the distributive property, we factor out $2a^2b^2$ and get

$$\begin{aligned} &= 2a^2b^2(2^2a^2 - 2 \cdot 3ab + 3^2b^2) \\ &= 2a^2b^2(4a^2 - 6ab + 9b^2). \end{aligned}$$

EXERCISES 4.2

For Exercises 1–10 find the GCF of the expressions. See Examples 1–3.

1. x^4, x^7, x^{11} **2.** y^5, y^3, y^9

3. x^2y^3, xy^2, x^2y **4.** a^4b^3, a^2b^2, ab^2

5. $15a^4, 9a^3, 12a^5$ **6.** $16x^3, 20x^2, 44x$

7. $12a^2b^3, 18ab^2, 24a^3b$ **8.** $20x^3y^2, 35x^2y^3, 15x^4y^5$

9. $54a^3b^2c^4, 72a^2b^3c^3, 180a^4b^2c^2$ **10.** $36x^2y^5z^3, 56xy^3z^2, 60x^3yz^3$

In Exercises 11–40, remove the GCF from the polynomial. See Examples 4 and 5.

11. $7x + 14$ **12.** $10y + 30$

13. $4x - 8y$ **14.** $3q - 6p$

15. $xy + xz$ **16.** $pq - pt$

17. $6ab - 20ac$ **18.** $3t^2 - 27tq$

19. $10x^2 + 15xy$ **20.** $7a^2b + 56ab^2$

21. $18w^2 - 24wy$ **22.** $30p^2 - 45pq$

23. $3m^3np - 3m^2n^2$ **24.** $2x^3 + 2xy^2$

25. $3a^2 - 6ab - 6ab^2$ **26.** $r^2st + 5rs^2t - 2rst^2$

27. $12x^3 - 6x^2y - 12xy^2$ **28.** $8c^4 - 20c^3 - 44c^2$

29. $4x^3y^2 + 8xy - 4xy^3$ **30.** $3p^3q - 15pq + 3pq^3$

31. $30x^5 - 5x^3 + 15x^8$ **32.** $14y^3 + 21y^2 + 28y^4$

33. $xy^5 - 3xy^3 - 6xy^2$ **34.** $y^4z + 4y^3z^2 - 7y^2z$

35. $30s^2t^5 + 60s^3t^4 - 150s^4t^3$ **36.** $21a^3b - 14a^2b^2 + 7ab^3$

37. $p^2qr^2 + pq^2r^2 + p^2q^2r$ **38.** $x^2y^2z - x^2yz^2 - xy^2z^2$

39. $12m^3n - 15m^2n^3 - 6m^2n^4$ **40.** $77a^2b - 22ab + 99a^2b^2$

Factoring the Difference of Two Squares 4.3

In Sections 4.1 and 4.2, we were concerned with identifying greatest common integer factors and greatest common monomial factors. In this section we advance our study of factoring polynomials by considering a special type of binomial, the difference of two squares. In Section 3.5 we saw that the difference of two squares was the product of the sum and the difference of two monomials. For example,

$$(x + 3)(x - 3) = x^2 - 9.$$

In Chapter 3 we began with the indicated product on the left and obtained the expression on the right. In this section we begin with the expression on the right and obtain the indicated product on the left. We say that $(x + 3)(x - 3)$ is the factored form of $x^2 - 9$. This example illustrates that the difference of two squares may be factored into the product of the sum and difference of the terms being squared. More generally, the difference between the square of x and the square of y, $x^2 - y^2$, will factor into the product of the sum $x + y$ and the difference $x - y$. That is,

$$x^2 - y^2 = (x + y)(x - y).$$

Example 1. Factor $x^2 - 4$.

Solution: Note that $x^2 - 4$ can be written as the difference of the square of x and the square of 2. That is, $x^2 - 4 = (x)^2 - (2)^2$. If we let $y = 2$ in the above formula, we see that $(x)^2 - (2)^2$ factors into $(x + 2)(x - 2)$; that is,

$$x^2 - 4 = (x)^2 - (2)^2 = (x + 2)(x - 2).$$

Example 2. Factor $9a^2 - 49$.

Solution: We see that $9a^2$ is the square of $3a$, $9a^2 = (3a)^2$, and that 49 is the square of 7, $49 = 7^2$. Thus $9a^2 - 49$ can be written as the difference of two squares as follows:

$$9a^2 - 49 = (3a)^2 - (7)^2.$$

As a result, we write $9a^2 - 49$ in factored form as

$$9a^2 - 49 = (3a)^2 - (7)^2 = (3a + 7)(3a - 7).$$

Recognizing an expression as the difference of two squares takes practice. Read the next four examples carefully before trying the exercises.

Example 3. Factor $9r^2 - s^2t^2$.

Solution: Note that $9r^2 - s^2t^2$ is the difference of the square of $3r$ and the square of st. Therefore,

$$\begin{aligned} 9r^2 - s^2t^2 &= (3r)^2 - (st)^2 \\ &= (3r + st)(3r - st). \end{aligned}$$

The next example illustrates that *the first step in factoring an expression is to remove the greatest common factor* (GCF) from the expression.

Example 4. Factor $y^2b - 25b$.

Solution: First remove the GCF of b. Thus $y^2b - 25b = b(y^2 - 25)$. Note that $y^2 - 25$ is the difference of the square of y and the square of 5. Therefore, by our formula for factoring the difference of two squares,

$$y^2 - 25 = (y)^2 - (5)^2 = (y + 5)(y - 5).$$

Thus

$$y^2b - 25b = b(y^2 - 25) = b(y - 5)(y + 5).$$

Example 5. Factor $3m^2 - 108n^2$

Solution: We begin by removing the greatest common factor of 3 from each term and getting $3(m^2 - 36n^2)$. The factor $m^2 - 36n^2$ is the difference of the square of m and the square of $6n$. Therefore,

$$\begin{aligned} 3m^2 - 108n^2 &= 3(m^2 - 36n^2) \\ &= 3[(m^2 - (6n)^2] \\ &= 3[(m + 6n)(m - 6n)] \\ &= 3(m + 6n)(m - 6n). \end{aligned}$$

Example 6. Factor $x^4 - 16y^4$.

Solution: Since $x^4 - 16y^4 = (x^2)^2 - (4y^2)^2$ it is the difference of the square of x^2 and the square of $4y^2$. Therefore,

$$x^4 - 16y^4 = (x)^2 - (4y^2)^2 = (x^2 + 4y^2)(x^2 - 4y^2).$$

However, the second factor, $x^2 - 4y^2$, is also the difference of two squares, the square of x and the square of $2y$. Therefore,

$$\begin{aligned} x^4 - 16y^4 &= (x^2 + 4y^2)(x^2 - 4y^2) \\ &= (x^2 + 4y^2)[(x)^2 - (2y)^2] \\ &= (x^2 + 4y^2)(x + 2y)(x - 2y). \end{aligned}$$

EXERCISES 4.3

For Exercises 1–20, factor the expression. See Examples 1–3.

1. $x^2 - 16$ **2.** $a^2 - 1$ **3.** $y^2 - 81$ **4.** $m^2 - 121$

5. $25 - b^2$ **6.** $9x^2 - 1$ **7.** $36y^2 - 49$ **8.** $4m^2 - 121$

9. $z^2 - 25$ **10.** $9y^2 - 25$ **11.** $x^2 - 64$ **12.** $m^2n^2 - 100$

13. $25y^2 - 144$ **14.** $9a^2 - 400$ **15.** $4x^2 - 169$ **16.** $36x^2 - 49$

17. $x^2 - y^2z^2$ **18.** $p^2 - q^2r^2$ **19.** $49t^2 - 81r^2$ **20.** $64a^2 - 121b^2$

For Exercises 21–40, first remove the GCF and then factor. See Examples 4–6.

21. $5x^2 - 20$	**22.** $ax^2 - 64a$	**23.** $3y^2 - 75$	**24.** $4z^2 - 100$
25. $4a^2 - 400$	**26.** $11y^2 - 176$	**27.** $x^3y^2 - 121xy^2$	**28.** $a^2b^3 - 64a^2b$
29. $7xy^2 - 343x$	**30.** $10z^2 - 1000$	**31.** $a^2x^2 - a^2b^2$	**32.** $3x^4 - 3y^4$
33. $3t^3 - 27r^2t$	**34.** $5y^3 - 125yz^2$	**35.** $20t^4 - 80t^2$	**36.** $5t^5 - 80t^3$
37. $2z^4 - 162$	**38.** $5x^3 - 45x$	**39.** $3(m - n)^2 - 75$	**40.** $7(a + b)^2 - 63$

4.4 Factoring Perfect Square Trinomials

In Section 3.5 it was shown that

$$(a + b)(a + b) = (a + b)^2 = a^2 + 2ab + b^2$$

and

$$(a - b)(a - b) = (a - b)^2 = a^2 - 2ab + b^2.$$

Because the trinomials $a^2 + 2ab + b^2$ and $a^2 - 2ab + b^2$ were obtained by squaring binomials, we call them *perfect square trinomials*. Furthermore, we say that $(a + b)^2$ and $(a - b)^2$ are the factored forms of $a^2 + 2ab + b^2$ and $a^2 - 2ab + b^2$, respectively.

We emphasize the relationship between a perfect square trinomial and its factored form with the following statement.

$$\text{Perfect square trinomial} = (\text{binomial})^2$$

$$a^2 + 2ab + b^2 = (a + b)^2$$

and

$$a^2 - 2ab + b^2 = (a - b)^2.$$

Notice that in a perfect square trinomial

1. The middle term is twice the product of the two terms of the binomial.
2. The sign of the middle term of the trinomial is the same as the sign in the binomial.

Example 1. Factor the perfect square trinomial $x^2 + 10x + 25$.

Solution: If we look at the patterns of the formulas above and compare them to the trinomial $x^2 + 10x + 25$, we see that

$$\begin{array}{llr} \text{Formula:} & a^2 + 2ab + b^2 = (a + b)^2 & \\ & \downarrow \quad\; \swarrow\searrow \quad \downarrow \quad\;\; \downarrow & \\ \text{Trinomial:} & x^2 + 2(5)x + (5)^2 = (x + 5)^2 & \text{since } 10x = 2(5)x \end{array}$$

Thus

$$x^2 + 10x + 25 = (x + 5)^2.$$

Example 2. Factor $x^2 - 20x + 100$.

Solution: First we see that $x^2 - 20x + 100$ is a perfect square trinomial since $x^2 - 20x + 100 = x^2 - 2(10 \cdot x) + 10^2$. Next we compare

$$x^2 - 2(x \cdot 10) + 10^2 \qquad [\text{notice that } 2(10 \cdot x) = 2(x \cdot 10)]$$

with the formula

$$a^2 - 2ab + b^2 = (a - b)^2.$$

As a result, we associate x with a, 10 with b, and $2(10 \cdot x) = 2(x \cdot 10)$ with $2ab$. Therefore, we can factor $x^2 - 2(10 \cdot x) + 10^2$ as $(x - 10)^2$. Thus

$$x^2 - 20x + 100 = (x - 10)^2.$$

Example 3. Factor $8xy^2 + 8xy + 2x$.

Solution: First remove the GCF of $2x$. Therefore,

$$8xy^2 + 8xy + 2x = 2x(4y^2 + 4y + 1).$$

The second factor is a perfect square trinomial and is the square of $(2y + 1)$. Thus

$$\begin{aligned} 8xy^2 + 8xy + 2x &= 2x(4y^2 + 4y + 1) \\ &= 2x(2y + 1)(2y + 1) \\ &= 2x(2y + 1)^2. \end{aligned}$$

Example 4. Factor $9z^2 - 30zq + 25q^2$.

Solution: Associating $9z^2 = (3z)^2$ with a^2 and $25q^2 = (5q)^2$ with b^2, we may write

$$\begin{aligned} 9z^2 - 30zq + 25q^2 &= (3z)^2 - 2(3z \cdot 5q) + (5q)^2 \\ &= (3z - 5q)(3z - 5q) \\ &= (3z - 5q)^2. \end{aligned}$$

The following exercises will give you practice at recognizing and factoring perfect square trinomials.

EXERCISES 4.4

For Exercises 1–10, identify which expressions are perfect square trinomials.

1. $y^2 + 6y + 12$ **2.** $x^2 + 8x + 16$ **3.** $x^2 + 50x + 100$

4. $z^2 - 18z + 81$ **5.** $4r^2 - 20r + 25$ **6.** $a^2 - 14ab + 49b^2$

7. $m^2 + 2mn + n^2$ **8.** $16x^2 - 48xy + 36y^2$ **9.** $9a^2 + 48ab + 64b^2$

10. $16y^2 - 72yz + 81z^2$

For Exercises 11–30, factor the polynomial. See Examples 1 and 2.

11. $x^2 - 2x + 1$ **12.** $a^2 + 8a + 16$ **13.** $x^2 + 4x + 4$

14. $b^2 - 6b + 9$ **15.** $x^2 + 14x + 49$ **16.** $z^2 + 12z + 36$

17. $x^2 + 6x + 9$ **18.** $a^2 - 34a + 289$ **19.** $m^2 - 22m + 121$

20. $a^2 + 24a + 144$ **21.** $x^2 + 18xy + 81y^2$ **22.** $z^2 - 30z + 225$

23. $4y^2 + 16y + 16$ **24.** $4a^2 - 20a + 25$ **25.** $9m^2 + 48m + 64$

26. $4m^2 + 12mn + 9n^2$ **27.** $49x^2 - 84x + 36$ **28.** $4z^2 - 24z + 36$

29. $a^4 - 16a^2 + 64$ **30.** $a^2b^2 - 2aby + y^2$

For Exercises 31–40, first factor out the GCF and then factor the remaining trinomial. See Examples 3 and 4.

31. $4m^2 + 120m + 900$ **32.** $5x^2 - 10x + 5$ **33.** $12m^2 - 60m + 75$

34. $2m^2 - 64m + 512$ **35.** $7a^4 + 84a^2 + 252$ **36.** $3t^4 + 6t^2 + 3$

37. $45z^2 + 240z + 320$ **38.** $2x^2 + 36x + 162$ **39.** $100y^2 + 120yz + 36z^2$

40. $100x^2 + 360xy + 324y^2$

4.5 Factoring Trinomials: Case I—with Leading Coefficient 1

General trinomials may be classified in several ways. In this section we examine those trinomials that are factorable and have a leading coefficient of 1. In particular, we shall be looking at trinomials with at least two binomial factors. Thus we consider polynomials such as

$$x^2 + 8x + 15.$$

In Section 3.5 we showed that

$$(x + 3)(x + 5) = x^2 + 8x + 15.$$

Therefore, $(x + 3)(x + 5)$ is the factored form of $x^2 + 8x + 15$.

We remind the reader that in Chapter 3 we started with the expression on the left and computed the product $x^2 + 8x + 15$. In this chapter we start with the expression on the right and obtain the factored form $(x + 3)(x + 5)$.

The factoring method that is developed in this section is an intelligent "trial and error" method. It requires a good understanding of how to multiply two binomials.

The following formula will help us understand the relationship between a trinomial with a leading coefficient of 1 and its factored form. In the following formula, b and d are integers.

The Product of Two Different Binomials	*The Resulting Trinomial*

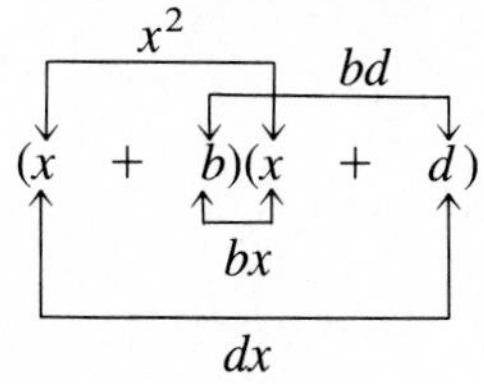

$$\overbrace{x^2}\ \ \overbrace{bd}$$

$$(x + b)(x + d) = x^2 + bx + dx + bd$$

$$\underbrace{bx}\ \ \underbrace{dx}$$

$$= x^2 + (b + d)x + bd$$

We say that the binomials are *factors* of the trinomial. We call $(x + b)(x + d)$ the *factored form* of the trinomial $x^2 + (b + d)x + bd$. We also say that we have *factored* the trinomial.

When seeking the binomial factors $(x + b)$ and $(x + d)$ of the trinomial it is important to note from the formula shown above that

1. The product of b and d is the constant term of the trinomial, bd.
2. The sum of b and d, $b + d$, is the coefficient of the middle term.

The following examples illustrate how to factor some trinomials.

Example 1. Factor $x^2 + 6x + 8$.

Solution: To factor $x^2 + 6x + 8$ as the product of two binomials of the form $(x + b)(x + d)$, we must find two integers b and d whose product is the constant term 8, and whose sum is 6, which is the coefficient of x. The possible pairs of positive factors of 8 are as follows; 1 and 8, 2 and 4. Since $2 \cdot 4 = 8$ and $2 + 4 = 6$, we choose $b = 2$ and $d = 4$; then

$$x^2 + 6x + 8 = (x + 2)(x + 4).$$

To check if we factored correctly, we multiply the factors $(x + 2)$ and $(x + 4)$.

$$\begin{aligned}(x + 2)(x + 4) &= x^2 + 4x + 2x + 8\\ &= x^2 + (4 + 2)x + 8\\ &= x^2 + 6x + 8.\end{aligned}$$

Since their product is the original polynomial, we have factored correctly.

Example 2. Factor $x^2 - 7x + 12$.

Solution: To write $x^2 - 7x + 12$ in the form $(x + b)(x + d)$, we need to find integers b and d such that $b \cdot d = 12$ and $b + d = -7$. Since the product of b and d is positive, b and d must have the same sign. But because the sum of b and d is negative, both b and d must be negative. Subject to these restrictions, we list in Table 4.1 the possible pairs of factors of 12 and their sum

Table 4.1

Factors b and d		Sum $b + d$
-1	-12	-13
-2	-6	-8
-3	-4	-7

Since $(-3)(-4) = +12$ and $(-3) + (-4) = -7$, we select $b = -3$ and $d = -4$. Thus

$$x^2 - 7x + 12 = (x - 3)(x - 4).$$

We leave it to the reader to check this factorization.

Example 3. Factor $-x^2 + 4x + 21$.

Solution: Since the coefficient of x^2 is -1, rather than $+1$, we factor out -1 from all three terms. That is,

$$-x^2 + 4x + 21 = -1(x^2 - 4x - 21).$$

The trinomial $x^2 - 4x - 21$ has a leading coefficient of 1 and we will now try to write it in factored form as $(x + b)(x + d)$. Thus we need to find integers b and d such that $b \cdot d = -21$ and $b + d = -4$. Since the constant term is negative, b and d must have opposite signs. Listing the possible pairs of factors of -21, we compute their sum as in Table 4.2.

Table 4.2

Factors b and d		Sum $b + d$
-1	$+21$	20
$+1$	-21	-20
-3	$+7$	$+4$
$+3$	-7	-4

Since $3(-7) = -21$ and $3 + (-7) = -4$, we may write

$$x^2 - 4x - 21 = (x + 3)(x - 7).$$

Thus

$$-x^2 + 4x + 21 = -1(x^2 - 4x - 21) = -1(x + 3)(x - 7).$$

In Example 3, if we had first tried the correct factors $b = 3$ and $d = -7$, we would not have continued to list the other factors. This is where "intelligent guessing" can be very helpful. For instance, in Example 3, the values $b = 1$ and $d = -21$ would be ignored because the absolute value of their sum is too large.

Example 4. Factor $2x^4 + 2x^3 - 60x^2$.

Solution: Note that since each term contains a GCF of $2x^2$, we begin by removing this common factor; thus we have

$$2x^4 + 2x^3 - 60x^2 = 2x^2(x^2 + x - 30).$$

The remaining trinomial factor has a leading coefficient of 1 and can be factored according to our formula

$$(x + b)(x + d) = x^2 + (b + d)x + bd.$$

Thus, to factor $x^2 + x - 30$, we seek integers b and d such that

$$b \cdot d = -30 \qquad \text{and} \qquad b + d = 1.$$

We note that since the constant term is negative, b and d must have opposite signs. Also, since the sum of b and d is 1 we can ignore factors such as 1 and -30, and -2 and 15. The correct choice is $b = -5$ and $d = 6$. This is because $(-5) \cdot 6 = -30$ and $-5 + 6 = 1$. Thus

$$x^2 + x - 30 = (x - 5)(x + 6)$$

and

$$\begin{aligned} 2x^4 + 2x^3 - 60x^2 &= 2x^2(x^2 + x - 30) \\ &= 2x^2(x - 5)(x + 6). \end{aligned}$$

When the signs of b and d are both changed, the sign of their sum is changed. For example, $(-5)(6) = -30$ and $-5 + 6 = 1$, while $(5)(-6) = -30$ but $5 + (-6) = -1$. This observation can sometimes be helpful in finding the correct factors of a trinomial.

Example 5. Factor $q^2 + 8qr - 33r^2$.

Solution: To factor the given trinomial as the product of two binomial factors, we need to find integers b and d such that

$$q^2 + 8qr - 33r^2 = (q + br)(q + dr)$$

or

$$q^2 + 8qr - 33r^2 = q^2 + (b + d)qr + (bd)r^2.$$

Therefore, we want two integers b and d with a product of -33 and a sum of 8. If we try $b = 3$ and $d = -11$, we get a sum of -8. By changing the signs of b and d, we get the correct factors. That is,

$$q^2 + 8qr - 33r^2 = (q - 3r)(q + 11r).$$

We conclude this section by observing that if we restrict ourselves to integer coefficients, then not every polynomial is factorable. For example, the polynomials

$$x^2 + x + 1, \qquad x^2 + 3x + 6, \qquad x^2 + 5x + 2$$

cannot be factored as $(x + b)(x + d)$ when b and d are integers. Readers should convince themselves of this by trying all the possible integer values for b and d. In this chapter we are restricting ourselves to polynomials that can be factored using only integers. In this section we considered only polynomials with factors $(x + b)$ and $(x + d)$, where b and d were integers.

EXERCISES 4.5

For Exercises 1–30, factor the trinomial. See Examples 1–4.

1. $x^2 + 5x + 4$
2. $x^2 + 11x + 18$
3. $x^2 + 9x + 14$
4. $x^2 + 7x + 10$
5. $z^2 + 3z + 2$
6. $y^2 + 9y + 20$
7. $x^2 - 5x + 6$
8. $t^2 - 8t + 12$
9. $z^2 - 5z + 4$
10. $m^2 - 16m + 15$
11. $t^2 - 11t + 24$
12. $x^2 - 31x + 30$
13. $y^2 - 5y - 14$
14. $s^2 - 4s - 45$
15. $m^2 - 3m - 40$
16. $x^2 - 19x - 20$
17. $y^2 - 5y - 24$
18. $x^2 - 6x - 27$
19. $x^2 + 4x - 21$
20. $y^2 + y - 20$
21. $x^2 + 7x - 30$
22. $x^2 + 8x - 9$
23. $x^2 + 2x - 15$
24. $y^2 + 9y - 22$
25. $z^2 + 11z + 30$
26. $t^2 - 10t + 16$
27. $x^2 - 15x + 54$
28. $x^2 - x - 42$
29. $x^2 + 5x - 36$
30. $x^2 + 16x + 28$

For Exercises 31–50, first remove the GCF and then factor. See Example 4.

31. $2x^2 - 2x - 4$
32. $2y^2 + 8y + 6$
33. $3z^2 - 24z + 45$
34. $3t^2 + 18t - 48$
35. $7y^2 + 77y + 210$
36. $11z^2 + 110z - 121$
37. $5x^2 + 15x - 200$
38. $13x^2 + 39x - 130$
39. $10t^2 + 40t - 210$
40. $6t^2 + 36t - 330$
41. $a^3 - a^2 - 20a$
42. $x^3 - 8x^2 + 16x$
43. $x^3 + 10x^2 - 39x$
44. $t^3 - 13t^2 + 40t$
45. $2y^3 - 10y^2 + 8y$
46. $3a^3 - 9a^2 - 120a$
47. $4xy^2 - 40xy + 100x$
48. $7x^2y + 7xy - 14y$
49. $3xyz^2 + 6xyz - 72xy$
50. $-5xy^3 - 30xy^2 - 40xy$

For Exercises 51–58, factor the trinomial. See Example 5.

51. $t^2 + 2ts - 35s^2$ **52.** $x^2 - 10xy + 16y^2$ **53.** $y^2 + 15yz + 56z^2$

54. $m^2 - 3mn - 28n^2$ **55.** $z^2 - 5zy - 36y^2$ **56.** $p^2 + 20pq + 100q^2$

57. $r^2 - 30rs + 81s^2$ **58.** $a^2 - 20ab + 64b^2$

For Exercises 59–62, first remove the GCF and then factor.

59. $5x^2 + 50xy + 105y^2$ **60.** $7a^2 - 49ab + 84b^2$

61. $x^3y + 8x^2y^2 + 7xy^3$ **62.** $m^2n + 12mn^2 - 45n^3$

4.6 Factoring Trinomials; Case II—with Leading Coefficient ≠ 1

Extending the methods developed in Section 4.5, we now discuss factoring a quadratic trinomial $Ax^2 + Bx + C$ as the product of two binomials when the leading coefficient A is an integer different from 1 or -1. An example of such a trinomial is

$$3x^2 - 7x - 6.$$

To express a general quadratic trinomial

$$Ax^2 + Bx + C \qquad \text{where } A, B, \text{ and } C \text{ are integers}$$

as the product of two binomials $(ax + b)$ and $(cx + d)$, we need to find integers a, b, c, and d so that

$$Ax^2 + Bx + C = (ax + b)(cx + d). \tag{1}$$

If we multiply out the right-hand side of Equation (1) using the FOIL Method, we see that

$$Ax^2 + Bx + C = \underset{\substack{\uparrow \\ \text{F}}}{(ac)}x^2 + \underset{\substack{\uparrow \\ (\text{O} + \text{I})}}{(ad + bc)}x + \underset{\substack{\uparrow \\ \text{L}}}{bd} \tag{2}$$

Equations (1) and (2) suggest the following procedure for factoring a general quadratic trinomial $Ax^2 + Bx + C$.

Write down possible binomial products of the form $(ax + b)(cx + d)$ with the aim of finding values of a, b, c, and d so that

1. The product of a and c equals the coefficient of the x^2 term. That is, $ac = A$.
2. The product of b and d equals the constant term. That is, $bd = C$.
3. The sum of ad and bc equals the coefficient of the x term. That is, $ad + bc = B$.

The following examples illustrate the suggestions outlined above for factoring *quadratic* trinomials of the form $Ax^2 + Bx + C$ when $A \neq 1$.

Example 1. Factor $3x^2 + 8x + 5$.

Solution: To factor $3x^2 + 8x + 5$ as the product of two binomials, we need to find integers a, b, c, and d so that

$$3x^2 + 8x + 5 = (ax + b)(cx + d). \tag{3}$$

We usually begin our factoring process by choosing two positive integers a and c whose product is 3. Therefore, we let $a = 3$ and $c = 1$ in (3). Thus, as a first step to factoring $3x^2 + 8x + 5$, we write

$$3x^2 + 8x + 5 = (\underset{\uparrow}{3x} + b)(\underset{\uparrow}{x} + d) \tag{4}$$

$$\qquad\qquad 3x^2$$

If we multiply the right-hand side of (4) using the FOIL method, we see that

$$3x^2 + 8x + 5 = 3x^2 + (3d + b)x + bd \qquad \text{FOIL} \tag{5}$$

From Equations (4) and (5) we observe that we need to find two integers b and d whose product $bd = 5$ (constant term) and such that $(3d + b) = 8$ is the coefficient of the x term. If we choose $b = 1$ and $d = 5$, we obtain

$$(3x + 1)(x + 5) = 3x^2 + 16x + 5 \qquad \textit{not correct}$$

$$x$$

$$15x$$

Since the only factors of 5 are 1 and 5, we try

$$(3x + 5)(x + 1) = 3x^2 + 8x + 5 \qquad \textit{correct}$$

$$5x$$

$$3x$$

Therefore,

$$3x^2 + 8x + 5 = (3x + 5)(x + 1).$$

Example 2. Factor $4x^2 - 13x + 9$.

Solution: We need to find integers a, b, c, and d so that

$$(ax + b)(cx + d) = 4x^2 - 13x + 9.$$

We select as possible values of a and c pairs of positive factors of 4 such as 4 and 1 or 2 and 2. Thus, as a first step in factoring $4x^2 - 13x + 9$, we would write

$$(4x + b)(x + d) = 4x^2 - 13x + 9 \tag{6}$$

or possibly

$$(2x + b)(2x + d) = 4x^2 - 13x + 9. \tag{7}$$

We note that only one of the above pair of binomial factors appearing in Equations (6) and (7) will lead to the correct answer.

Next we need to choose integers b and d so that when we multiply out either

$$(4x + b)(x + d) \quad \text{or} \quad (2x + b)(2x + d)$$

we get a negative middle term, $-13x$, and positive constant, 9. Since $bd = 9$, we should choose for b and d negative factors of 9 such as -1 and -9 or -3 and -3.

In Table 4.3 we list some of the possible binomial factors of $4x^2 - 13x + 9$ and check the middle term of their product to see which one yields a middle term of $-13x$.

Table 4.3

Products of Binomial Factors	Middle Term of the Product
$(2x - 1)(x - 9)$	$-19x$
$(2x - 9)(x - 1)$	$-11x$
$(2x - 3)(2x - 3)$	$-12x$
$(4x - 1)(x - 9)$	$-37x$
$(4x - 9)(x - 1)$	$-13x$

The last binomial factors in the list yield the correct middle term $-13x$. Therefore,

$$4x^2 - 13x + 9 = (4x - 9)(x - 1).$$

Example 2 indicates that factoring the general trinomial requires some intelligent guessing in order to get the correct combination of binomial factors. If we practice enough, we soon become very good at eliminating a lot of the possibilities and thus we will be able to factor a trinomial with just a few trials.

Example 3. Factor $15x^2 - 7x - 4$.

Solution: To factor $15x^2 - 7x - 4$ into the form $(ax + b)(cx + d)$, we consider:

1. The possible positive pairs of factors a and c of 15 are as follows; 15 and 1, 5 and 3.
2. The pairs of factors b and d of -4 are as follows; 4 and -1, -4 and 1, 2 and -2.

If we first try to factor $15x^2 - 7x - 4$ using $a = 15$ and $c = 1$, we get

$$15x^2 - 7x - 4 = (15x + b)(x + d)$$

or

$$15x^2 - 7x - \underset{\uparrow}{4} = 15x^2 + (15d + b)x + \underset{\uparrow}{bd} \qquad \text{FOIL}$$

But we would find that none of the possible factors of -4 that we could substitute for b and d yields the correct middle term $-7x$. Thus we try $a = 5$ and $c = 3$ and write

$$15x^2 - 7x - 4 = (5x + b)(3x + d).$$

If we try $b = 4$ and $d = -1$, we get

$$(5x + 4)(3x - 1) = 15x^2 + 7x - 1 \qquad \textit{incorrect}$$

$12x$

$-5x$

Thus we see that the middle term does not have the correct sign. However, if we switch signs in the binomials we obtain

$$(5x - 4)(3x + 1) = 15x^2 - 7x - 1 \qquad \textit{correct}$$

$-12x$

$5x$

Therefore, $15x^2 - 7x - 1 = (5x - 4)(3x + 1)$.

When factoring a polynomial, it is always wise to first remove any greatest common factor. It is also suggested to factor out -1 if the leading coefficient is negative. Thus the new leading coefficient will be positive and we may then keep a and c positive.

Example 4. Factor $-10x^2y - 16xy + 8y$.

Solution: First we remove the greatest common factor $-2y$. Thus we have

$$-10x^2y - 16xy + 8y = -2y(5x^2 + 8x - 4)$$

Next we factor $5x^2 + 8x - 4$ into the form $(ax + b)(cx + d)$. The positive factors of 5 are 5 and 1. Thus, to factor $5x^2 + 8x - 4$, we begin by writing

$$(5x + b)(x + d) = 5x^2 + 8x - 4$$

Next we need to choose integers b and d so that the product of the binomials yields a middle term of $8x$ and a constant term of -4. The factors of the constant term $bd = -4$ are as follows: 1 and -4, -1 and 4, and 2 and -2. If we try as possible factors of $5x^2 + 8x - 4$, the following binomial products

$$(5x + 1)(x - 4) \qquad (5x - 4)(x + 1)$$
$$(5x - 1)(x + 4) \qquad (5x - 1)(x + 4)$$

we find that we do not obtain the correct middle term $8x$. The only remaining factors of -4 are 2 and -2. If we try these for b and d in $(5x + b)(x + d)$, we obtain

$$(5x + 2)(x - 2) = 5x^2 - 8x - 4 \qquad \text{incorrect middle term}$$
$$(5x - 2)(x + 2) = 5x^2 + 8x - 4 \qquad \textit{correct} \text{ middle term}$$

The last pair of factors is the desired factorization of $5x^2 + 8x - 4$. As a result we have

$$-10x^2y - 16xy + 8y = -2y(5x^2 + 8x - 4) = -2y(5x - 2)(x + 2).$$

Example 5. Factor $3x^2 - 13xy - 10y^2$.

Solution: First we notice that there is no greatest common factor.

The possible positive factors of 3 are 3 and 1. Therefore, the binomial factors have the form

$$(3x + by)(x + dy) = 3x^2 + (3d + b)xy + bdy^2.$$

Thus we have to determine b and d so that

$$3x^2 + (3d + b)xy + bdy^2 = 3x^2 - 13xy - 10y^2.$$

The possible factors of -10, the coefficient of y^2, are as follows: 10 and -1, -10 and 1, 5 and -2, and -5 and 2. These represent possible pairs of values that we may substitute for b and d.

Next we need to choose values for b and d so that the middle term is $-13xy$. Instead of just listing all possible binomial factors, we try to select numbers that look like they might give us the desired middle term. In Table 4.4 we list some binomial factors that we might try before coming up with the correct factors.

Table 4.4

Products of Binomial Factors: $(3x + by)(x + dy)$	Middle Term of the Product: $(3d + b)xy$
$(3x + 10y)(x - y)$	$-3xy + 10xy = 7xy$
$(3x - y)(x + 10y)$	$30xy - xy = 29xy$
$(3x + 5y)(x - 2y)$	$-6xy + 5xy = -xy$
$(3x + 2y)(x - 5y)$	$-15xy + 2xy = -13xy$

Since the last pair of binomial factors yields the correct middle term $-13xy$, we have

$$3x^2 - 13xy - 10y^2 = (3x + 2y)(x - 5y).$$

In Example 5, if we had first tried the binomial factors $(3x + 2y)(x - 5y)$, we would not have bothered with the remaining possibilities. Thus there is an element of chance involved when we list possible binomial factors. If we think carefully and select the right numbers, our list will be short; otherwise, our list may be long.

EXERCISES 4.6

For Exercises 1–10, find the middle term of the product of the binomials.

1. $(x + 2)(2x - 3)$ **2.** $(x - 5)(3x - 2)$ **3.** $(x + 5)(3x - 2)$ **4.** $(x + 4)(5x - 2)$

5. $(x - 4)(5x + 2)$ **6.** $(2x + 3)(3x - 2)$ **7.** $(7x + 2)(5x - 3)$ **8.** $(6x - 5)(2x - 7)$

9. $(4x + 5)(3x + 4)$ **10.** $(9x - 8)(x - 1)$

For Exercises 11–40, factor the trinomial. See Examples 1–3.

11. $2x^2 + 11x + 5$ **12.** $2y^2 - 5y + 2$ **13.** $2x^2 + 15x + 7$ **14.** $2x^2 + 9x + 7$

15. $5x^2 + 16x + 3$ **16.** $3x^2 + 10x + 3$ **17.** $7x^2 + 74x - 33$ **18.** $7x^2 + 30x - 25$

19. $5z^2 - 53z - 22$ **20.** $2y^2 - 13y + 15$ **21.** $3a^2 + 8a + 4$ **22.** $5b^2 - 13b + 6$

23. $6t^2 + 13t + 6$ **24.** $6x^2 - 19x + 10$ **25.** $6t^2 - t - 35$ **26.** $6z^2 - z - 15$

27. $2x^2 + 7x - 15$ **28.** $3a^2 - 7a - 6$ **29.** $5x^2 + 26x + 5$ **30.** $6b^2 + 5b - 6$

31. $7y^2 - 10y - 8$ **32.** $3z^2 - 10z - 8$ **33.** $5a^2 - 12a + 4$ **34.** $4m^2 - 25m + 6$

35. $6y^2 + 5y - 56$ **36.** $2p^2 + 31p - 51$ **37.** $25y^2 + 10y - 8$ **38.** $10z^2 + 19z + 6$

39. $10z^2 + 11z - 6$ **40.** $10z^2 - 19z + 6$

For Exercises 41–60, factor the trinomial. See Examples 4 and 5.

41. $9x^2 + 33x - 60$ **42.** $6x^2 + 33x + 36$ **43.** $4x^2 - 14x + 6$

44. $6a^2 - 16a + 8$ **45.** $6z^2 + 15z + 6$ **46.** $9x^2 + 30x - 24$

47. $15y^2 + 35y - 30$ **48.** $2x^2 - 16x + 30$ **49.** $14y^3 - 20y^2 - 16y$

50. $30x^2y + 38xy - 20y$ **51.** $8x^3y + 24x^2y - 14xy$ **52.** $10a^3b + a^2b - 21ab$

53. $24p^3q + 42p^2q - 45pq$ **54.** $40y^3z + 10y^2z - 105yz$ **55.** $36x^4 + 21x^3 - 30x^2$

56. $24y^4 - 4y^3 - 48y^2$ **57.** $16x^2 + 32xy + 15y^2$ **58.** $15m^2 - 26mn - 24n^2$

59. $24y^3 + 26y^2z - 8yz^2$ **60.** $18a^3 + 3a^2b - 105ab^2$

4.7 Solving Quadratic Equations by Factoring

In previous sections we learned how to factor some quadratic expressions of the form $ax^2 + bx + c$. In this section we learn how to use factoring to solve *quadratic equations*.

> Any equation that can put in the form $ax^2 + bx + c = 0$ where a, b, and c are real numbers, with $a \neq 0$, is called a *quadratic equation*. This is called the *standard form* of a quadratic equation.

The following equations are examples of quadratic equations:

$$x^2 - 8x + 15 = 0, \qquad 2x^2 - 8 = 0, \qquad x^2 - 2x = 0, \qquad 2x^2 - x = 3.$$

The fundamental property of the real numbers that is used when solving equations by factoring is

Factor Property of Zero

> If a and b are real numbers and $a \cdot b = 0$, then $a = 0$ or $b = 0$.

To find the solution set of a quadratic equation such as

$$x^2 - 8x + 15 = 0,$$

we must find values of x that will satisfy the equation. To satisfy the equation above means that the value assigned to x will make the left side reduce to 0. When a quadratic equation is in standard form, such as $x^2 - 8x + 15 = 0$, it is not easy to see what values of x will satisfy the equation. However, if we factor $x^2 - 8x + 15$, we see that

$$x^2 - 8x + 15 = (x - 3)(x - 5) = 0.$$

But the product $(x - 3)(x - 5)$ can equal 0 only if $x - 3 = 0$ or $x - 5 = 0$ (factor property of zero). Therefore, we set each factor equal to 0 and solve the resulting equations for x, as follows:

$$\begin{aligned} x - 3 &= 0 \\ x &= 3 \end{aligned} \qquad \text{or} \qquad \begin{aligned} x - 5 &= 0 \\ x &= 5. \end{aligned}$$

We check to see that 3 and 5 satisfy the original quadratic equation by substituting 3 and 5 into $x^2 - 8x + 15 = 0$.

CHECK:

$$\begin{aligned} 3^2 - 8(3) + 15 &= 0 \\ 9 - 24 + 15 &= 0 \\ 0 &= 0. \quad \text{true} \end{aligned} \qquad \begin{aligned} 5^2 - 8(5) + 15 &= 0 \\ 25 - 40 + 15 &= 0 \\ 0 &= 0. \quad \text{true} \end{aligned}$$

Thus the solution set of the quadratic equation $x^2 - 8x + 15 = 0$ is $\{3, 5\}$. Any member of the solution set is called a *root* of the equation.

In general, to solve a quadratic equation by factoring, we follow four steps:

1. Collect all terms on one side of the equality sign, usually the left side.
2. Factor the expression as completely as possible.
3. Set each factor that contains an unknown equal to 0 and solve the resulting equations (factor property of zero).
4. Check the answers from step 3 by substituting them into the original equation.

The following examples illustrate the foregoing procedure.

Example 1. Solve the equation $2x^2 - x = 3$.

Solution: First we collect all terms on the left side of the equation, by adding -3 to both sides of $2x^2 - x = 3$. Thus we obtain

$$2x^2 - x - 3 = 0.$$

Next we factor the expression $2x^2 - x - 3$ and rewrite our equation as follows:

$$2x^2 - x - 3 = (2x - 3)(x + 1) = 0.$$

Setting each factor equal to 0 and solve the resulting equations, we get

$$\begin{aligned} 2x - 3 &= 0 \\ 2x &= 3 \\ x &= \frac{3}{2} \end{aligned} \qquad \text{or} \qquad \begin{aligned} x + 1 &= 0 \\ x &= -1. \end{aligned}$$

NOTE. Each of these values of x will make one factor equal to 0. Substituting $\frac{3}{2}$ and -1 for x in the original equation would confirm that they satisfy the given equation. Therefore, the solution set of $2x^2 - x = 3$ is $\left\{\frac{3}{2}, -1\right\}$.

Example 2. Solve the equation $13x = 3x^2 - 10$.

Solution: First collect all terms on one side of the equality sign. If we subtract $13x$ from both sides, we obtain $0 = 3x^2 - 13x - 10$. We then use the symmetric property of equality to transform the equation into standard form. That is,

$$3x^2 - 13x - 10 = 0.$$

Factoring $3x^2 - 13x - 10$, we obtain $3x^2 - 13x - 10 = (3x + 2)(x - 5) = 0$. Next we set each factor equal to zero and solve the resulting equations.

$$\begin{aligned} 3x + 2 &= 0 \\ 3x &= -2 \\ x &= -\frac{2}{3}. \end{aligned} \qquad \begin{aligned} x - 5 &= 0 \\ x &= 5. \end{aligned}$$

The reader can check that $-\frac{2}{3}$ and 5 are the roots of the quadratic equation by substituting $-\frac{2}{3}$ and 5 into the original equation. Therefore, the solution set of $13x = 3x^2 - 10$ is $\left\{-\frac{2}{3}, 5\right\}$.

Example 3. Solve $3t^2 - 27t = 0$.

Solution

$$3t^2 - 27t = 0.$$

Factoring the left side of the equation, we see that

$$3t(t - 9) = 0.$$

Thus $3t = 0$ or $t - 9 = 0$. Solving these, we get

$$t = 0 \qquad \text{or} \qquad t = 9.$$

We see that the solution set is $\{0, 9\}$. The check is simple and is therefore omitted.

Example 4. Solve the equation $5x(2x + 5) = 15$.

Solution: We rewrite the equation in standard form.

$$\begin{aligned} 5x(2x + 5) &= 15 \\ 10x^2 + 25x &= 15 \\ 10x^2 + 25x - 15 &= 0. \end{aligned}$$

Dividing each side of the last equation by 5, we obtain

$$2x^2 + 5x - 3 = 0.$$

Next we factor and set each factor that contains an unknown equal to zero. That is,

$$\begin{aligned} 2x^2 + 5x - 3 &= 0 \\ (2x - 1)(x + 3) &= 0. \end{aligned}$$

Therefore, $2x - 1 = 0$ or $x + 3 = 0$. Solving each equation for x, we get $x = \frac{1}{2}$ or $x = -3$. The solution set to the original equation is $\left\{\frac{1}{2}, -3\right\}$.

Example 5. Solve the equation $x^3 - 2x^2 - 15x = 0$.

Solution: The expression $x^3 - 2x^2 - 15x = 0$ is not a quadratic equation but is called a cubic equation because of the x^3 term. However, removing the common factor x reveals that the other factor, $x^2 - 2x - 15$, is a quadratic. With that as our first step we can then proceed as before. Factoring the expression $x^3 - 2x^2 - 15x$ completely, we obtain $x(x^2 - 2x - 15) = x(x + 3)(x - 5) = 0$. Setting each factor that contains an unknown equal to 0 and solving the resulting equations yields

$$\begin{array}{lll} x = 0 & x + 3 = 0 & x - 5 = 0 \\ & x = -3 & x = 5. \end{array}$$

Substituting 0, −3, and 5 into the original equation, $x^3 - 2x^2 - 15x = 0$ confirms that these are the desired solutions to the given equation.
Therefore, the solution set of $x^3 - 2x^2 - 15x = 0$ is $\{0, -3, 5\}$.

All the examples and exercises in this section may be solved by using the techniques of factoring that we have already mastered. In Chapter 8 we shall learn how to solve equations that do not factor as easily as the equations studied in this chapter.

EXERCISES 4.7

For Exercises 1–10, use the factor property of zero to find the solution set.

1. $(x - 3)(x - 1) = 0$ **2.** $(x - 2)(x + 5) = 0$ **3.** $y(y - 7) = 0$

4. $z(z - 5) = 0$ **5.** $(x - \frac{1}{2})(x + \frac{2}{3}) = 0$ **6.** $(5y + 7)(2y - 1) = 0$

7. $0 = (3x - 5)(x + 4)$ **8.** $0 = (t + 9)(2t - 3)$ **9.** $3x(x + 2)(2x - 1) = 0$

10. $(y + 7)(y - 11)(2y + 5) = 0$

For Exercises 11–46, solve the equation. See Examples 1–3.

11. $x^2 - 9 = 0$ **12.** $x^2 - 16 = 0$ **13.** $y^2 - 100 = 0$

14. $z^2 - 144 = 0$ **15.** $p^2 - 8p + 16 = 0$ **16.** $w^2 + 18w + 81 = 0$

17. $x^2 + 12x + 36 = 0$ **18.** $y^2 - 14y + 49 = 0$ **19.** $x^2 + 25 = 10x$

20. $y^2 + 1 = 2y$ **21.** $t^2 = 12 + t$ **22.** $28 = x^2 + 3x$

23. $-7y - 10 = y^2$ **24.** $2z - z^2 = -24$ **25.** $z^2 = 4z + 12$

26. $y^2 - 2y = 3$ **27.** $y^2 + y = 42$ **28.** $w^2 = 5w + 24$

29. $x^2 + 28 = -16x$ **30.** $z^2 + 50 = 15z$ **31.** $10x^2 - 7x + 1 = 0$

32. $6y^2 - 5y + 1 = 0$ **33.** $3p^2 - 7p - 6 = 0$ **34.** $2w^2 + 7w - 15 = 0$

35. $2z^2 + 15 = z$ **36.** $4x^2 = 25x - 6$ **37.** $8y^2 + 14y - 15 = 0$

38. $12z^2 + 8z - 15 = 0$ **39.** $9w^2 = 12w - 4$ **40.** $20y - 4 = 25y^2$

41. $80x^2 - 45 = 0$ **42.** $18z^2 - 50 = 0$ **43.** $2w^2 = 16w - 30$

44. $4x^2 = 14x - 6$ **45.** $48 - 24x = -3x^2$ **46.** $245 + 70y = -5y^2$

For Exercises 47–54, write the quadratic equation in standard form and then solve. See Example 4.

47. $2z(6z - 5) = 2$ **48.** $x(3x - 13) = 10$ **49.** $4y(4y + 2) - 15 = 0$

50. $(y - 4)(y + 2) = 7$ **51.** $(2x + 3)(x - 2) = 9$ **52.** $4y(4y + 9) + 18 = 0$

53. $(3x + 2)(x - 1) = 22$ **54.** $(x - 3)^2 = 4$

For Exercises 55–60, solve the cubic equation. See Example 5.

55. $x^3 - 64x = 0$ **56.** $y^3 - 121y = 0$ **57.** $x^3 - 3x^2 + 2x = 0$

58. $x^3 + 6x^2 + 8x = 0$ **59.** $6x^3 + 7x^2 - 3x = 0$ **60.** $6x^3 - 13x^2 - 5x = 0$

Applications 4.8

In this section we examine some ways of applying mathematics to a variety of problems. The application of mathematics usually requires the translation of a problem from a verbal or written form to a symbolic or algebraic form. The reader should review at this time the steps outlined in Section 2.6 for solving applied problems. In addition to those steps we would suggest that (1) for problems dealing with rectangles, triangles, and so on, make a sketch of the figure and label it, where appropriate, with variables for the unknown quantities; and (2) check that the solution to the equation makes sense as a solution to the applied (word) problem.

We now examine verbal problems whose solution involves formulating and solving quadratic equations.

Example 1. A contractor accepts a job to pave a rectangular parking area that is to cover 1200 square feet. If the parking area is designed to be 10 feet longer than it is wide, find its dimensions.

Solution: We choose convenient variables to represent the length, say L, and the width, say W. We are told that L is 10 feet longer than W (Figure 4.1). The verb "is" translates into "=," and "10 more" means that L is larger than W by 10. Therefore, $L = W + 10$.

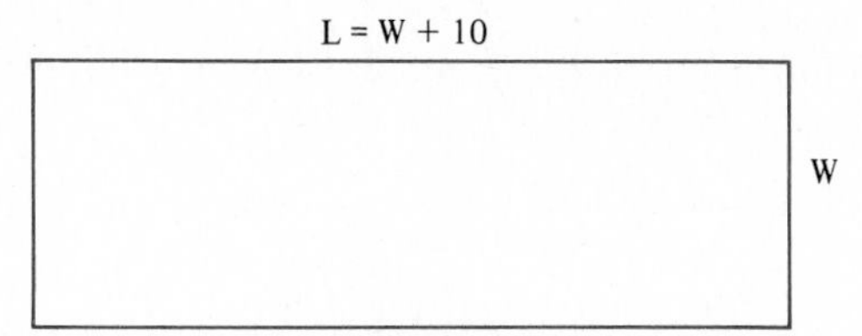

Figure 4.1

Since the formula for the area of a rectangle is length (L) times width (W), we write the algebraic statement: $L \cdot W = 1200$ or $(W + 10)W = 1200$, since $L = W + 10$. Thus

$$(W + 10)W = 1200$$

$$W^2 + 10W = 1200 \qquad \text{distributive property}$$

$$W^2 + 10W - 1200 = 0 \qquad \text{collect all terms on one side of the equation}$$

$$(W + 40)(W - 30) = 0 \qquad \text{factor}$$

$$W + 40 = 0 \qquad W - 30 = 0 \qquad \text{set each factor equal to zero}$$

$$W = -40 \qquad W = 30. \qquad \text{solve for } W$$

The solution set of the equation is $\{-40, 30\}$.

However, the only meaningful answer is $W = 30$, since the width must be a positive number. In applied problems dealing with physical quantities, we must be careful to select only those roots that are valid values of the variables. Therefore, the width $W = 30$ and since $L = W + 10$, $L = 40$. We check these values by noting that $L \cdot W = 40 \cdot 30 = 1200$.

Example 2. The product of two consecutive integers is 29 more than their sum. Find the integers.

Solution: If we let x represent the first integer, then the next integer, being 1 greater, may be represented as $x + 1$. Their product is $x(x + 1)$ and their sum is $x + (x + 1) = 2x + 1$. Since the product *is* (=) 29 *more* (+) than the *sum*, we may write

$$\text{product} = \text{sum} + 29.$$

Therefore,

$x(x + 1) = (2x + 1) + 29$	
$x^2 + x = 2x + 30$	simplify
$x^2 - x - 30 = 0$	collect all terms on one side of the equation
$(x + 5)(x - 6) = 0$	factor
$x + 5 = 0 \qquad x - 6 = 0$	set each factor equation to zero
$x = -5 \qquad x = 6.$	solve for x

Since the integers can be positive or negative, we have two choices for the consecutive integers, -5 and -4 or 6 and 7.

CHECK:

$$(-5)(-4) = (-5) + (-4) + 29 \quad \text{and} \quad (6)(7) = 6 + 7 + 29$$
$$20 = -9 + 29 \qquad\qquad 42 = 42. \quad \text{true}$$
$$20 = 20. \quad \text{true}$$

Thus both pairs -5 and -4, and 6 and 7, are solutions to the problem.

When an object is projected vertically upward, its distance, h, above the ground is approximately $h = -16t^2 + vt$, where v is the initial velocity with which it is projected, and t is the time elapsed from its release.

Example 3. A small model rocket is launched vertically upward from the ground with an initial velocity of 64 feet per second. If we substitute $v = 64$ into the formula $h = -16t^2 + vt$, the height of the rocket above ground after t seconds is given by the formula

$$h = -16t^2 + 64t.$$

How much time will elapse (a) before it *first* reaches a height, h, of 48 feet; and (b) before it returns to the ground?

Solution

(a) Since $h = 48$, we substitute in the formula and obtain the quadratic equation

$$48 = -16t^2 + 64t.$$

To solve this equation, we put it into standard form,

$$16t^2 - 64t + 48 = 0.$$

Next we get

$$\begin{aligned} t^2 - 4t + 3 &= 0 && \text{divide both sides by 16} \\ (t - 1)(t - 3) &= 0 && \text{factor} \\ t - 1 = 0 \quad \text{or} \quad t - 3 &= 0 && \text{set each factor equal to 0} \\ t = 1 \quad \text{or} \quad t &= 3. && \text{solve each equation} \end{aligned}$$

Therefore, we see that it takes 1 second after launch to rise to a height of 48 feet (Figure 4.2). The other possible answer $t = 3$ means that on the way back down, after 3 seconds, it will again be 48 feet above the ground.

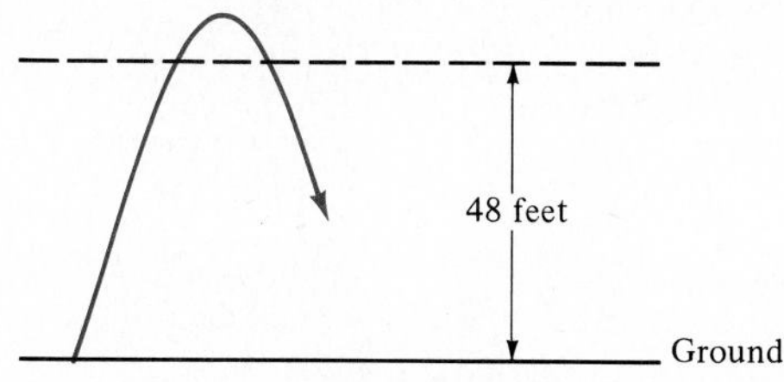

Figure 4.2

(b) When it returns to the ground, its height above the ground will be 0. Therefore, $h = 0$. From the formula we have

$$\begin{aligned} 0 &= -16t^2 + 64t \\ 0 &= t^2 - 4t && \text{divide both sides by } -16 \\ 0 &= t(t - 4) && \text{factor} \\ t = 0 \quad \text{or} \quad t - 4 &= 0 && \text{set each factor equal to 0} \\ t = 0 \quad \text{or} \quad t &= 4 && \text{solve each equation} \end{aligned}$$

The model rocket takes 4 seconds after launch to return to the ground. Notice that $t = 0$ means that at the time of launch $t = 0$, it is 0 feet above the ground.

Example 4. Find two consecutive odd integers such that the square of the first exceeds the second by 40.

Solution: Consecutive odd integers, like consecutive even integers, are two apart. Thus 3, 5, 7, and so on, are consecutive odd integers. If x represents the first of the two odd integers, then $x + 2$ represents the second odd integer. "The square of the first *exceeds* the second by 40" means that the first integer squared is 40 *larger* than the second odd integer. Since x represents the first integer and $x + 2$ represents the second integer, the *difference* between the square of the first integer and the second integer will equal 40. Thus

$$x^2 - (x + 2) = 40.$$

We solve this equation as follows:

$$x^2 - x - 2 = 40$$
$$x^2 - x - 42 = 0$$
$$(x + 6)(x - 7) = 0$$
$$x + 6 = 0 \qquad x - 7 = 0$$
$$x = -6 \qquad x = 7.$$

Since 7 is the only odd integer in the solution set, the two odd integers must be 7 and 9.

CHECK:

$$7^2 - 9 = 40$$
$$49 - 9 = 40$$
$$40 = 40.$$

EXERCISES 4.8

Solve the following.

For Exercises 1–4, see Example 1.

1. The length of a rectangle is 6 more than the width. If the area of the rectangle is 55 square inches, find the length and width.

2. The length of a rectangle is 3 more than twice the width. If the area of the rectangle is 44 square inches, find the length and width.

3. A rectangle is three times as long as it is wide. If the area of the rectangle is 75 square inches, find the length and width.

4. The length of a rectangle is 1 more than twice its width. If the area of the rectangle is 105 square inches, find the dimensions of the rectangle.

For Exercises 5–8, recall that the area of a triangle is given by the formula in the figure on page 151.

5. The height of a triangle is half of the length of its base. If the area is 64 square inches, find the length of the base and the height of the triangle.

6. The height of a triangle is one-third of the length of its base. If the area of the triangle is 54 square inches, find the length of the base and the height of the triangle.

7. One leg of a right triangle is 3 more than twice the other leg. If the area of the triangle is 22 square inches, find the length of the two legs. (HINT: Either leg can be considered an altitude, h.)

8. One leg of a right triangle is 9 feet longer than the other leg. If the area of the triangle is 56 square feet, find the length of the two legs. (HINT: Either leg may be considered an altitude, h.)

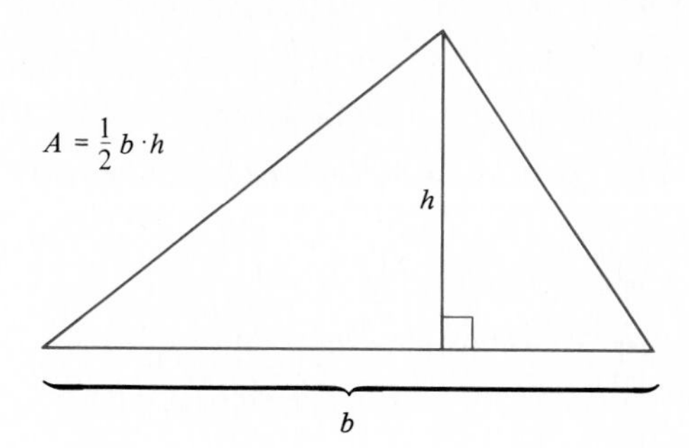

For Exercises 9–12, see Example 2.

9. Find two consecutive integers with a product that is 55 more than their sum.

10. Find two consecutive integers with a product that is 155 more than their sum.

11. Find two consecutive integers with a product that is 17 more than five times their sum.

12. Find two consecutive integers that have a product that is 10 more than 10 times their sum.

For Exercises 13–18, see Example 3.

13. A small model rocket is launched vertically upward from the ground. If the initial velocity of the rocket is 96 feet per second, how much time will elapse before the rocket **(a)** first reaches a height of 128 feet; **(b)** returns to the ground?

14. A small model rocket is launched vertically upward from the ground. The initial velocity of the rocket is 80 feet per second. How much time will elapse before the rocket **(a)** first reaches a height of 100 feet; **(b)** returns to the ground?

15. An arrow is shot vertically upward from ground level with an initial velocity of 112 feet per second. How much time will elapse before the arrow **(a)** first reaches a height of 160 feet; **(b)** returns to the ground?

16. An arrow is shot vertically upward from ground level with an initial velocity of 128 feet per second. How much time will elapse before the arrow **(a)** first reaches a height of 192 feet; **(b)** returns to the ground?

17. A man throws a ball vertically upward from the ground with an initial velocity of 72 feet per second. How much time will elapse before the ball **(a)** first reaches a height of 32 feet; **(b)** returns to the ground?

18. An object is projected vertically upward from the ground with a velocity of 144 feet per second. How much time will elapse before the object **(a)** first reaches a height of 288 feet; **(b)** returns to the ground?

For Exercises 19–22, see Example 4.

19. Find two consecutive odd integers such that their product is 37 less than 10 times their sum.

20. Find two consecutive odd integers with a product that is 1 less than 13 times their sum.

21. Find three consecutive odd integers such that the product of the first and last integers is 36 more than the sum of all three integers.

22. Find three consecutive odd integers with the property that the square of the middle integer is seven times the sum of all three integers.

For Exercises 23–27, Example 4 may be helpful.

23. Find two consecutive even integers with the property that their product is 14 more than their sum.

24. Find two consecutive even integers with the property that the square of the first plus the square of the second is 12 more than their product.

25. Find three consecutive positive integers whose squares sum to 194.

26. Find four consecutive positive integers whose squares have the sum of 86.

27. Find four consecutive even integers that are all positive such that the product of the first and last integers is 12 more than the sum of all four integers.

Key Words

Common factor
Composite number
Difference of two squares
Factor
Factor completely
Greatest common factor
Perfect square trinomial
Prime number
Quadratic equation
Remove or factor out
Root

CHAPTER 4 TEST

In Problems 1 and 2, find all the positive integer factors of the number.

1. 27 **2.** 29

3. Find the greatest common factor of 15, 24, and 30.

4. Find the GCF of $24x^2y$, $36xy^2$, and $12x^3y^2$.

For Problems 5 and 6, factor by removing the GCF.

5. $2x^4 - 6x^2y$ **6.** $9x^2y^3 + 12x^3y^2$

In Problems 7–16, factor the polynomial.

7. $x^2 - 25$ **8.** $9x^2 - 16$ **9.** $y^2 + 14y + 49$ **10.** $y^2 - 20y + 100$

11. $4x^2 + 32x + 64$ **12.** $x^4 + 12x^2 + 36$ **13.** $x^2 - 11x + 28$ **14.** $2x^2 + 7x + 5$

15. $3z^2 - 16z + 20$ **16.** $5w^3 - 18w^2 - 8w$

In Problems 17–20, solve the equation.

17. $x^2 - 49 = 0$ **18.** $x^2 + 10x + 25 = 0$

19. $y^3 - 4y^2 - 21y = 0$ **20.** $3x(x + 10) = 10x - 12$

For Problems 21 and 22, find the integers with the given properties.

21. Find two consecutive integers with a product that is 11 more than three times their sum.

22. Find three consecutive odd integers such that the sum of their squares is 8 more than seven times their sum.

CHAPTER 5

Rational Expressions

Electrical resistance is what makes toasters and irons hot. Computing the resistance, R, of an electrical circuit that has two parallel resistances R_1 and R_2 involves rational equations such as $\frac{1}{R} = \frac{1}{R_1} + \frac{1}{R_2}$.

Source: *PAR-NYC © 1983*

Thus far in our study of algebra we have been working with numbers and polynomials. In this chapter we study rational expressions and use them to solve practical problems from business, engineering, and the sciences. The successful solution of many problems in this chapter is dependent on your ability to factor polynomials.

5.1 Reduction of Rational Expressions

In Chapter 1, after we studied the integers, we learned about the rational numbers. Recall that a rational number is the quotient, $\frac{p}{q}$, of two integers with $q \neq 0$. For example,

$$\frac{2}{3}, \quad \frac{-5}{7}, \quad \frac{17}{5}, \quad \frac{4}{1}$$

are all rational numbers.

A *rational expression*, which is also called a fraction, is the quotient, $\frac{P}{Q}$, of two polynomials with $Q \neq 0$. For example,

$$\frac{2x + 1}{3x - 7}, \quad \frac{x^2 + 3x - 4}{7x}, \quad \frac{3xy + 2y^5}{x^2 + y^2}$$

are all rational expressions. Notice that since any integer is a polynomial, every rational number is also a rational expression.

The value of a rational expression is determined by the values of the variables in the expression. For example, $\frac{7x}{x - 2}$ has the value 21 when $x = 3$ because if we replace x with 3 we get

$$\frac{7(3)}{3 - 2} = \frac{21}{1} = 21.$$

Similarly, when $x = -1$,

$$\frac{7(-1)}{-1 - 2} = \frac{-7}{-3} = \frac{7}{3}.$$

Notice, however, that $\frac{7x}{x - 2}$ is undefined when $x = 2$ because

$$\frac{7(2)}{2 - 2} = \frac{14}{0}$$

and division by zero is undefined. A rational expression is not defined when its denominator has a value of zero.

Example 1. For what values of the variables is each rational expression undefined?

(a) $\dfrac{2x + 6}{3x - 9}$. (b) $\dfrac{y - 4}{y^2 - 2y}$. (c) $\dfrac{3z - 12}{z^2 + 1}$.

Solution: Each of these rational expressions is defined except when its denominator is zero. Thus, to find where they are undefined, we set each denominator equal to zero and solve for the variable.

(a) The denominator is $3x - 9$. Thus we solve $3x - 9 = 0$, or

$$3x = 9$$
$$x = 3.$$

Hence $\dfrac{2x + 6}{3x - 9}$ is undefined when $x = 3$.

(b) Setting the denominator, $y^2 - 2y$, equal to zero and factoring, we get

$$y^2 - 2y = 0$$
$$y(y - 2) = 0.$$

Hence $y = 0$ or $y - 2 = 0$. But $y - 2 = 0$ when $y = 2$. Thus the expression $\dfrac{y - 4}{y^2 - 2y}$ is undefined when $y = 0$ or $y = 2$.

(c) The denominator is $z^2 + 1$. But $z^2 \geq 0$ for all real numbers, thus $z^2 + 1 \geq 1$ for all real numbers. Therefore, $z^2 + 1$ is never zero. We conclude that $\dfrac{3z - 12}{z^2 + 1}$ is defined for all real numbers z.

The rational number $\dfrac{6}{10}$ may be rewritten as

$$\frac{3}{5}, \quad \frac{9}{15}, \quad \frac{18}{30}, \quad \text{or} \quad \frac{51}{85}.$$

We say that $\dfrac{3}{5}$ is the *reduced form* of these expressions. This is because there is no factor (other than ± 1) that is common to both the numerator and the denominator. Notice that $\dfrac{18}{30}$ is not *in reduced form* because 6 is a common factor of 18 and 30. That is,

$$\frac{18}{30} = \frac{6 \cdot 3}{6 \cdot 5}.$$

The process of removing all factors that are common to both the numerator and the denominator of a rational expression is called *reducing to lowest terms*. To reduce a rational expression to lowest terms, we first factor both the numerator and the denominator. Next we *remove any common factors* by applying the *Fundamental Property of Rational Expressions*.

Fundamental Property of Rational Expressions

> If P, Q, and D are polynomials with $Q \neq 0$ and $D \neq 0$, then
> $$\frac{P \cdot D}{Q \cdot D} = \frac{P}{Q}.$$

In this rule, the statements $Q \neq 0$ and $D \neq 0$ restrict the equality,

$$\frac{P \cdot D}{Q \cdot D} = \frac{P}{Q},$$

to those values of the variables that do not make either Q or D equal to zero. We now use the rule to reduce

$$\frac{5x^2 - 15x}{8x - 24}$$

to lowest terms. We begin by factoring the numerator and denominator. Thus

$$\frac{5x^2 - 15x}{8x - 24} = \frac{5x(x - 3)}{8(x - 3)}.$$

Since $(x - 3)$ is a factor in the numerator and the denominator, we may use the Fundamental Property of Rational Expressions and remove it. Thus we see that

$$\frac{5x^2 - 15x}{8x - 24} = \frac{5x\cancel{(x - 3)}}{8\cancel{(x - 3)}} = \frac{5x}{8}.$$

The placing of slash marks through the common factors indicates the application of the Fundamental Property of Rational Expressions. It is very important to note that

$$\frac{5x^2 - 15x}{8x - 24} = \frac{5x}{8}$$

only when the common factor $x - 3 \neq 0$: that is, only when $x \neq 3$. This is because $\frac{5x}{8}$ is defined for all values of x, but $\frac{5x^2 - 15x}{8x - 24}$ is undefined when $x = 3$.

When using the Fundamental Property of Rational Expressions, care should be taken to identify the values of the variables for which the denominator is equal to zero. The application of the Fundamental Property of Rational Expressions is sometimes referred to as "dividing the numerator and denominator by a common factor."

Example 2. Reduce $\frac{30y^4z^2}{24y^3z^5}$ to lowest terms.

Solution: Note that the denominator of this expression is zero when $y = 0$ or $z = 0$. Thus the rest of this solution assumes that $y \neq 0$ and $z \neq 0$.

$$\frac{30y^4z^2}{24y^3z^5} = \frac{\cancel{2} \cdot \cancel{3} \cdot 5 \cdot \cancel{y^3} \cdot y \cdot \cancel{z^2}}{\cancel{2} \cdot \cancel{3} \cdot 4 \cdot \cancel{y^3} \cdot z^3 \cdot \cancel{z^2}}$$

$$= \frac{5y}{4z^3}.$$

We conclude that the reduced form of

$$\frac{30y^4z^2}{24y^3z^5} \quad \text{is} \quad \frac{5y}{4z^3}.$$

Example 3. Reduce $\dfrac{5x^2 + 15}{7x^2 + 21}$ to lowest terms.

Solution: Factoring the numerator and denominator, we see that

$$\frac{5x^2 + 15}{7x^2 + 21} = \frac{5\cancel{(x^2 + 3)}}{7\cancel{(x^2 + 3)}}$$

$$= \frac{5}{7}.$$

Since $7x^2 + 21 = 7(x^2 + 3)$, we see that the denominator is never equal to zero. This is because $7 \neq 0$ and for all x, $x^2 \geq 0$, thus $x^2 + 3 \geq 3$. It follows that

$$\frac{5x^2 + 15}{7x^2 + 21} = \frac{5}{7}$$

for all real numbers x.

For the rest of this chapter we assume that the polynomial in the denominator of a rational expression is not evaluated at any number that would make it equal to zero. Thus we will usually not mention the restrictions that are implicit in the examples and exercises.

Example 4. Reduce $\dfrac{p^2 - 9}{p^2 + 6p + 9}$ to lowest terms.

Solution: Factoring both the numerator and denominator and then removing all common factors results in

$$\frac{p^2 - 9}{p^2 + 6p + 9} = \frac{(p - 3)\cancel{(p + 3)}}{(p + 3)\cancel{(p + 3)}} = \frac{p - 3}{p + 3}.$$

Example 5. Reduce $\frac{a - b}{b - a}$ to lowest terms.

Solution: Recall that we are looking for expressions that are factors of both the numerator and denominator. Notice that

$$\frac{a - b}{b - a} = \frac{1\cancel{(a - b)}}{-1\cancel{(a - b)}}$$

$$= \frac{1}{-1}$$

$$= -1.$$

Example 6. Reduce $\frac{x^3 + 6x^2 + 8x}{x^3 + 7x^2 + 10x}$ to lowest terms.

Solution: If we first factor the numerator and denominator and then remove any common factors, the result is

$$\frac{x^3 + 6x^2 + 8x}{x^3 + 7x^2 + 10x} = \frac{x(x^2 + 6x + 8)}{x(x^2 + 7x + 10)}$$

$$= \frac{\cancel{x(x + 2)}(x + 4)}{\cancel{x(x + 2)}(x + 5)}$$

$$= \frac{x + 4}{x + 5}.$$

EXERCISES 5.1

For Exercises 1–20, find the values if any, for which the rational expression is not defined. See Example 1.

1. $\frac{3z}{4z^2}$ **2.** $\frac{7y}{11y^3}$ **3.** $\frac{2x^2}{5x^3}$ **4.** $\frac{3w^2}{7w^3}$

5. $\frac{4}{7x}$ **6.** $\frac{5}{6y}$ **7.** $\frac{3z}{2}$ **8.** $\frac{5x}{8}$

9. $\frac{5}{p - 3}$ **10.** $\frac{4}{x - 2}$ **11.** $\frac{13}{z + 1}$ **12.** $\frac{3}{w + 4}$

13. $\frac{2x}{x - 5}$ **14.** $\frac{5w}{w - 2}$ **15.** $\frac{3x + 12}{x^2 - 6x + 8}$ **16.** $\frac{3x + 6}{x^2 - 5x + 4}$

17. $\frac{7q}{q^2 + 2q + 1}$ **18.** $\frac{8y}{y^2 + 3y + 2}$ **19.** $\frac{2x + 10}{x^2 + 7x + 10}$ **20.** $\frac{4y + 8}{y^2 + 8y + 12}$

For Exercises 21–50, reduce the expression to lowest terms. See Examples 2–6.

21. $\dfrac{26}{39}$ **22.** $\dfrac{84}{126}$ **23.** $\dfrac{3p}{7p}$

24. $\dfrac{6x^2}{12x^2}$ **25.** $\dfrac{5y^2}{20y}$ **26.** $\dfrac{8z^3}{72z^5}$

27. $\dfrac{-25x}{5x^3}$ **28.** $\dfrac{27q^5}{-81q^3}$ **29.** $\dfrac{14(x-3)}{28(x-3)}$

30. $\dfrac{35(m+7)}{14(m+7)}$ **31.** $\dfrac{-12(y+5)}{4(y+5)}$ **32.** $\dfrac{-9(z-11)}{36(z-11)}$

33. $\dfrac{15p-30}{20p-40}$ **34.** $\dfrac{14q+70}{21q+105}$ **35.** $\dfrac{27y+54}{9y+18}$

36. $\dfrac{24x-48}{4x-8}$ **37.** $\dfrac{x+1}{x^2+2x+1}$ **38.** $\dfrac{w-1}{w^2-2w+1}$

39. $\dfrac{y-4}{y^2-8y+16}$ **40.** $\dfrac{z+3}{z^2+6z+9}$ **41.** $\dfrac{p^2-p-2}{p^2+3p+2}$

42. $\dfrac{q^2+q-30}{q^2-11q+30}$ **43.** $\dfrac{x^2-4x-21}{x^2-3x-28}$ **44.** $\dfrac{z^2-7z+10}{z^2+4z-12}$

45. $\dfrac{p+q}{-p-q}$ **46.** $\dfrac{y-2}{2-y}$ **47.** $\dfrac{x^3+4x^2-32x}{x^3+x^2-20x}$

48. $\dfrac{q^3-q^2-6q}{q^3+6q^2+8q}$ **49.** $\dfrac{2y^3+16y^2+24y}{2y^3-8y^2+24y}$ **50.** $\dfrac{5z^3+45z^2+100z}{5z^3+5z^2+60z}$

Products and Quotients of Rational Expressions 5.2

We know that to find the product of two fractions, we multiply numerator by numerator and denominator by denominator. For example,

$$\frac{3}{4}\cdot\frac{5}{7}=\frac{3\cdot 5}{4\cdot 7}=\frac{15}{28}.$$

The same method is used to find the product of two rational expressions.

Product of Two Rational Expressions

> If P, Q, R, and S are polynomials with $Q \neq 0$ and $S \neq 0$, then
>
> $$\frac{P}{Q}\cdot\frac{R}{S}=\frac{P\cdot R}{Q\cdot S}.$$

When finding the product of two rational expressions, it is common practice to reduce the answer to lowest terms. For example, to multiply $\frac{y^3}{15}$ and $\frac{20}{y^2}$ we proceed as follows:

$$\begin{aligned}\frac{y^3}{15} \cdot \frac{20}{y^2} &= \frac{20y^3}{15y^2}\\ &= \frac{4 \cdot \cancel{5} \cdot \cancel{y^2} \cdot y}{3 \cdot \cancel{5} \cdot \cancel{y^2}}\\ &= \frac{4y}{3}.\end{aligned}$$

Often we remove all common factors that will appear in the final product before multiplying numerator by numerator and denominator by denominator. For example,

$$\begin{aligned}\frac{x^5}{20} \cdot \frac{24}{x^7} &= \frac{\cancel{x^5}}{\cancel{4} \cdot 5} \cdot \frac{\cancel{4} \cdot 6}{\cancel{x^5} \cdot x^2}\\ &= \frac{6}{5x^2}.\end{aligned}$$

This latter method is particularly advantageous when the rational expressions become more complicated.

Example 1. Find the product of $\frac{(a+b)^3}{15c^4}$ and $\frac{25c}{(a+b)^2}$.

Solution

$$\frac{(a+b)^3}{15c^4} \cdot \frac{25c}{(a+b)^2}$$ form the product

$$= \frac{\cancel{(a+b)^2} \cdot (a+b)}{3 \cdot \cancel{5} \cdot \cancel{c} \cdot c^3} \cdot \frac{5 \cdot \cancel{5} \cdot \cancel{c}}{\cancel{(a+b)^2}}$$ factor the numerators and denominators and then remove all common factors

$$= \frac{(a+b)}{3c^3} \cdot \frac{5}{1}$$ multiply the remaining factors

$$= \frac{5(a+b)}{3c^3} = \frac{5a+5b}{3c^3}.$$

Thus

$$\frac{(a+b)^3}{15c^4} \cdot \frac{25c}{(a+b)^4} = \frac{5a+5b}{3c^3}.$$

Example 2. Find the product of $\dfrac{x^2 - x - 2}{x^2 - 9}$ and $\dfrac{x^2 - 6x + 9}{x^2 + 6x + 5}$.

Solution

$$\frac{x^2 - x - 2}{x^2 - 9} \cdot \frac{x^2 - 6x + 9}{x^2 + 6x + 5}$$ form the product

$$= \frac{(x - 2)\cancel{(x + 1)}}{\cancel{(x - 3)}(x + 3)} \cdot \frac{\cancel{(x - 3)}(x - 3)}{\cancel{(x + 1)}(x + 5)}$$ factor and remove all common factors

$$= \frac{(x - 2)(x - 3)}{(x + 3)(x + 5)}$$ multiply the remaining factors

$$= \frac{x^2 - 5x + 6}{x^2 + 8x + 15}.$$

Thus

$$\frac{x^2 - x - 2}{x^2 - 9} \cdot \frac{x^2 - 6x + 9}{x^2 + 6x + 5} = \frac{x^2 - 5x + 6}{x^2 + 8x + 15}.$$

We know that to find the quotient of two fractions, we multiply the dividend by the reciprocal of the divisor. For example,

$$\frac{3}{7} \div \frac{2}{11} = \frac{3}{7} \cdot \frac{11}{2} = \frac{33}{14}.$$

The same method is used to find the quotient of two rational expressions.

Quotient of Two Rational Expressions

If P, Q, R, and S are polynomials with $Q \neq 0$, $R \neq 0$ and $S \neq 0$, then

$$\frac{P}{Q} \div \frac{R}{S} = \frac{P}{Q} \cdot \frac{S}{R}.$$

Thus division of rational expressions is performed in terms of multiplication of rational expressions.

Example 3. Find the quotient $\frac{3a}{a+2} \div \frac{5a^2}{(a+2)^2}$.

Solution

$$\frac{3a}{a+2} \div \frac{5a^2}{(a+2)^2}$$

$$= \frac{3a}{a+2} \cdot \frac{(a+2)^2}{5a^2}$$ rewrite the division problem as a multiplication problem

$$= \frac{3\cancel{a}}{\cancel{a+2}} \cdot \frac{\cancel{(a+2)}(a+2)}{5 \cdot a \cdot \cancel{a}}$$ remove all common factors

$$= \frac{3(a+2)}{5a} = \frac{3a+6}{5a}.$$ multiply the remaining factors

Example 4. Divide $\frac{x^2 - 5x + 4}{x^2 + 10x + 25} \div \frac{x^2 - 2x - 8}{x^2 + 7x + 10}$.

Solution

$$\frac{x^2 - 5x + 4}{x^2 + 10x + 25} \div \frac{x^2 - 2x - 8}{x^2 + 7x + 10}$$

$$= \frac{x^2 - 5x + 4}{x^2 + 10x + 25} \cdot \frac{x^2 + 7x + 10}{x^2 - 2x - 8}$$ change the division problem to a multiplication problem

$$= \frac{\cancel{(x-4)}(x-1)}{(x+5)\cancel{(x+5)}} \cdot \frac{\cancel{(x+5)}\cancel{(x+2)}}{\cancel{(x-4)}\cancel{(x+2)}}$$ factor and remove all common factors

$$= \frac{(x-1)}{(x+5)} = \frac{x-1}{x+5}.$$ multiply the remaining factors

From our work in the first two sections of this chapter, it should now be clear that the ability to factor polynomials is an important skill when working with rational expressions.

EXERCISES 5.2

For Exercises 1–12, find the product. Reduce your answer to lowest terms. See Example 1.

1. $\frac{3a^2}{35} \cdot \frac{14}{6a}$
2. $\frac{8x^5}{9x} \cdot \frac{3}{4x^3}$
3. $\frac{10x^2}{7y^3} \cdot \frac{21y^2}{5x^2}$
4. $\frac{13m^3}{12n^2} \cdot \frac{6n}{26m^3}$
5. $\frac{9z^4}{2w} \cdot \frac{4w^2}{3z^3}$
6. $\frac{81x^4}{11y^5} \cdot \frac{121y^4}{9x^3}$

7. $\frac{(x+y)^2}{18z} \cdot \frac{16z^2}{(x+y)^3}$

8. $\frac{(c+d)}{15a^4} \cdot \frac{25a^5}{(c+d)^5}$

9. $\frac{4(p+q)^3}{9r^3} \cdot \frac{3r^2}{8(p+q)}$

10. $\frac{7(m+n)^5}{16} \cdot \frac{24}{14(m+n)^6}$

11. $\frac{11c^4}{(a+b)^5} \cdot \frac{(a+b)^4}{22c^3}$

12. $\frac{36x^2}{9(y+1)^3} \cdot \frac{2(y+1)^2}{48x}$

For Exercises 13–26, find the product. Reduce your answer to lowest terms. See Example 2.

13. $\frac{(x+1)(x-2)}{(x+7)} \cdot \frac{(x+7)^2}{(x-1)(x-2)}$

14. $\frac{(a+6)^2(a+1)}{(a+5)} \cdot \frac{(a+5)}{(a+6)(a+1)}$

15. $\frac{3x+12}{7x} \cdot \frac{8x}{2x+8}$

16. $\frac{5a-15}{3a^2} \cdot \frac{14a^2}{7a-21}$

17. $\frac{4y+12}{6y-12} \cdot \frac{5y-10}{3y+9}$

18. $\frac{9c+27}{2c-14} \cdot \frac{3c-21}{11c+33}$

19. $\frac{x^2+7x+12}{x^2-4x+4} \cdot \frac{x^2-4}{x^2+6x+8}$

20. $\frac{a^2-7a+12}{a^2+4a+4} \cdot \frac{a^2-4}{a^2-6a+8}$

21. $\frac{y^3-25y}{y^2-3y-10} \cdot \frac{y^2+y-2}{y^3+5y^2}$

22. $\frac{b^4-b^2}{b^2+3b+2} \cdot \frac{b^2+4b+4}{b^2-b}$

23. $\frac{x^2-4x-21}{x^2-3x-28} \cdot \frac{x^2+6x+8}{x^2+10x+21}$

24. $\frac{c^2+c-2}{c^2+5c+6} \cdot \frac{c^2+7c+12}{c^2+7c-8}$

25. $\frac{2z^2-5z-3}{2z^2+z-1} \cdot \frac{2z^2+3z-2}{2z^2+5z+2}$

26. $\frac{3w^2-2w-1}{2w^2-w-1} \cdot \frac{2w^2-3w-2}{3w^2-5w-2}$

For Exercises 27–34, find the quotient. Reduce your answer to lowest terms. See Example 3.

27. $\frac{9y^3}{2} \div \frac{3y}{4}$

28. $\frac{5m^2}{6} \div \frac{25m^3}{18}$

29. $\frac{27}{4x^3} \div \frac{9}{2x}$

30. $\frac{4}{7c^5} \div \frac{2}{14c^6}$

31. $\frac{1}{(y+1)^3} \div \frac{3}{(y+1)}$

32. $\frac{2}{(a-2)} \div \frac{1}{(a-2)^2}$

33. $\frac{3b}{(b-4)^5} \div \frac{9b^2}{(b-4)^3}$

34. $\frac{5w^2}{(w+1)} \div \frac{w}{(w+1)^3}$

For Exercises 35–46, find the quotient. Reduce your answer to lowest terms. See Example 4.

35. $\frac{2x+10}{x-3} \div \frac{x+5}{3x-9}$

36. $\frac{a+7}{5a+15} \div \frac{4a+28}{a+3}$

37. $\frac{3y+12}{2y-10} \div \frac{5y+20}{3y-15}$

38. $\frac{7b+14}{2b-14} \div \frac{3b+6}{3b-21}$

39. $\frac{5a+25}{3a-6} \div \frac{a^2-25}{a^2-4}$

40. $\frac{z^2-16}{z^2-1} \div \frac{6z+24}{5z-5}$

41. $\dfrac{x^2 - 3x + 2}{x^2 + x - 6} \div \dfrac{x^2 + 2x - 3}{x^2 + 4x + 3}$

42. $\dfrac{p^2 + 8p + 7}{p^2 - 2p - 3} \div \dfrac{p^2 + 3p + 2}{p^2 - p - 6}$

43. $\dfrac{18 - 2c^2}{c^2 + 4c + 3} \div \dfrac{15 - 20c + 5c^2}{c^2 - 1}$

44. $\dfrac{2b^2 - b - 3}{b^2 - 3b + 2} \div \dfrac{10b^2 - 5b - 15}{3b^2 + 3b - 6}$

45. $\dfrac{3w^2 - 4w + 1}{2w^2 - w - 1} \div \dfrac{3w^2 - 16w + 5}{2w^2 + 3w + 1}$

46. $\dfrac{5a^2 - 2a + 3}{6a^2 + 13a + 6} \div \dfrac{5a^2 + 2a - 3}{2a^2 + 5a + 3}$

For Exercises 47–50, perform the operations indicated. Reduce your answer to lowest terms.

47. $\left(\dfrac{x^2 - x - 2}{x^2 + 5x} \cdot \dfrac{2x^2}{x^2 - 3x + 2}\right) \div \dfrac{x + 1}{x + 5}$

48. $\left(\dfrac{a^2 + 7a + 10}{7a^3 - 21a^2} \cdot \dfrac{a^3}{a^2 + a - 2}\right) \div \dfrac{a + 5}{a - 1}$

49. $\left(\dfrac{2z^2 + 4z - 30}{3z + 3} \div \dfrac{z^2 - 4z + 3}{z^2 + 2z + 1}\right) \cdot \dfrac{3z - 3}{7z + 35}$

50. $\left(\dfrac{4q^2 - 4}{q^2 + 5q + 6} \div \dfrac{6q^3 + 6q^2}{q^2 + q - 2}\right) \cdot \dfrac{3q^3}{q^2 - 2q + 1}$

5.3 Addition and Subtraction of Rational Expressions

When we add two fractions that have the same denominator, we add the numerators and place this sum over the common denominator. Thus

$$\frac{a}{c} + \frac{b}{c} = \frac{a + b}{c}.$$

We use the same procedure to add two rational expressions that have a common denominator.

Addition of Two Rational Expressions

If P, Q, and R are polynomials with $Q \neq 0$, then

$$\frac{P}{Q} + \frac{R}{Q} = \frac{P + R}{Q}.$$

Example 1. Add

(a) $\dfrac{7}{x}$ and $\dfrac{2}{x}$. (b) $\dfrac{x}{x + 5}$ and $\dfrac{x - 3}{x + 5}$.

Solution: Since in each part both rational expressions have the same denominator, we simply add the numerators and keep the common denominator.

(a) $\dfrac{7}{x} + \dfrac{2}{x} = \dfrac{7+2}{x} = \dfrac{9}{x}$.

(b) $\dfrac{x}{x+5} + \dfrac{x-3}{x+5} = \dfrac{x+x-3}{x+5} = \dfrac{2x-3}{x+5}$.

As the next example illustrates, after adding two rational expressions, it may be possible to reduce the answer.

Example 2. Find each sum.

(a) $\dfrac{8}{3y} + \dfrac{10}{3y}$. (b) $\dfrac{5x+11}{2x+3} + \dfrac{x-2}{2x+3}$.

Solution

(a) $\dfrac{8}{3y} + \dfrac{10}{3y} = \dfrac{18}{3y} = \dfrac{\cancel{3} \cdot 6}{\cancel{3} \cdot y} = \dfrac{6}{y}$.

(b) $\dfrac{5x+11}{2x+3} + \dfrac{x-2}{2x+3} = \dfrac{6x+9}{2x+3}$

$= \dfrac{3\cancel{(2x+3)}}{\cancel{(2x+3)}}$

$= 3.$

Before we can add two rational expressions, they must have the same denominator. Thus before we can add

$$\frac{7}{15x^3} + \frac{3}{20x},$$

we must rewrite each rational expression so that both have the same denominator. We usually find the *least common denominator* using the following procedure.

Step 1. Factor each denominator completely.

Step 2. Form the product of all the distinct factors from step 1.

Step 3. Raise each factor to the largest power that it has in step 1.

Applying these steps to $\dfrac{7}{15x^3} + \dfrac{3}{20x}$, we get

Step 1. $15x^3 = 3 \cdot 5 \cdot x^3$ and $20x = 2^2 \cdot 5 \cdot x$.

Step 2. $2 \cdot 3 \cdot 5 \cdot x$.

Step 3. $2^2 \cdot 3 \cdot 5 \cdot x^3 = 4 \cdot 3 \cdot 5 \cdot x^3 = 60x^3$.

Thus the least common denominator is $60x^3$.

The Fundamental Property of Rational Expressions states that

$$\frac{P \cdot D}{Q \cdot D} = \frac{P}{Q} \quad \text{or, equivalently,} \quad \frac{P}{Q} = \frac{P \cdot D}{Q \cdot D},$$

where P, Q, and D are polynomials with $Q \neq 0$ and $D \neq 0$. We will use this property to rewrite

$$\frac{7}{15x^3} + \frac{3}{20x}$$

as the sum of two rational expressions, each with the same denominator. To do this, we multiply the numerator and denominator of each term by the factors of the least common denominator that are not factors of its denominator. Since $2^2 = 4$ is the factor of $60x^3$ that is not a factor of $15x^3$, we multiply the numerator and denominator of $\frac{7}{15x^3}$ by 4. Similarly, we multiply the numerator and denominator of $\frac{3}{20x}$ by $3x^2$. Thus

$$\frac{7}{15x^3} + \frac{3}{20x} = \frac{4}{4} \cdot \frac{7}{15x^3} + \frac{3x^2}{3x^2} \cdot \frac{3}{20x}$$

$$= \frac{28}{60x^3} + \frac{9x^2}{60x^3}.$$

Notice that we replaced each fraction with a fraction that has a denominator of $60x^3$. Once we have written the fractions in our problem with the least common denominator, we add them as in Examples 1 and 2. Thus

$$\frac{7}{15x^3} + \frac{3}{20x} = \frac{28}{60x^3} + \frac{9x^2}{60x^3}$$

$$= \frac{9x^2 + 28}{60x^3}.$$

It is customary to use *LCD* to denote the least common denominator.

Example 3. Find the sum of $\frac{7}{6x - 4}$ and $\frac{2x}{3x - 2}$.

Solution: First we factor each denominator. The denominator of the first expression is

$$6x - 4 = 2(3x - 2).$$

The denominator of the second expression, $3x - 2$, is not factorable. Next we form the LCD, $2(3x - 2)$, and rewrite each rational expression with the LCD as the denominator. Thus

$$\frac{7}{6x-4} + \frac{2x}{3x-2} = \frac{7}{2(3x-2)} + \frac{2x}{(3x-2)}$$

$$= \frac{7}{2(3x-2)} + \frac{2}{2} \cdot \frac{2x}{(3x-2)}$$

$$= \frac{7}{6x-4} + \frac{4x}{6x-4}$$

$$= \frac{7+4x}{6x-4} = \frac{4x+7}{6x-4}.$$

Therefore,

$$\frac{7}{6x-4} + \frac{2x}{3x-2} = \frac{4x+7}{6x-4}.$$

As mentioned previously, after adding or subtracting two rational expressions, it may be possible to reduce the answer. The next two examples demonstrate this statement.

Example 4. Subtract $\dfrac{2x}{x^2-9} - \dfrac{1}{x-3}$.

Solution: First we factor each denominator. The denominator of the first expression is $x^2 - 9 = (x-3)(x+3)$; the second denominator, $x - 3$, is not factorable. Thus the LCD is $(x-3)(x+3)$. Next we rewrite the problem as

$$\frac{2x}{x^2-9} - \frac{1}{x-3} = \frac{2x}{(x-3)(x+3)} - \frac{1(x+3)}{(x-3)(x+3)}$$

$$= \frac{2x-(x+3)}{(x-3)(x+3)}$$

$$= \frac{2x-x-3}{(x-3)(x+3)}$$

$$= \frac{\cancel{x-3}}{\cancel{(x-3)}(x+3)}$$

$$= \frac{1}{x+3}.$$

Example 5. Find $\frac{2}{3} + \frac{x}{x+1} - \frac{2}{x}$.

Solution: First we multiply the numerator and denominator of each term by an appropriate polynomial so that all the terms have the same denominator of $3x(x + 1)$. Thus we get

$$\frac{2}{3} \cdot \frac{x(x+1)}{x(x+1)} + \frac{x}{x+1} \cdot \frac{3x}{3x} - \frac{2}{x} \cdot \frac{3(x+1)}{3(x+1)}$$

$$= \frac{2x^2 + 2x}{3x(x+1)} + \frac{3x^2}{3x(x+1)} - \frac{6x+6}{3x(x+1)}$$

$$= \frac{5x^2 - 4x - 6}{3x(x+1)}.$$

Since the numerator cannot be factored, we conclude that

$$\frac{2}{3} + \frac{x}{x+1} - \frac{2}{x} = \frac{5x^2 - 4x - 6}{3x^2 + 3x}.$$

EXERCISES 5.3

Find the sum or difference. When possible, reduce your answer to lowest terms.

For Exercises 1–12, see Examples 1 and 2.

1. $\frac{5}{x} + \frac{3}{x}$ **2.** $\frac{7}{y} + \frac{2}{y}$ **3.** $\frac{4}{b} - \frac{1}{b}$

4. $\frac{6}{b} - \frac{2}{b}$ **5.** $\frac{7}{2z} + \frac{3}{2z}$ **6.** $\frac{11}{3z} - \frac{2}{3z}$

7. $\frac{4}{2x-1} + \frac{3}{2x-1}$ **8.** $\frac{5}{a+7} + \frac{6}{a+7}$ **9.** $\frac{13}{y+4} - \frac{2}{y+4}$

10. $\frac{17}{6c+5} - \frac{11}{6c+5}$ **11.** $\frac{3x+1}{x+1} + \frac{2}{x+1}$ **12.** $\frac{5y+9}{2y+3} - \frac{y+3}{2y+3}$

For Exercises 13–50, see Examples 3–5.

13. $\frac{2}{a} + \frac{a}{5}$ **14.** $\frac{3}{x} + \frac{x}{4}$

15. $\frac{5}{2} - \frac{2}{x}$ **16.** $\frac{3}{7} - \frac{1}{y}$

17. $\dfrac{b}{7} + \dfrac{1}{b}$

18. $\dfrac{z}{4} + \dfrac{3}{z}$

19. $\dfrac{4}{x+2} + \dfrac{1}{x}$

20. $\dfrac{3}{x-1} + \dfrac{2}{x}$

21. $\dfrac{7}{a+5} - \dfrac{2}{a}$

22. $\dfrac{5}{c-7} - \dfrac{3}{c}$

23. $\dfrac{z-1}{z(z+3)} + \dfrac{5}{z}$

24. $\dfrac{a+2}{a(a-5)} + \dfrac{3}{a}$

25. $\dfrac{p-2}{p(p+1)} - \dfrac{2}{p}$

26. $\dfrac{m+3}{m(m-1)} - \dfrac{1}{m}$

27. $\dfrac{x+5}{(x-1)(x+2)} + \dfrac{7}{(x+2)}$

28. $\dfrac{y-3}{(y+1)(y-2)} + \dfrac{2}{(y+1)}$

29. $\dfrac{a+1}{(a+2)(a+4)} - \dfrac{1}{(a+2)}$

30. $\dfrac{b+2}{(b+3)(b-1)} - \dfrac{2}{(b+3)}$

31. $\dfrac{8}{x^2-4} + \dfrac{x+1}{x+2}$

32. $\dfrac{7}{x^2-16} + \dfrac{x-1}{x-4}$

33. $\dfrac{11}{y^2-9} - \dfrac{10y+1}{y+3}$

34. $\dfrac{4}{z^2-25} - \dfrac{z+2}{z+5}$

35. $\dfrac{z-1}{z^2-36} + \dfrac{z+2}{z+6}$

36. $\dfrac{w+2}{w^2-49} + \dfrac{w-1}{w+7}$

37. $\dfrac{5}{a+1} + \dfrac{1}{a-1}$

38. $\dfrac{7}{b+3} + \dfrac{4}{b-3}$

39. $\dfrac{10}{x-1} - \dfrac{4}{x+1}$

40. $\dfrac{6}{y-2} - \dfrac{3}{y+2}$

41. $\dfrac{z}{z+4} + \dfrac{z+1}{z-4}$

42. $\dfrac{y}{y+3} + \dfrac{y-1}{y-3}$

43. $\dfrac{p}{p-1} - \dfrac{p-2}{p+1}$

44. $\dfrac{q}{q+5} - \dfrac{q+1}{q-5}$

45. $\dfrac{x}{x^2-1} + \dfrac{x-1}{x^2+3x+2}$

46. $\dfrac{y+1}{y^2+5y+6} + \dfrac{y}{y^2-4}$

47. $\dfrac{a+1}{a^2-9} - \dfrac{a+1}{a^2+5a+6}$

48. $\dfrac{b+3}{b^2+8b+16} - \dfrac{b-3}{b^2-16}$

49. $\dfrac{1}{5} + \dfrac{x}{x-1} - \dfrac{2x}{x+1}$

50. $\dfrac{2}{7} - \dfrac{y}{y+3} + \dfrac{y}{y-3}$

5.4 Complex Fractions

A *complex fraction* is a rational expression with a numerator or denominator that contains a fraction or fractions. Thus

$$\frac{5}{\frac{3}{4}}, \quad \frac{\frac{1}{4}}{\frac{3}{5}}, \quad \frac{3 + \frac{1}{x}}{x}, \quad \frac{\frac{2y}{3}}{y - \frac{1}{3}}, \quad \frac{\frac{z}{z-1}}{\frac{3z}{z+1}}$$

are all examples of complex fractions. Such expressions are often difficult to interpret. In this section we present two methods for rewriting complex fractions.

A complex fraction is *simplified* when it has been rewritten as a rational expression that does not contain fractions in either the numerator or the denominator and it has been reduced to lowest terms. There are two methods that are most often used to simplify complex fractions. They are

Method 1. Treat the complex fraction as a division problem and multiply the numerator by the reciprocal of the denominator of the complex fraction.

Method 2. Multiply the numerator and the denominator by the LCD of all the fractions in both the numerator and the denominator.

These methods will be demonstrated using the complex fractions given above.

Example 1. Simplify $\frac{\frac{1}{4}}{\frac{3}{5}}$.

Solution

Method 1. Consider this as a division problem and multiply the numerator by the reciprocal of the denominator. Thus

$$\frac{\frac{1}{4}}{\frac{3}{5}} = \frac{1}{4} \div \frac{3}{5} = \frac{1}{4} \cdot \frac{5}{3} = \frac{5}{12}.$$

Method 2. Multiply the numerator and denominator by the LCD of the fractions in the numerator and denominator. The LCD of $\frac{1}{4}$ and $\frac{3}{5}$ is 20. Thus

$$\frac{\frac{1}{4}}{\frac{3}{5}} = \frac{20 \cdot \frac{1}{4}}{20 \cdot \frac{3}{5}} = \frac{5}{12}.$$

Example 2. Simplify $\frac{3 + \frac{1}{x}}{x}$.

Solution

Method 1. First add

$$3 + \frac{1}{x} = \frac{3x}{x} + \frac{1}{x} = \frac{3x + 1}{x}.$$

Also note that $x = \dfrac{x}{1}$. Thus

$$\frac{3 + \frac{1}{x}}{x} = \frac{\frac{3x + 1}{x}}{\frac{x}{1}} = \frac{3x + 1}{x} \cdot \frac{1}{x}$$

$$= \frac{3x + 1}{x^2}$$

Method 2. The LCD of the fractions in the numerator and denominator is x. Thus

$$\frac{3 + \frac{1}{x}}{x} = \frac{x\left(3 + \frac{1}{x}\right)}{x \cdot x} = \frac{3x + 1}{x^2}.$$

Note in Example 2 that method 2 is more efficient than method 1. When either the numerator or denominator of a complex fraction contains the sum or difference of fractions with unequal denominators, method 2 is preferred to method 1; otherwise, method 1 is preferred.

Example 3. Simplify $\dfrac{\frac{2y}{3}}{y - \frac{1}{3}}$.

Solution: Because the denominator, $y - \frac{1}{3}$, is the difference of two rational expressions with unequal denominators, we will use method 2. Since the LCD is 3, we have

$$\frac{\frac{2y}{3}}{y - \frac{1}{3}} = \frac{3\left(\frac{2y}{3}\right)}{3(y - \frac{1}{3})}$$

$$= \frac{2y}{3y - 1}.$$

The next example demonstrates the importance of factoring when simplifying complex fractions.

Example 4. Simplify

$$\frac{\dfrac{z-1}{z^2+3z+2}}{\dfrac{z^2+3z-4}{z^2+5z+6}}.$$

Solution: Method 1 will be more efficient for this problem. Multiplying the numerator by the reciprocal of the denominator and factoring, we get

$$\frac{z-1}{z^2+3z+2}\cdot\frac{z^2+5z+6}{z^2+3z-4} = \frac{z-1}{(z+1)(z+2)}\cdot\frac{(z+2)(z+3)}{(z-1)(z+4)}$$

$$= \frac{z+3}{(z+1)(z+4)}$$

$$= \frac{z+3}{z^2+5z+4}.$$

EXERCISES 5.4

For Exercises 1–30, simplify the complex fraction. Remember the comment that precedes Example 3 and refer to Examples 3 and 4.

1. $\dfrac{\frac{7}{9}}{\frac{1}{5}}$ **2.** $\dfrac{\frac{2}{7}}{\frac{3}{8}}$ **3.** $\dfrac{\dfrac{x}{y}}{\dfrac{x^2}{y^3}}$

4. $\dfrac{\dfrac{c^2}{d}}{\dfrac{c}{d^3}}$ **5.** $\dfrac{\dfrac{5w^3}{z^2}}{\dfrac{10w}{z}}$ **6.** $\dfrac{\dfrac{7p^4}{18q}}{\dfrac{14p}{9q^3}}$

7. $\dfrac{1+\dfrac{1}{a}}{3a}$ **8.** $\dfrac{2+\dfrac{3}{b}}{b}$ **9.** $\dfrac{3}{1-\dfrac{5}{c}}$

10. $\dfrac{1}{2-\dfrac{3}{d}}$ **11.** $\dfrac{a-\dfrac{1}{b}}{b-\dfrac{1}{a}}$ **12.** $\dfrac{z+\dfrac{1}{y}}{y+\dfrac{1}{z}}$

13. $\dfrac{2x + \dfrac{1}{3x}}{x - \dfrac{1}{2x}}$ **14.** $\dfrac{3y - \dfrac{1}{2y}}{y + \dfrac{1}{3y}}$ **15.** $\dfrac{\dfrac{(a + 2)}{(a - 1)(a - 3)}}{\dfrac{(a + 1)(a + 2)}{(a - 3)}}$

16. $\dfrac{\dfrac{(x - 5)}{(x + 2)(x + 3)}}{\dfrac{(x - 1)(x - 5)}{(x + 3)}}$ **17.** $\dfrac{\dfrac{(b - 3)}{b^2 - 4}}{\dfrac{b^2 - 9}{b + 2}}$ **18.** $\dfrac{\dfrac{y + 5}{y^2 - 9}}{\dfrac{y^2 - 25}{y - 3}}$

19. $\dfrac{\dfrac{x^2 - 1}{x + 6}}{\dfrac{x + 1}{x^2 - 36}}$ **20.** $\dfrac{\dfrac{w^2 - 64}{w - 9}}{\dfrac{w - 8}{w^2 - 81}}$ **21.** $\dfrac{\dfrac{1}{ab}}{\dfrac{1}{a} + \dfrac{1}{b}}$

22. $\dfrac{\dfrac{1}{c} - \dfrac{1}{d}}{\dfrac{1}{cd}}$ **23.** $\dfrac{\dfrac{1}{x} + \dfrac{1}{xy}}{\dfrac{1}{x} + \dfrac{1}{y}}$ **24.** $\dfrac{\dfrac{1}{x} - \dfrac{1}{y}}{\dfrac{1}{xy} - \dfrac{1}{x}}$

25. $\dfrac{1 - \dfrac{3}{a + 5}}{\dfrac{3}{5 + a} - 1}$ **26.** $\dfrac{\dfrac{2}{b + 3} - 5}{5 - \dfrac{2}{3 + b}}$ **27.** $\dfrac{\dfrac{x + 1}{x^2 + 4x + 3}}{\dfrac{x^2 + 5x + 4}{x + 3}}$

28. $\dfrac{\dfrac{z - 3}{z^2 + 5z + 6}}{\dfrac{z - 3}{z^2 + 4z + 3}}$ **29.** $\dfrac{\dfrac{a^2 - a - 12}{a^2 - 6a + 5}}{\dfrac{a^2 + 7a + 12}{a^2 - 1}}$ **30.** $\dfrac{\dfrac{y^2 - 9}{y^2 + 4y + 4}}{\dfrac{y^2 - 6y + 9}{y^2 - 4}}$

Equations with Rational Expressions 5.5

To solve an equation that contains *rational expressions,* we multiply each side of the equation by *the LCD of the rational expressions* in the equation. However, if the LCD contains a variable, the resulting equation may have solutions that are not solutions to the original problem. Such solutions are called *extraneous solutions*. They can be identified because they will fail to satisfy the original equation. The process of multiplying each side of an equation by the LCD is called *clearing fractions*. We illustrate the preceding discussion in the following examples.

Example 1. Solve $\dfrac{7}{x - 1} + \dfrac{1}{2} = 1$.

Solution: We clear fractions by multiplying both sides of the equation by the LCD $2(x - 1)$. The result of this multiplication is

$$2(x-1)\left(\frac{7}{x-1} + \frac{1}{2}\right) = 2(x-1)(1)$$

$$2\cancel{(x-1)} \cdot \frac{7}{\cancel{(x-1)}} + \cancel{2}(x-1)\frac{1}{\cancel{2}} = 2(x-1)$$

$$\begin{aligned} 14 + (x-1) &= 2x - 2 \\ x + 13 &= 2x - 2 \\ x - 2x &= -2 - 13 \\ -x &= -15 \quad \text{or} \quad x = 15. \end{aligned}$$

The reader should confirm by a check that 15 is the solution to the original equation.

Example 2. Solve $\dfrac{x-3}{x-2} = -3 - \dfrac{1}{x-2}$.

Solution: We clear fractions, multiplying both sides of the equation by the LCD, $(x - 2)$. This results in the following equation:

$$\cancel{(x-2)}\frac{x-3}{\cancel{(x-2)}} = (x-2)(-3) - \cancel{(x-2)}\frac{1}{\cancel{(x-2)}}$$

$$x - 3 = -3x + 6 - 1.$$

Collecting like terms, we get

$$4x = 8 \qquad \text{or} \qquad x = 2.$$

We check the proposed solution 2 by substituting it in the original equation.

CHECK:

$$\text{Let } x = 2: \quad \frac{x-3}{x-2} = -3 - \frac{1}{x-2}$$

$$\frac{2-3}{2-2} = -3 - \frac{1}{2-2}$$

$$\frac{-1}{0} = -3 - \frac{1}{0}.$$

But division by zero is not defined. Thus 2 is not a solution to the original equation. This is an example of an extraneous solution. Since 2 is the only possible solution and it is extraneous, the original equation has no solution.

Example 2 indicates that *we should always check our proposed solutions to an equation when we have multiplied both sides of an equation by an expression that involves a variable.*

Example 3. Solve $\frac{4}{x} + 1 = \frac{5}{x^2}$.

Solution: The LCD is x^2. Multiplying both sides of the original equation by x^2 results in the following equation:

$$x^2\left(\frac{4}{x} + 1\right) = x^2 \cdot \frac{5}{x^2}$$

$$4x + x^2 = 5$$
$$x^2 + 4x - 5 = 0.$$

The last equation is a quadratic equation that we solve by factoring. Thus

$$(x + 5)(x - 1) = 0.$$

Therefore, -5 and 1 are both possible solutions. A check would confirm that they both satisfy the original equation.

Example 4. Solve $\frac{y}{y - 5} + \frac{5}{y^2 - 25} = \frac{-1}{y + 5}$.

Solution: The factors of $y^2 - 25$ are $y - 5$ and $y + 5$; therefore, the LCD of $y^2 - 25 = (y - 5)(y + 5)$, $(y - 5)$, and $(y + 5)$ is $(y - 5)(y + 5)$. If we multiply each side of the equation by this LCD, we get

$$(y - 5)(y + 5)\left[\frac{y}{y - 5} + \frac{5}{(y - 5)(y + 5)}\right] = (y - 5)(y + 5)\frac{-1}{(y + 5)}$$

$$\cancel{(y - 5)}(y + 5)\frac{y}{\cancel{(y - 5)}} + \cancel{(y - 5)}\cancel{(y + 5)}\frac{5}{\cancel{(y - 5)}\cancel{(y + 5)}}$$
$$= (y - 5)\cancel{(y + 5)}\frac{-1}{\cancel{(y + 5)}}$$

$$(y + 5)y + 5 = (y - 5)(-1)$$
$$y^2 + 5y + 5 = -y + 5$$
$$y^2 + 6y = 0$$
$$y(y + 6) = 0.$$

Using the Factor Property of Zero, we see that $y = 0$ or $y + 6 = 0$. Thus 0 and -6 are both candidates for solutions to the original equation. The reader should check that each of these candidates satisfies the original equation.

Example 5. Solve $\frac{x + 1}{x - 2} + \frac{1}{x - 1} = \frac{3}{x^2 - 3x + 2}$.

Solution: The factors of $x^2 - 3x + 2$ are $(x - 2)$ and $(x - 1)$. Thus the LCD is $(x - 2)(x - 1)$. We multiply each side of the equation by this LCD and solve the resulting equation.

$$\cancel{(x-2)}(x-1)\frac{x+1}{\cancel{x-2}} + (x-2)\cancel{(x-1)}\frac{1}{\cancel{x-1}}$$
$$= \cancel{(x-2)}\cancel{(x-1)}\frac{3}{\cancel{(x-2)}\cancel{(x-1)}}$$

$$\begin{aligned}(x-1)(x+1) + (x-2) &= 3\\ x^2 - 1 + x - 2 &= 3\\ x^2 + x - 6 &= 0\\ (x+3)(x-2) &= 0.\end{aligned}$$

Thus -3 and 2 are possible solutions. A check would confirm that -3 satisfies the original equation. Let us see what happens when $x = 2$.

CHECK: Let $x = 2$:

$$\frac{x+1}{x-2} + \frac{1}{x-1} = \frac{3}{x^2 - 3x + 2}$$

$$\frac{2+1}{2-2} + \frac{1}{2-1} = \frac{3}{2^2 - 3(2) + 2}$$

$$\frac{3}{0} + \frac{1}{1} = \frac{3}{0}.$$

Since division by 0 is undefined, 2 is not a solution to the original equation. Therefore, the only solution to the original equation is $x = -3$.

Example 6. Solve

$$\frac{5}{2z+2} = \frac{2}{z+2} + \frac{1}{z^2 + 3z + 2}.$$

Solution: The denominators may be factored or grouped as follows:

$$2z + 2 = 2(z+1), \qquad z + 2 = (z+2), \qquad z^2 + 3z + 2 = (z+2)(z+1).$$

Therefore, the LCD is $2(z+2)(z+1)$. If we multiply each term of the equation by this LCD, we obtain

$$2(z+2)(z+1)\frac{5}{2z+2}$$
$$= 2(z+2)(z+1)\frac{2}{z+2} + 2(z+2)(z+1)\frac{1}{z^2 + 3z + 2}$$

$$\cancel{2}(z+2)\cancel{(z+1)}\frac{5}{2\cancel{(z+1)}}$$
$$= 2\cancel{(z+2)}(z+1)\frac{2}{\cancel{z+2}} + 2\cancel{(z+2)}\cancel{(z+1)}\frac{1}{\cancel{(z+2)}\cancel{(z+1)}}$$

$$\begin{aligned} 5(z + 2) &= 4(z + 1) + 2 \\ 5z + 10 &= 4z + 4 + 2 \\ z &= -4. \end{aligned}$$

It is easy to check that -4 is a solution to the original equation.

EXERCISES 5.5

For Exercises 1–14, solve the equation and check your answer. See Example 1.

1. $\frac{x}{5} + \frac{3}{2} = \frac{1}{2}$

2. $\frac{y}{2} - \frac{2}{3} = \frac{4}{3}$

3. $\frac{z}{7} + \frac{1}{3} = 1$

4. $\frac{p}{6} + \frac{2}{3} = 1$

5. $\frac{2}{x} = \frac{5}{3} + 1$

6. $\frac{3}{p} = \frac{7}{2} - 1$

7. $\frac{3}{y} + \frac{1}{2y} = \frac{3}{8}$

8. $\frac{7}{2q} - \frac{2}{q} = \frac{1}{9}$

9. $\frac{4}{x} - 2 = \frac{2}{5}$

10. $\frac{1}{w} + 5 = \frac{8}{3}$

11. $\frac{3}{t} + \frac{7}{2} = 5$

12. $\frac{2}{p} - \frac{1}{5} = 2$

13. $\frac{5}{q} - \frac{1}{3} = \frac{1}{2}$

14. $\frac{6}{z} + \frac{2}{5} = \frac{4}{7}$

For Exercises 15–50, solve the equation and check your answer. Identify any extraneous solutions. See Examples 2–6.

15. $\frac{1}{x - 1} + \frac{1}{(x - 1)(x + 2)} = \frac{-1}{x + 2}$

16. $\frac{-1}{y + 2} + \frac{1}{(y + 2)(y + 3)} = \frac{1}{y + 3}$

17. $\frac{1}{p + 3} + \frac{1}{p - 4} = \frac{7}{p^2 - p - 12}$

18. $\frac{1}{q - 5} + \frac{1}{q + 2} = \frac{7}{q^2 - 3q - 10}$

19. $\frac{2}{2z - 1} + \frac{1}{z + 1} = \frac{13}{2z^2 + z - 1}$

20. $\frac{2}{2x + 1} - \frac{5}{3x - 2} = \frac{-1}{6x^2 - x - 2}$

21. $\frac{x - 1}{x - 2} + 1 = \frac{1}{x - 2}$

22. $\frac{x + 2}{x - 5} + 1 = \frac{7}{x - 5}$

23. $\frac{5x + 3}{x + 2} + 2 = \frac{-7}{x + 2}$

24. $\frac{2p + 6}{p + 4} + 1 = \frac{-2}{p + 4}$

25. $\frac{2q - 5}{q - 11} + 2 = \frac{17}{q - 11}$

26. $\frac{w - 5}{w + 5} + \frac{2}{5} = \frac{-10}{w + 5}$

27. $\frac{7}{2} - \frac{3(x + 3)}{x + 2} = \frac{-1}{x + 2}$

28. $\frac{z + 5}{z + 6} = \frac{3}{4} - \frac{1}{2(z + 6)}$

29. $2 - \frac{6}{w + 5} = \frac{w + 1}{w + 5}$

30. $\frac{1}{3} + \frac{y - 1}{y + 2} = \frac{5}{y + 2}$

31. $\frac{x + 1}{2} - \frac{1}{x} = \frac{5}{3}$

32. $\frac{q - 2}{3} - \frac{3}{q} = 2$

33. $\frac{-4}{y} + 1 = \frac{5}{y^2}$

34. $\frac{-7}{x} - 1 = \frac{10}{x^2}$

35. $2 + \frac{1}{x} = \frac{1}{x^2}$

36. $3 - \frac{7}{x} = \frac{-2}{x^2}$

37. $\frac{y}{y-4} + \frac{4}{y^2-16} = \frac{-1}{y+4}$

38. $\frac{6}{w^2-36} + \frac{w}{w-6} = \frac{-1}{w+6}$

39. $\frac{z}{z+2} - \frac{2}{z^2-4} = \frac{1}{z-2}$

40. $\frac{p}{p+1} - \frac{7}{p^2-1} = \frac{1}{p-1}$

41. $\frac{2}{(y+4)(y-3)} + \frac{1}{(y-3)(y+1)} = \frac{2}{(y+4)(y+1)}$

42. $\frac{4}{(z-5)(z-4)} - \frac{3}{(z-4)(z-3)} = \frac{2}{(z-3)(z-5)}$

43. $\frac{p}{p^2+3p-10} + \frac{6}{p^2+7p+10} = \frac{p}{p^2-4}$

44. $\frac{w}{w^2-4w-21} + \frac{-7}{w^2-10w+21} = \frac{w}{w^2-9}$

45. $\frac{3}{2z+5} = \frac{1}{z+3} + \frac{-2}{z^2+2z-3}$

46. $\frac{-2}{x+5} = \frac{4}{3x+1} + \frac{-6}{x^2+4x-5}$

47. $\frac{5z}{z^2+2z-3} = \frac{2}{z^2+z-2} + \frac{3z}{z^2+5z+6}$

48. $\frac{3x}{x^2+5x+6} = \frac{4}{x^2-4} - \frac{x}{x^2+x-6}$

49. $\frac{3x}{x^2+9x+20} + \frac{x+2}{x^2+7x+12} = \frac{6x}{x^2+8x+15}$

50. $\frac{3w-6}{w^2+w-2} + \frac{w+4}{w^2+2w-3} = \frac{3w+18}{w^2+5w+6}$

5.6 Ratios and Proportions

If there are 2 Democrats and 5 Republicans on a city council, we say that the *ratio* of Democrats to Republicans is 2 to 5. The ratio 2 to 5 is also written as 2:5 or more commonly as the fraction $\frac{2}{5}$.

> The ratio of the number a to the number b ($b \neq 0$) is the quotient $\frac{a}{b}$.

Ratios are often encountered in our daily experiences. Thus we talk about the ratio of oil to gas for the proper fuel mixture for a moped or the ratio of male students to female students at a college we might attend. We usually express a ratio in lowest terms.

Example 1. A medical technician counts 40 white blood cells and 280 red blood cells in a sample she is testing.
(a) What is the ratio of white blood cells to red blood cells?
(b) What is the ratio of red blood cells to white blood cells?
(c) What is the ratio of red blood cells to the total of blood cells that are either red or white?

Solution

(a) $\frac{40}{280}$ or $\frac{1}{7}$, read "1 to 7."

(b) $\frac{280}{40}$ or $\frac{7}{1}$, read "7 to 1."

(c) The total of blood cells that are either red or white is $280 + 40 = 320$. Therefore, the ratio of red blood cells to this total is

$$\frac{280}{320} = \frac{7}{8},$$

read "7 to 8."

As in Example 1(b), it is common practice to leave the 1 in the denominator of a fraction when it represents a ratio.

Example 2. Daniel's height is 42 inches. His sister Jennifer is 4 feet 6 inches tall. What is the ratio of Daniel's height to Jennifer's height?

Solution: First we express both heights in the same unit of measure. If we convert Jennifer's height to inches, the result is 54. Therefore, the ratio we seek is

$$\frac{42}{54} = \frac{7}{9}.$$

Example 3. What is the ratio of the months of the year with a total of at most 30 days to the months of the year with more than 30 days?

Solution: Remember the nursery rhyme, "Thirty days hath September, April, June, and November. All the rest . . . ." Notice that there are 4 months with exactly 30 days. But February also has fewer than 30 days; thus the ratio we seek is $\frac{5}{7}$.

We use a ratio to compare two quantities. When two ratios are equal, they form a *proportion*. We express this symbolically with the following statement.

> A proportion is a statement of equality about two ratios. If the equal ratios are $\frac{a}{b}$ and $\frac{c}{d}$, then
>
> $$\frac{a}{b} = \frac{c}{d}$$
>
> is called a proportion.

Each of the four quantities a, b, c, and d in the proportion $\frac{a}{b} = \frac{c}{d}$ is called a *term of the proportion*. The quantities a and d are called the *extremes* of the proportion. The quantities b and c are called the *means* of the proportion. For example, in the proportion

$$\frac{5}{7} = \frac{15}{21}$$

the extremes are 5 and 21, and the means are 7 and 15.

If we multiply each side of the proportion $\frac{a}{b} = \frac{c}{d}$ by the quantity bd, the result is

$$\cancel{b}d \cdot \frac{a}{\cancel{b}} = b\cancel{d} \cdot \frac{c}{\cancel{d}}$$

$$ad = bc.$$

Thus we see that in any proportion, the product of the means equals the product of the extremes.

Fundamental Property of Proportions

> If $\frac{a}{b} = \frac{c}{d}$, then $ad = bc$.

The Fundamental Property of Proportions is used to solve a proportion when one of its terms is unknown.

Example 4. Solve the proportion $\frac{x}{28} = \frac{3}{7}$ for x.

Solution: Using the Fundamental Property of Proportions, we may write

$$7x = 3 \cdot 28 = 84.$$

Thus

$$x = 12.$$

It is easy to check that $x = 12$ is the correct solution.

In Example 2 we converted feet to inches in order to have a common unit of measure. Sometimes it is not desirable or possible to have a common unit of measure. For example, we may want the ratio of students to computer terminals. As an example, where a common unit of measure is not desirable consider the mining of gold. In order to get an ingot of 1 cubic foot of solid gold, it is sometimes necessary to remove a cubic mile of earth! The ratio, 1 cubic foot to 1 cubic mile, is easier to comprehend than

$$\frac{1}{147{,}197{,}950{,}000},$$

which is the same ratio expressed in terms of a common unit of measure, cubic feet. The denominator in the last ratio is equal to $(5280)^3$, where 5280 is the number of feet in a mile.

If one ratio of a proportion compares different quantities, then the other ratio must also compare the same types of quantities. For example, if one ratio compares the temperature in Celsius to the altitude in feet, then the other ratio must also compare temperature in Celsius to altitude in feet.

Example 5. An owner of a 40-pound dog is given a prescription for the dog. The instructions state that the dog is to be given 180 milliliters of the medicine each morning. If the ratio of dosage to the weight of the animal is the same for all dogs, what is the correct dosage for a 70-pound dog?

Solution: Let x be the dosage in milliliters for the 70-pound dog. We know that

$$\frac{40}{180} = \frac{70}{x}.$$

But reducing the left-hand side of the equation to lowest terms, we see that

$$\frac{40}{180} = \frac{2}{9}.$$

Therefore, we need to solve

$$\frac{2}{9} = \frac{70}{x}$$

for x. Since the product of the means equals the product of the extremes, we know that

$$2x = 9 \cdot 70$$
$$2x = 630$$
$$x = 315.$$

The correct dosage is 315 milliliters.

Example 6. Solve for x in the following proportion:

$$\frac{x}{5} = \frac{4}{x + 1}.$$

Solution: Applying the Fundamental Property of Proportions, we get

$$x(x + 1) = 5 \cdot 4$$

or

$$x^2 + x = 20$$
$$x^2 + x - 20 = 0.$$

We solve this quadratic equation by factoring the left-hand side. Thus we have $(x + 5)(x - 4) = 0$. From this we know that $x = -5$ or $x = 4$.

A check would reveal that both -5 and 4 are solutions to the original proportion.

Example 7. A pilot is flying a small plane from Allentown to Philipsburg, Pennsylvania. On his map the two airports are 22.5 inches apart. He knows that on the map 3 inches represents 20 miles.

(a) What is the actual distance in miles between the two airports?

(b) If the plane cruises at 120 miles per hour and he has enough fuel to cruise for $2\frac{1}{2}$ hours, should he refuel at Philipsburg before returning to Allentown?

Solution

(a) From the information given, we may write the proportion

$$\frac{3}{20} = \frac{22.5}{x},$$

where x is the actual distance between the two airports. Applying the Fundamental Property of Proportions, we get

$$3x = 20 \cdot 22.5$$
$$3x = 450$$
$$x = 150.$$

Therefore, the airports are 150 miles apart.

(b) Recall that distance = rate × time. Let t represent how long it takes to fly 150 miles. Thus

$$150 = 120 \times t$$

or

$$120t = 150$$

$$t = 1\frac{1}{4} \text{ hours.}$$

Therefore, it will take $2\left(1\frac{1}{4}\right) = 2\frac{1}{2}$ hours to make the entire round-trip flight. Unless he likes to live dangerously, he should refuel at Philipsburg.

EXERCISES 5.6

For Exercises 1–6, write the ratio as a fraction. Reduce each fraction to lowest terms. See Example 1.

1. 4 to 7 **2.** 3 to 8 **3.** 6 to 18 **4.** 18 to 6 **5.** 35 to 7 **6.** 7 to 35

For Exercises 7–16, express both quantities in a common unit and write the ratio as a fraction reduced to lowest terms. See Example 2.

7. 8 inches to 1 yard
8. 27 inches to 2 yards
9. 20 minutes to 2 hours
10. 6 minutes to 1 hour
11. 60 cents to 3 dollars
12. 75 cents to 5 dollars
13. 40 hours to 1 week
14. 12 hours to 1 week
15. 5 feet to 10 yards
16. 8 feet to 4 yards

For Exercises 17–40, solve the proportion. See Examples 4 and 6.

17. $\frac{y}{11} = \frac{8}{44}$ **18.** $\frac{z}{5} = \frac{9}{15}$ **19.** $\frac{3}{x} = \frac{9}{21}$ **20.** $\frac{4}{p} = \frac{24}{42}$

21. $\frac{8}{40} = \frac{q}{25}$ **22.** $\frac{32}{40} = \frac{x}{10}$ **23.** $\frac{8}{5} = \frac{40}{y}$ **24.** $\frac{9}{7} = \frac{81}{x}$

25. $\frac{1}{3} = \frac{z}{4}$ **26.** $\frac{3}{5} = \frac{2}{x}$ **27.** $\frac{r}{11} = \frac{7}{50}$ **28.** $\frac{s}{8} = \frac{5}{16}$

29. $\frac{x}{3} = \frac{12}{x}$ **30.** $\frac{8}{z} = \frac{z}{2}$ **31.** $\frac{32}{r} = \frac{r}{2}$ **32.** $\frac{p}{5} = \frac{20}{p}$

33. $\frac{y}{4} = \frac{5}{y-1}$ **34.** $\frac{z}{2} = \frac{21}{z+1}$ **35.** $\frac{x+3}{4} = \frac{7}{x}$ **36.** $\frac{p-3}{7} = \frac{4}{p}$

37. $\dfrac{y+2}{4} = \dfrac{2}{y}$ **38.** $\dfrac{x-3}{9} = \dfrac{2}{x}$ **39.** $\dfrac{1}{x+1} = \dfrac{x-1}{3}$ **40.** $\dfrac{3}{p-2} = \dfrac{p+3}{2}$

Problems 41 and 42 are based on the following table, which gives the units of vitamin A for four food items. See Example 3.

Food	Units of Vitamin A per 100 grams
Raw apples	90
Clams	100
Grape jam	10
Wild rice	2

41. Find the ratio of the vitamin A content of 100 grams of clams to the vitamin A content of 100 grams of raw apples.

42. Find the ratio of the vitamin A content of 400 grams of clams to the vitamin A content of 200 grams of raw apples plus 500 grams of wild rice.

Solve Exercises 43–50 by setting up and solving a proportion. See Examples 5 and 7.

43. On a map, 3 inches represents 100 miles. If two towns are 7.5 inches apart on the map, how far apart are they in actual miles?

44. In a photograph of a painting, the image of an object is $1\frac{1}{2}$ inches long. In the painting, the same object is 9 inches long. In the photograph, how long is the image of an object that is 12 inches long in the painting?

45. Fifty pounds of fertilizer will cover 2200 square feet. How many pounds are needed to cover 8000 square feet?

46. A two-cycle engine requires an oil-to-gas ratio, in ounces, of 1 to 30. How many ounces of oil should be added to 5 gallons of gas? (HINT: One gallon equals 128 ounces.)

47. The adult dosage for a new liquid cold medicine is based on the weight of the patient. A 120-pound person is instructed to take 1.5 fluid ounces of the liquid every 6 hours.
(a) What is the dosage for a 200-pound person?
(b) How much of the medicine is needed to supply a 200-pound person for 1 week?

48. If 2 pounds of coffee will make 160 cups of coffee, how many pounds of coffee are needed to make 1000 cups of coffee?

49. The ratio of male to female students in a local college is 5 to 4. If there are 900 female students at the college, what is the total enrollment?

50. A new diet suggests that you drink 8 ounces of tomato juice for every 90 pounds of body weight. How many ounces of tomato juice should a 225-pound person drink?

Applications and Word Problems 5.7

Formulas that involve rational expressions are frequently used in engineering, business, and the sciences. In this section we discuss some of these formulas as well as some word problems that involve rational expressions.

Example 1. If the same number is subtracted from both the numerator and the denominator of $\frac{7}{5}$, the result is the reduced fraction $\frac{3}{2}$. Find the number.

Solution: Let x be the number we subtract from both the numerator and the denominator; then

$$\frac{7-x}{5-x} = \frac{3}{2}.$$

We can clear fractions by multiplying each side of the equation by $2(5 - x)$.

$$2\cancel{(5-x)}\frac{7-x}{\cancel{5-x}} = \cancel{2}(5-x)\frac{3}{\cancel{2}}$$

$$\begin{aligned} 2(7-x) &= (5-x)3 \\ 14 - 2x &= 15 - 3x \\ x &= 1. \end{aligned}$$

CHECK:

Let $x = 1$ in $\frac{7-x}{5-x}$; then $\frac{7-1}{5-1} = \frac{6}{4} = \frac{3}{2}$.

Thus 1 is the correct answer.

Example 2. The sum of a number and its reciprocal is $\frac{10}{3}$, find the number.

Solution: Let x be the number. Thus

$$x + \frac{1}{x} = \frac{10}{3}.$$

Multiply each side of the equation by $3x$ in order to clear fractions.

$$3x\left(x + \frac{1}{x}\right) = 3x\left(\frac{10}{3}\right)$$

$$3x \cdot x + 3x\frac{1}{x} = 3x \cdot \frac{10}{3}$$

$$3x^2 + 3 = 10x.$$

The result is a quadratic equation that we solve by factoring.

$$3x^2 - 10x + 3 = 0$$
$$(3x - 1)(x - 3) = 0.$$

Thus $3x - 1 = 0$ or $x - 3 = 0$. Therefore, $x = \frac{1}{3}$ and $x = 3$ are both possible solutions to the original problem.

CHECK:

$$\text{Let } x = \frac{1}{3}; \qquad \text{then } x + \frac{1}{x} = \frac{1}{3} + 3 = \frac{10}{3}.$$

$$\text{Let } x = 3; \qquad \text{then } x + \frac{1}{x} = 3 + \frac{1}{3} = \frac{10}{3}.$$

We conclude that both 3 and $\frac{1}{3}$ are solutions to the problem.

Example 3. A few miles south of the Delaware Water Gap there is a long stretch of open water on the Delaware River with a steady current of 2 miles per hour. Maria and Carlos paddled their canoe 4 miles upstream and then returned to their point of departure in $1\frac{1}{2}$ hours. How fast can they paddle in still water?

Solution: We use the formula $d = rt$ (distance = rate · time). Let x be their rate of paddling in still water. Because the rate of the current is 2 miles per hour, their rate downstream is $x + 2$ miles per hour and their rate upstream is $x - 2$ miles per hour. If we solve $d = rt$ for t, the result is $t = d/r$. Substituting 4 for d, we may write

$$\text{time upstream} = \frac{\text{distance upstream}}{\text{rate upstream}} = \frac{4}{x - 2}$$

$$\text{time downstream} = \frac{\text{distance downstream}}{\text{rate downstream}} = \frac{4}{x + 2}.$$

But

$$\begin{pmatrix}\text{time}\\ \text{upstream}\end{pmatrix} + \begin{pmatrix}\text{time}\\ \text{downstream}\end{pmatrix} = \begin{pmatrix}\text{total}\\ \text{time}\end{pmatrix}$$

or

$$\frac{4}{x - 2} + \frac{4}{x + 2} = \frac{3}{2}. \qquad \left(1\frac{1}{2} = \frac{3}{2}\right)$$

We clear fractions in this equation by multiplying each term by $2(x - 2)(x + 2)$.

Thus we have

$$2\cancel{(x-2)}(x+2)\frac{4}{\cancel{x-2}} + 2(x-2)\cancel{(x+2)}\frac{4}{\cancel{x+2}} = \cancel{2}(x-2)(x+2)\frac{3}{\cancel{2}}$$

or

$$\begin{aligned} 8(x+2) + 8(x-2) &= 3(x-2)(x+2) \\ 8x + 16 + 8x - 16 &= 3(x^2 - 4) \\ 16x &= 3x^2 - 12 \\ 3x^2 - 16x - 12 &= 0. \end{aligned}$$

We solve this quadratic equation by factoring and then setting each factor equal to 0. Thus

$$3x^2 - 16x - 12 = 0$$

or

$$(3x + 2)(x - 6) = 0.$$

Therefore,

$$3x + 2 = 0 \qquad \text{or} \qquad x - 6 = 0.$$

It follows that

$$x = -\frac{2}{3} \qquad \text{or} \qquad x = 6.$$

Since their rate of paddling is a positive number, $x = 6$ is the only possible solution. Thus their rate in still water is 6 miles per hour.

Most car owners are very much aware of the concept of *depreciation*. For example, if you purchase a new car for $8000 in January and it is only worth $5800 in June, we say that it has depreciated $2200 in 6 months. Some items, such as automobiles, furniture, and appliances, depreciate very rapidly immediately after purchase. During the rest of their period of usefulness, they depreciate at a much slower rate. Other items, such as tools and large manufacturing equipment, depreciate at a more regular rate over their period of usefulness.

For proper long-range planning, accounting purposes, and tax benefits, it is important for people in business to plan for the depreciation of their equipment. There are several ways of calculating the depreciation of equipment.

The next example discusses regular or linear depreciation. With this method, an item that has an assumed 10-year period of usefulness and originally cost $75,000 will depreciate

$$\frac{\$75{,}000}{10} = \$7500$$

each year for 10 years. If an item that originally cost C dollars is going to be depreciated over y years, the value V after k years is given by the formula

$$V = C\left(1 - \frac{k}{y}\right).$$

Notice that this formula involves the rational expression $\frac{k}{y}$.

Example 4. If a $30,000 printing press is to be depreciated for 20 years, what will be its value after 8 years?

Solution: For this item $C = \$30{,}000$, $y = 20$, and $k = 8$. Thus

$$\begin{aligned} V &= C\left(1 - \frac{k}{y}\right) \\ &= 30{,}000\left(1 - \frac{8}{20}\right) \\ &= 30{,}000\left(\frac{12}{20}\right) \\ &= 30{,}000\left(\frac{3}{5}\right) \\ &= 18{,}000. \end{aligned}$$

Thus at the end of 8 years, the value of the printing press is $18,000. It has depreciated $12,000 in 8 years.

Example 5. A small tractor that is used in furniture warehouses originally cost $18,000. It is estimated that the tractor will be worth $10,500 five years after purchase. Use the formula for linear depreciation to determine the number of years over which the tractor will be depreciated.

Solution: For the tractor $C = \$18{,}000$ and $V = 10{,}500$ when $k = 5$. Thus

$$10{,}500 = 18{,}000\left(1 - \frac{5}{y}\right).$$

We can clear fractions if we multiply each side of this equation by y. The result of this multiplication is

$$\begin{aligned} 10{,}500y &= 18{,}000y\left(1 - \frac{5}{y}\right) \\ &= 18{,}000y - 90{,}000. \end{aligned}$$

Thus

$$7500y = 90{,}000 \qquad \text{or} \qquad y = 12.$$

The tractor is depreciated over 12 years.

If you had toast for breakfast or used an iron to touch up some clothing last night, you were aided by electrical *resistance*. Resistance is a property of every metallic device when an electric current flows through the device. In toasters and irons, the result of resistance is heat. Resistance is measured in ohms. The symbol for a resistor is —/\/\/\/— . Figure 5.1 shows two resistors that are connected in parallel in part of an electric circuit. When two resistors are connected in parallel, the total resistance in that part of the circuit is computed using the formula

$$\frac{1}{R} = \frac{1}{R_1} + \frac{1}{R_2}.$$

In the formula, R is the total resistance in ohms and R_1 and R_2 are the resistances in ohms of the first and second resistors, respectively.

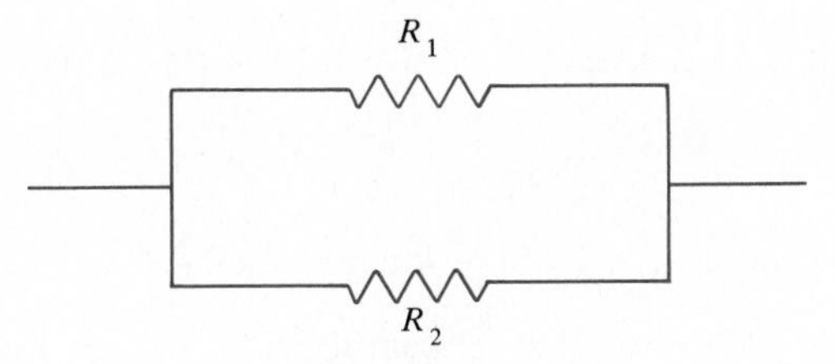

Figure 5.1

Example 6. An 8-ohm resistor is connected in parallel to a 6-ohm resistor. What is the total resistance in that part of the circuit?

Solution: Let $R_1 = 8$ and $R_2 = 6$; then

$$\frac{1}{R} = \frac{1}{R_2} + \frac{1}{R_2}$$

$$\frac{1}{R} = \frac{1}{8} + \frac{1}{6}.$$

We can clear fractions by multiplying both sides of the above equation by $24R$.

$$24R\frac{1}{R} = 24R\left(\frac{1}{8} + \frac{1}{6}\right)$$

$$24 = 3R + 4R$$

$$24 = 7R.$$

Thus

$$R = \frac{24}{7} \quad \text{or} \quad 3\frac{3}{7} \text{ ohms.}$$

Therefore, a single resistor of $3\frac{3}{7}$ ohms is equal to two resistors of 6 ohms and 8 ohms when the two resistors are connected in parallel.

Example 7. The water tank on a forest service fire truck can be filled from a hose in 2 hours. It takes 3 hours to fill the same tank with a smaller hose.
(a) How long will it take to fill the tank if both hoses are used?
(b) In 1 hour, how much of the tank will be filled by both hoses?
(c) What fraction of the tanks' capacity is filled by each hose?

Solution

(a) Let x denote the number of hours that it takes to fill the tank with both hoses. We will solve this problem by considering how much of the tank is filled in 1 hour. Observe that:

(1) Since it takes 2 hours for the first hose to fill the tank, in 1 hour it will fill $\frac{1}{2}$ of the tank.

(2) The second hose will fill $\frac{1}{3}$ of the tank in 1 hour.

(3) Since the two hoses together can fill the tank in x hours, in 1 hour they will fill $\frac{1}{x}$ of the tank.

Thus for 1 hour we know that

$$\frac{1}{2} + \frac{1}{3} = \frac{1}{x}$$

$$\begin{pmatrix}\text{amount of the}\\\text{tank filled by}\\\text{the first hose}\end{pmatrix} + \begin{pmatrix}\text{amount of the}\\\text{tank filled by}\\\text{the second hose}\end{pmatrix} = \begin{pmatrix}\text{amount of the}\\\text{tank filled by}\\\text{both hoses}\end{pmatrix}$$

We can clear fractions if we multiply each side of the equation by the LCD, $6x$.

$$6x\left(\frac{1}{2} + \frac{1}{3}\right) = 6x\frac{1}{x}$$

$$\begin{aligned} 3x + 2x &= 6 \\ 5x &= 6 \\ x &= \frac{6}{5}. \end{aligned}$$

Thus in $\frac{6}{5}$ hours, the two hoses will fill the tank. That is, it will take

$$\frac{6}{5}(60) = 72 \text{ minutes}$$

for both hoses to fill the tank.

(b) Since it takes $\frac{6}{5}$ hours for both hoses to fill the tank, in 1 hour they will fill

$$\frac{1}{\frac{6}{5}} = \frac{5}{6}$$

of the tank.

(c) The first hose fills the tank at the rate of $\frac{1}{2}$ tank per hour. Thus in $\frac{6}{5}$ hours it will fill

$$\frac{1}{2} \cdot \frac{6}{5} = \frac{3}{5}$$

of the tank. Similarly, the second hose will fill

$$\frac{1}{3} \cdot \frac{6}{5} = \frac{2}{5}$$

of the tank.

EXERCISES 5.7

Solve the following exercises.

For Exercises 1–6, see Examples 1 and 2.

1. If the same number is added to both the numerator and the denominator of $\frac{7}{10}$, the result is the reduced fraction $\frac{3}{4}$. Find the number.
2. If the same number is subtracted from both the numerator and the denominator of $\frac{5}{8}$, the result is $\frac{2}{3}$. Find the number.
3. The denominator of a certain fraction is 4 more than the numerator. If 1 is added to the numerator and subtracted from the denominator, the result equals $\frac{2}{3}$. Find the original fraction.
4. The numerator of a certain fraction is 4 more than the denominator. If 1 is subtracted from the numerator and added to the denominator, the result is $\frac{5}{4}$. Find the original fraction.
5. A number plus twice the reciprocal of the number equals $\frac{9}{2}$; find the number.
6. A number plus three times the reciprocal of the number equals $\frac{13}{2}$; find the number.

For Exercises 7–10, see Example 3.

7. An excursion boat makes $1\frac{1}{2}$ hour tours on a river. Each tour goes 6 miles upstream and then returns to its starting point. If the river has a steady current of 3 miles per hour, what is the rate of the boat in still water?

8. A tugboat can make a trip 9 miles upstream and return in 1 hour and 36 minutes. If the rate of the boat in still water is 12 miles per hour, what is the rate of the river's current? (HINT: 1 hour and 36 minutes $= \frac{8}{5}$ hours.)

9. A small airplane flies northwest for 100 miles against a headwind of 10 miles per hour and then returns with the wind to its starting point. If the round trip takes 1 hour and 50 minutes, what is the plane's speed in still air? (HINT: 1 hour and 50 minutes $= 1\frac{5}{6}$ hours.)

10. An airplane that has a cruising speed of 135 miles per hour in still air flies 200 miles southwest directly against a steady headwind. It then returns with the wind to its point of departure. If the entire trip takes 3 hours, what is the rate of the wind in miles per hour?

For Exercises 11–16, refer to Examples 4 and 5.

11. Use the formula given in Example 4 to find the value of a \$50,000 item after 9 years if the item is being depreciated for 20 years.

12. A drill press that costs \$120,000 is being depreciated for 20 years according to the formula given in Example 4. What is the value of the drill press after 16 years?

13. An industrial extruder that manufactures automobile parts originally cost \$480,000. It is being depreciated over 20 years using the formula given in Example 4. When will it be worth \$144,000?

14. An industrial blender that originally cost \$38,000 is being depreciated over 20 years using the formula given in Example 4. When will it be worth \$15,200?

15. A company uses the formula given in Example 4 when figuring depreciation on its equipment. The company wants to depreciate an item that originally cost \$75,000. Experience indicates that after 9 years the item will be valued at \$30,000. Using this information, determine over how many years the item should be depreciated.

16. It is estimated that at the end of 5 years, a drill press that originally cost \$96,000 will have a value of \$56,000. Use the formula given in Example 4 to determine the number of years over which the drill press will be depreciated.

For Exercises 17–20, refer to Example 6.

17. A resistor of 10 ohms is connected in parallel with a resistor of 12 ohms. What single resistance is equal to these two resistances?

18. Two resistors of 9 and 15 ohms, respectively, are connected in parallel. What single resistance is equal to these resistances?

19. An 8-ohm resistor is to be connected in parallel to another resistor. If the total resistance in the circuit is to be 6 ohms, at how many ohms should the second resistor be rated?

20. A 10-ohm resistor is to be connected in a circuit parallel to another resistor. If the total resistance in the circuit is to be 2.5 ohms, at how many ohms should the second resistor be rated?

For Exercises 21–25, see Example 7.

21. One pipe can fill a swimming pool in 6 hours. A hose can fill the same pool in 10 hours.
(a) How long will it take to fill the pool if both the pipe and the hose are used?
(b) In 1 hour, how much of the pool is filled if both the hose and the pipe are used?
(c) What fraction of the pool is filled by the pipe?

22. The spray tank on an arborist's truck can be filled by a hose in 1 hour and 20 minutes. A larger hose can fill the tank in 48 minutes.
(a) How long will it take for both hoses to fill the tank?
(b) In 10 minutes, how much of the tank is filled by both hoses?
(c) What fraction of the tanks capacity is filled by each hose?
(HINT: Change all the times to either minutes or hours.)

23. Christina can assemble a display case in 40 minutes. It takes Daniel 1 hour to assemble the same display case. If they work together, how long will it take to assemble the display case? (HINT: Change all the times to either minutes or hours.)

24. It takes 5 minutes for the drain pipe to empty a full sink of water. A faucet can fill the empty sink in 4 minutes. If the drain is left open, how long will it take to fill an empty sink?

25. A small computer can complete a payroll in 6 hours. A larger and newer computer can complete the same payroll in 4 hours.
(a) If the two computers can work on the payroll at the same time, how long will the job take?
(b) What fraction of the payroll is processed by each computer?

Key Words

Clearing fractions
Complex fraction
Extraneous solution
Extremes of a proportion
Fundamental property of proportions
Fundamental property of rational expressions
Least common denominator (LCD)
Means of a proportion
Ratio
Rational expression
Reduce to lowest terms
Reduced form
Removing common factors
Simplified

CHAPTER 5 TEST

For Problems 1 and 2, determine what values of the variable will make the rational expression undefined.

1. $\dfrac{17}{z - 5}$ **2.** $\dfrac{3x + 1}{x^2 - 5x + 6}$

For Problems 3–6, reduce the expression to lowest terms.

3. $\dfrac{180}{336}$ **4.** $\dfrac{75x^2}{45x^3}$ **5.** $\dfrac{y + 2}{y^2 - 4}$ **6.** $\dfrac{z^2 + 4z - 21}{z^2 - 7z + 12}$

For Problems 7–12, perform the multiplication or division indicated. Reduce your answer to lowest terms.

7. $\dfrac{7x^2}{15} \cdot \dfrac{6}{28x^4}$ **8.** $\dfrac{12x^3}{6(x+3)^2} \cdot \dfrac{(x+3)}{4x^2}$ **9.** $\dfrac{7y+14}{2y-2} \cdot \dfrac{4y-4}{3y+6}$

10. $\dfrac{3z^4}{10} \div \dfrac{6z}{5}$ **11.** $\dfrac{9x^2}{(x-3)^3} \div \dfrac{3x}{(x-3)^5}$ **12.** $\dfrac{y^2+2y-3}{y^2+4y+3} \div \dfrac{y^2-3y+2}{y^2+y-6}$

For Problems 13–16, find the sum or difference indicated. Reduce your answer to lowest terms.

13. $\dfrac{3}{x} + \dfrac{x}{2}$ **14.** $\dfrac{5}{y} - \dfrac{3y}{2}$

15. $\dfrac{w+1}{w^2-1} + \dfrac{w-1}{w-1}$ **16.** $\dfrac{x-3}{x+2} - \dfrac{x}{x^2-2x-8}$

For Problems 17 and 18, simplify the complex fraction.

17. $\dfrac{x+\dfrac{1}{3}}{7-\dfrac{2}{x}}$ **18.** $\dfrac{\dfrac{y+3}{y^2+3y-10}}{\dfrac{y+3}{y^2+7y+10}}$

For Problems 19–22, solve the equation and check your answer. Identify any extraneous solutions.

19. $\dfrac{x}{3} - \dfrac{7}{4} = \dfrac{5}{4}$ **20.** $\dfrac{3}{10} + \dfrac{13}{5y} = -1$

21. $\dfrac{z}{z-3} - \dfrac{6}{z^2-9} = \dfrac{1}{z+3}$ **22.** $\dfrac{w}{w+1} = \dfrac{3}{w^2-w-2} + \dfrac{2w-1}{w-2}$

23. Express 2 feet 6 inches and 5 yards in a common unit of length and then write them as a ratio reduced to lowest terms.

24. Identify the means and the extremes in the proportion

$$\frac{7}{x} = \frac{x-2}{5}.$$

Solve Problems 25–28.

25. On a map, 3 inches represents 50 miles. If two interchanges are 4.5 inches apart on the map, how far apart are they in actual miles?

26. The ratio of female students to male students at a local community college is 5 to 7. If there are 4000 female students, what is the total student population?

27. A 6-ohm resistor is connected parallel to a 12-ohm resistor as shown in the figure on page 195. What single resistance is equal to these two resistors? Recall that if two resistors are in parallel, then

$$\frac{1}{R} = \frac{1}{R_1} + \frac{1}{R_2}$$

when R is the total resistance.

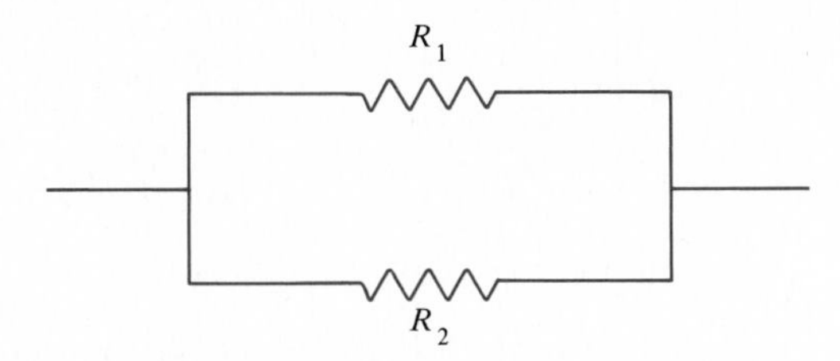

28. The container for a chemical bath can be filled by a faucet in 2.5 minutes. The drain pipe can empty the container in 3 minutes. If the drain is accidentally left open while the faucet is left on, how long will it take to fill the container?

CHAPTER 6
Linear Equations in Two Variables and Graphing

CHAPTER 6

Linear equations in two variables are helpful for solving problems involving two unknown quantities. For instance, an ice cream store sells only cones and sundaes. On a busy summer day the owner knew the total cash receipts and the total amount of cones and sundaes sold but not how many of each kind. The solution to a simple system of linear equations will tell him how many cones and sundaes he sold so that he can properly restock his inventory.

Thus far in our study of algebra, we have discussed solutions of equations in *one variable* such as

$$3x + 2 = 3 \qquad \text{first-degree equation}$$

and

$$x^2 + x - 6 = 0. \qquad \text{quadratic equation}$$

In this chapter we study first-degree equations and inequalities in *two variables* and their corresponding solution sets. In order to obtain a ''geometric picture'' or graph of the solution set of a first-degree equation in two variables, we shall introduce the rectangular (or Cartesian) coordinate system.

6.1 Rectangular Coordinate System

In Chapter 1 we studied the real number line. Recall that each point on the number line has a real number attached to it called its coordinate. Also, each real number is associated with a point called the graph of the real number. In Chapter 2 we used the number line to graph inequalities in one variable. In a similar way, the study of equations in two variables is aided by associating to each point in the plane an ordered pair of real numbers, such as (2, 3). To do this, we use two real number lines, one horizontal and the other one vertical, intersecting at right angles at their origins (or zero points), as indicated in Figure 6.1.

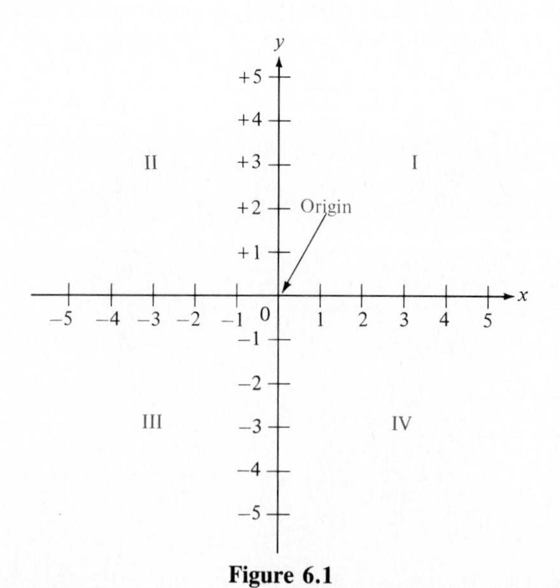

Figure 6.1

The horizontal line is called the *x-axis*, and the vertical line is called the *y-axis*. The positive side of the x-axis extends to the right and the negative side extends to the left.

The y-axis has the positive side going upward and the negative side downward. The two axes form a *rectangular* or *Cartesian coordinate system*. The point where the two axes intersect is called the origin of the coordinate system.

The coordinate system separates the plane into four regions called quadrants. The quadrants are labeled with Roman numerals counterclockwise as shown in Figure 6.1. By convention, points lying on either the x-axis or y-axis do not belong to any of the quadrants.

Let us now take a particular pair of numbers, say 2 and 3, written in order as (2, 3). Starting at the origin, the first number, 2, represents how far along the x-axis we move in a positive direction. Again starting at the origin, the second number, 3, represents how far we move vertically along the y-axis. At 2 on the x-axis, draw a line parallel to the y-axis. At 3 on the y-axis draw a line parallel to the x-axis. These two lines intersect at the point, P, as shown in Figure 6.2. In this way we have located the point associated with the ordered pair of numbers (2, 3). To indicate this association between point P and the ordered pair (2, 3), we write $P(2, 3)$ as shown in Figure 6.2.

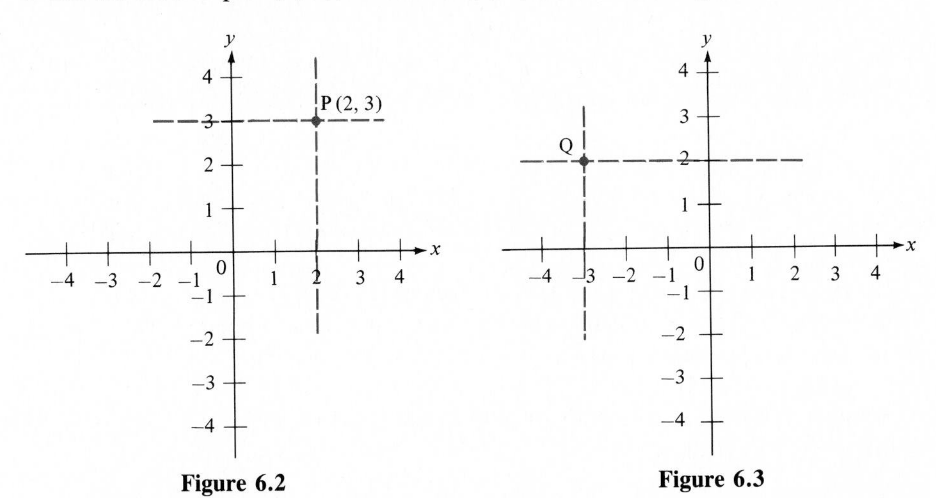

Figure 6.2 **Figure 6.3**

To find the ordered pair of numbers associated with a given point Q in a rectangular coordinate system such as in Figure 6.3, we draw a horizontal line and a vertical line through the point Q. One of these lines intersects the x-axis at -3 and the other intersects the y-axis at 2. The ordered pair of numbers that locate the point Q in the plane is $(-3, 2)$. Notice that in the ordered pairs of numbers (2, 3) and $(-3, 2)$, the value associated with the x-axis appears first in the parentheses, and the value associated with the y-axis appears second.

In general, given a rectangular coordinate system, a pair of numbers (x, y) determines a point in the plane, and for each point in the plane, we can determine the pair of numbers (x, y) associated with it.

The pair of numbers (x, y) is called an *ordered pair*. The first number (the x-value) is called the *x-coordinate* or *abscissa* of the corresponding point. The second number y is called the *y-coordinate* or *ordinate* of the point. The two numbers together are called the *coordinates* of the point. The point is called the *graph* of the ordered pair (x, y).

Locating a point with specific coordinates, such as (2, 3), is called *plotting* or *graphing* the point.

In Figure 6.4 we have graphed four ordered pairs, one for each quadrant. The ordered pairs are (3, 2), (−2, 4), (−3, −2), and (4, −1).

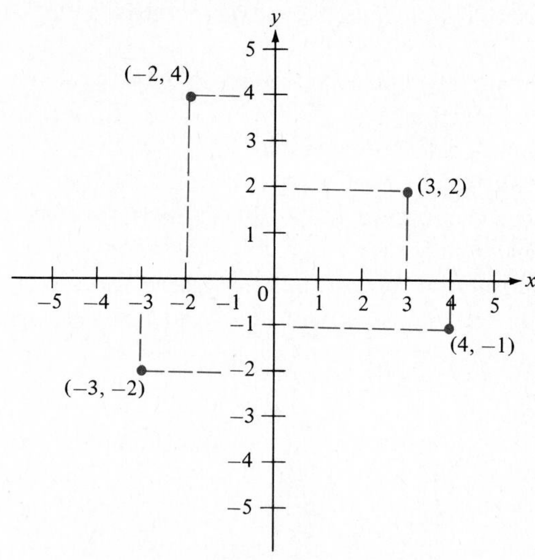

Figure 6.4

We now outline an equivalent way of graphing an ordered pair of numbers (x, y) without drawing horizontal and vertical lines. We see from Figure 6.4 that a pair of coordinates (a, b) can be thought of as directions for locating a point P. The next example illustrates this method.

We note first that we frequently identify a point P with its pair of coordinates and speak of the "point (a, b)."

Example 1. Plot the points with the following coordinates on a rectangular coordinate system.

$$(-4, 2). \qquad (4, 0). \qquad (-2, -4). \qquad \left(\frac{3}{2}, 3\right).$$

Solution: To locate the point (−4, 2), we start at the origin and go to −4 on the x-axis. Then go up 2 units to the point with coordinates (−4, 2). See Figure 6.5.

To find (4, 0) we go to +4 on the x-axis. Since the y-coordinate is 0, we "move" 0 units in a vertical direction. Proceeding in a similar manner, we locate (−2, −4) and $\left(\frac{3}{2}, 3\right)$.

The result of graphing all four of these points is shown in Figure 6.5.

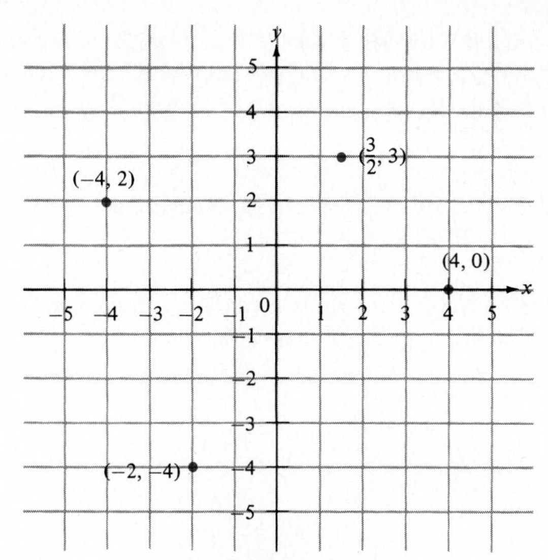

Figure 6.5

Example 2. Find the coordinates of points P, Q, R, and S shown in the rectangular coordinate system in Figure 6.6.

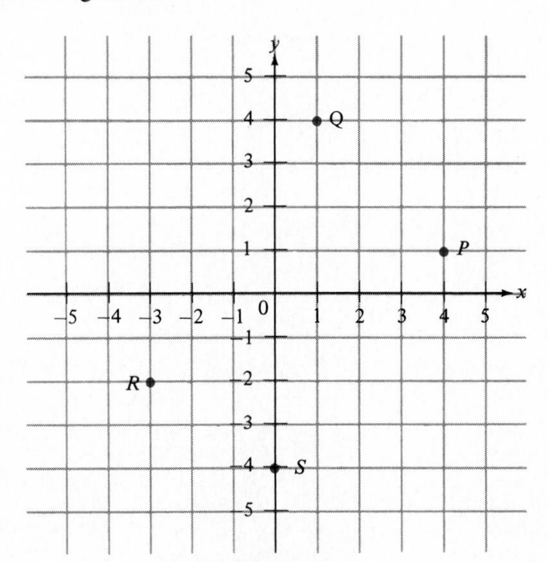

Figure 6.6

Solution: Referring to the rectangular coordinate system shown in Figure 6.6 we see that the coordinates of the points are

$$P(4, 1), \quad Q(1, 4), \quad R(-3, -2), \quad \text{and} \quad S(0, -4).$$

Notice in Example 2 that the ordered pairs (4, 1) and (1, 4) are associated with two different points, P and Q, respectively. Thus the order in which the numbers appear in the parentheses is important.

EXERCISES 6.1

For Exercises 1–8, find the coordinates of points labeled with the letters A to H in the rectangular coordinate system shown. Also state what quadrant, if any, the point lies in. See Example 2.

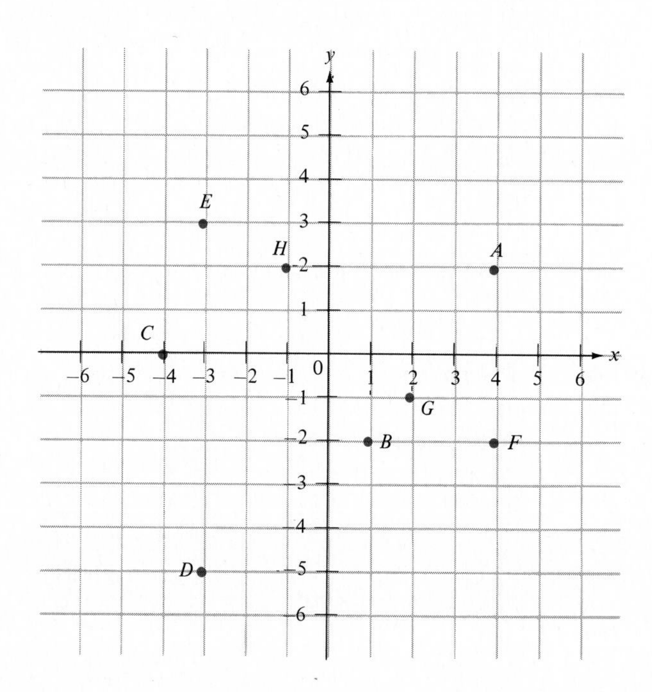

1. *A* **2.** *B* **3.** *C* **4.** *D* **5.** *E* **6.** *F* **7.** *G* **8.** *H*

For Exercises 9–26, plot the points in a rectangular coordinate system. See Example 1.

9. (1, 4) **10.** (−2, 3) **11.** (4, 1) **12.** (−3, −4)
13. (3, −2) **14.** (0, −5) **15.** (2, 0) **16.** (−2, −2)
17. (1, −3) **18.** (−5, 2) **19.** $(\frac{3}{2}, 2)$ **20.** (−4, 1)
21. (0, 3) **22.** (−3, 0) **23.** (0, 0) **24.** (0, 2)
25. (−3.5, −1.5) **26.** (2.5, −4)

Linear Equations in Two Variables 6.2

A *linear* (or first-degree) *equation in two variables,* x and y, is an equation that can be put in the form

$$Ax + By = C,$$

where A, B, and C are real numbers (A and B cannot both be 0).

Examples of linear equations in two variables are

$$x + y = 5, \qquad 5x - y = 3, \qquad y = -2x + 1, \qquad 2x + 3y - 5 = 0.$$

A *solution* of a linear equation $Ax + By = C$ is an ordered pair of numbers (a, b) such that when x is replaced by a, and y is replaced by b, the result is a true statement.

If (a, b) is a solution to an equation in two variables, x and y, then the ordered pair (a, b) is said to *satisfy* the equation. The set of all ordered pairs (a, b) that satisfy the equation is called the *solution set* of the equation.

Let us consider a particular linear equation such as $2x + y = 10$. The ordered pair (3, 4) satisfies this equation because when we replace x with 3 and y with 4 in the equation $2x + y = 10$, we obtain

$$\begin{aligned} 2(3) + 4 &= 10 \\ 6 + 4 &= 10 \\ 10 &= 10. \qquad \text{true} \end{aligned}$$

It is important to notice that although (3, 4) is a solution to $2x + y = 10$, the ordered pair (4, 3) is not a solution to this equation. To verify this, we let $x = 4$ and $y = 3$ in $2x + y = 10$ and obtain

$$\begin{aligned} 2(4) + 3 &= 10 \\ 8 + 3 &= 10 \\ 11 &= 10. \qquad \text{false} \end{aligned}$$

So when specifying solutions to most linear equations, remember that the order in which two numbers are written in parentheses is important. The next example illustrates this remark.

Example 1. Determine which of the ordered pairs (2, 1), (1, 2), and $\left(\frac{2}{3}, 3\right)$ are solutions to the equation $3x + 2y = 8$.

Solution: To see if (2, 1) is a solution, we replace x with 2 and y with 1 in the equation $3x + 2y = 8$. Thus we obtain

$$\begin{aligned} 3(2) + 2(1) &= 8 \\ 6 + 2 &= 8 \\ 8 &= 8. \qquad \text{true} \end{aligned}$$

Therefore, (2, 1) is a solution.

For the ordered pair (1, 2) we let $x = 1$ and $y = 2$ in $3x + 2y = 8$. As a result, we see that

$$\begin{aligned} 3(1) + 2(2) &= 8 \\ 3 + 4 &= 8 \\ 7 &= 8. \qquad \text{false} \end{aligned}$$

Therefore, (1, 2) is *not* a solution.

Finally, for $\left(\frac{2}{3}, 3\right)$ we substitute $x = \frac{2}{3}$ and $y = 3$ in the equation. The result is

$$\begin{aligned} 3\left(\frac{2}{3}\right) + 2(3) &= 8 \\ 2 + 6 &= 8 \\ 8 &= 8. \qquad \text{true} \end{aligned}$$

Therefore, $\left(\frac{2}{3}, 3\right)$ is a solution.

Summarizing, we see that (2, 1) and $\left(\frac{2}{3}, 3\right)$ are solutions to the equation $3x + 2y = 8$, but (1, 2) is *not* a solution.

We saw before that (3, 4) was a solution to the equation $2x + y = 10$. It is natural to ask "How do we find other solutions to the equation $2x + y = 10$?" The answer is simple. We substitute a value for one of the variables in the equation $2x + y = 10$ and solve the resulting equation for the other variable. For example, if we were to choose a value for y, say $y = 2$, we need to find a value for x so that $(x, 2)$ is a solution to $2x + y = 10$.

As a first step to finding such an x-value, we replace y with 2 in $2x + y = 10$. Thus we obtain a first-degree equation in *one* variable,

$$2x + 2 = 10.$$

We then use the methods outlined in Chapter 2 for solving equations in one variable in order to find x. Proceeding in this way, we see that

$$\begin{aligned} 2x + 2 + (-2) &= 10 + (-2) && \text{add } -2 \text{ to both sides} \\ 2x &= 8 && \text{addition of signed numbers} \\ x &= 4. && \text{divide both sides by 2} \end{aligned}$$

Therefore, in the equation $2x + y = 10$, *x has to be 4* when y is 2. Thus the ordered pair (4, 2) is another solution to the equation $2x + y = 10$.

To find still another solution to $2x + y = 10$, we can choose a value for x first, rather than y, and then find the corresponding y-value. For example, if we let $x = 5$ in the equation $2x + y = 10$, we obtain

$$\begin{aligned} 2(5) + y &= 10 \\ 10 + y &= 10. \end{aligned}$$

Solving this equation in one variable for y, we see that

$$(-10) + 10 + y = (-10) + 10 \qquad \text{add } -10 \text{ to both sides}$$
$$y = 0. \qquad \text{addition of signed numbers}$$

Therefore, since $y = 0$ when $x = 5$, another solution to the equation $2x + y = 10$ is $(5, 0)$.

The procedures used above for finding solutions to the equation $2x + y = 10$ suggest the following steps for finding solutions to any linear equation $Ax + By = C$.

1. Choose any value for one of the variables, either x or y.
2. Substitute the chosen value into the equation $Ax + By = C$.
3. Solve the resulting equation for the other variable.
4. Write the values obtained as an ordered pair.

The steps just outlined show that an equation in two variables, $Ax + By = C$, has an infinite number of solutions. This is in contrast to a first-degree equation in one variable, which usually has only one solution.

Example 2. For $5x - y = 3$, find the value of x or y so that the resulting ordered pairs will be a solution to the equation.
(a) $(2, y)$. (b) $(x, -2)$. (c) $(0, y)$.

Solution

(a) **Step 1.** The first number appearing in $(2, y)$ represents the x-value. Thus let $x = 2$.

Step 2. Replace x with 2 in $5x - y = 3$. The result is

$$5(2) - y = 3.$$

Step 3. Solve the equation for y.

$$\begin{aligned} 5(2) - y &= 3 \\ 10 - y &= 3 \\ -y &= -7 \\ y &= 7. \end{aligned}$$

Step 4. A solution is $(2, 7)$.

(b) **Step 1.** For $(x, -2)$ we see that the y-value is -2.

Step 2. Replace y with -2 in $5x - y = 3$. Thus we obtain the equation

$$5x - (-2) = 3 \qquad \text{or} \qquad 5x + 2 = 3.$$

Step 3. Solve for x.

$$\begin{aligned} 5x + 2 &= 3 \\ 5x &= 1 \\ x &= \frac{1}{5}. \end{aligned}$$

Step 4. A solution is $\left(\frac{1}{5}, -2\right)$.

(c) To solve this problem, we use steps 1–4 without writing each step separately. Thus, to find y so that $(0, y)$ is a solution of $5x - y = 3$, we replace x with 0 in the equation. The resulting equation to be solved for y is

$$5(0) - y = 3.$$

Solving for y, we have

$$\begin{aligned} -y &= 3 \\ y &= -3. \end{aligned}$$

A solution is the ordered pair $(0, -3)$.

In the following examples, we shall solve linear equations in two variables using steps 1 to 4 without writing out each step separately.

Example 3. Find four ordered pairs in the solution set of

$$2x + y = 5.$$

Solution: To find the ordered pairs, we make up a table assigning a value to either x or y so that the resulting equation in one variable appears easy to solve.
To start with, we assign the following values to the indicated variables.

x	0		1	
y		0		-1

For $x = 0$: We substitute 0 for x in $2x + y = 5$. The resulting equation is

$$2(0) + y = 5. \qquad \text{an equation in one variable}$$

Solving for y, we obtain

$$\begin{aligned} 0 + y &= 5 \\ y &= 5. \end{aligned}$$

Therefore, a solution to $2x + y = 5$ is $(0, 5)$.
For $y = 0$: Substituting 0 for y in the given equation results in

$$\begin{aligned} 2x + 0 &= 5 \\ 2x &= 5. \end{aligned}$$

Solving for x, we get

$$x = \frac{5}{2}.$$

A solution is $\left(\frac{5}{2}, 0\right)$.

For $x = 1$: Substituting 1 for x, we get

$$\begin{aligned} 2(1) + y &= 5 \\ 2 + y &= 5 \\ y &= 3. \end{aligned}$$

A solution is (1, 3).

For $y = -1$: First replace y with -1 and obtain

$$\begin{aligned} 2x + (-1) &= 5 \\ 2x - 1 &= 5. \end{aligned}$$

Solving for x, we obtain $x = 3$.

$$\begin{aligned} 2x &= 6 \\ x &= 3. \end{aligned}$$

Another solution is $(3, -1)$.

Our completed table is as follows:

x	0	$\frac{5}{2}$	1	3
y	5	0	3	-1

Thus four ordered pairs that are solutions to $2x + y = 5$ are (0, 5); $\left(\frac{5}{2}, 0\right)$; (1, 3); and $(3, -1)$.

Example 4. Find four ordered pairs that satisfy the equation $y = 3x + 1$.

Solution: For this equation it is convenient to set up a table as we did in Example 3. However, because of the way this equation is written, it is easier to assign values to x, and then determine the corresponding values of y as shown below.

x	0	1	-1	2
y				

For $x = 0$: $y = 3 \cdot 0 + 1$, $y = 1$.

For $x = 1$: $y = 3 \cdot 1 + 1$, $y = 4$.

For $x = -1$: $y = 3(-1) + 1$
$y = -3 + 1$
$y = -2.$

For $x = 2$: $y = 3(2) + 1$
$y = 6 + 1$
$y = 7.$

Our completed table is as follows:

x	0	1	-1	2
y	1	4	-2	7

The four ordered pairs satisfying $y = 3x + 1$ are

$(0, 1)$; $(1, 4)$; $(-1, -2)$; $(2, 7)$.

EXERCISES 6.2

For Exercises 1–8, determine which of the ordered pairs are solutions of the equation. See Example 1.

1. $x + y = 8$; $(5, 3)$, $(7, 2)$, $(-1, 9)$
2. $y - x = 7$; $(1, 8)$, $(8, 1)$, $(9, 2)$
3. $2x - 3y = 9$; $(0, -3)$, $(2, -\frac{5}{3})$, $(6, 1)$
4. $4x + y = -6$; $(-2, 2)$, $(0, -6)$, $(2, -3)$
5. $6x + 4y = 12$; $(0, 2)$ $(3, 0))$, $(1, 1)$
6. $3x + 2y = 6$; $(2, 0)$, $(0, 3)$, $(1, \frac{3}{2})$
7. $y = 2x + 3$; $(1, 5)$, $(5, 1)$, $(0, 3)$
8. $y = -\frac{1}{2}x + 4$; $(4, 0)$, $(4, 2)$, $(2, 3)$

For Exercises 9–24, find the values of x or y so that the resulting ordered pairs are solutions to the equation. See Example 2.

9. $3x - 2y = 8$; $(x, 0)$, $(0, y)$, $(x, -1)$
10. $5x - y = 2$; $(x, 0)$, $(0, y)$, $(2, y)$
11. $3x + 4y = 1$; $(x, 0)$, $(x, -2)$, $(0, y)$
12. $y = 3x$; $(0, y)$, $(3, y)$, $(-4, y)$
13. $y = -4x$; $(0, y)$, $(x, 0)$, $(1, y)$
14. $y = \frac{1}{2}x + 2$; $(0, y)$, $(x, 0)$, $(2, y)$
15. $y = -\frac{1}{3}x - 1$; $(0, y)$, $(x, 0)$, $(3, y)$
16. $y = 0 \cdot x + 2$; $(2, y)$, $(-1, y)$, $(4, y)$
17. $y = 0 \cdot x - 1$; $(4, y)$, $(5, y)$, $(9, y)$
18. $x + 0 \cdot y = 3$; $(x, 3)$, $(x, -4)$, $(x, 15)$
19. $x + 0 \cdot y = -2$; $(x, 1)$, $(x, 6)$, $(x, 27)$
20. $x - y = 5$; $(x, 4)$, $(2, y)$, $(-2, y)$

21. $3x + y = 7$; $(3, y)$, $(x, -8)$, $(1, y)$

22. $x + 5y = 10$; $(x, 1)$, $(-5, y)$, $(x, 3)$

23. $4x - y = 11$; $(x, 5)$, $(2, y)$, $(x, -3)$

24. $5x + 3y = 12$; $(x, -1)$, $(x, -11)$, $(-6, y)$

For Exercises 25–34, complete the table for the equation. See Example 3.

25. $y = 2x$

x	2		−4	0
y		8		

26. $y = -3x$

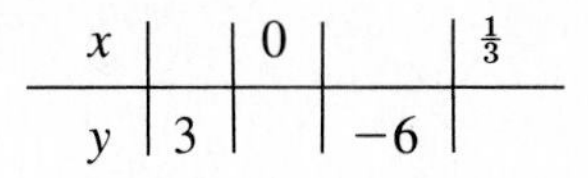

x		0		$\frac{1}{3}$
y	3		−6	

27. $y = \frac{1}{2}x$

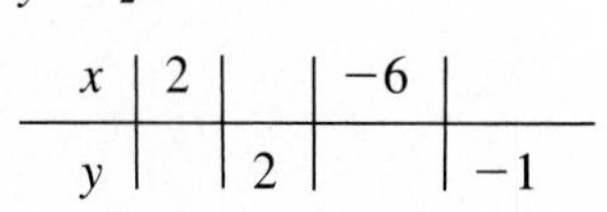

x	2		−6	
y		2		−1

28. $y = -\frac{1}{3}x$

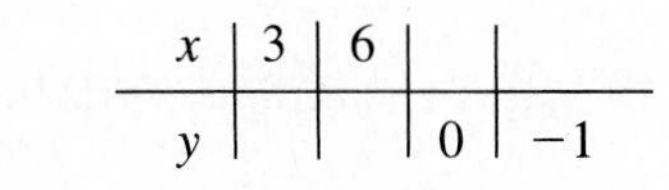

x	3	6		
y			0	−1

29. $y = 6x - 2$

x	0		1	2
y		0		

30. $y = 7x - 3$

x	0		1	−1
y		0		

31. $x + y = 9$

x	0		3	
y		0		5

32. $x - y = 2$

x	0		4	
y		0		3

33. $2x + 5y = 10$

x	0		4	
y		0		−2

34. $\frac{1}{3}x + \frac{1}{5}y = 1$

x	0			6
y		0	−10	

For Exercises 35–44, find four ordered pairs that satisfy the following equation. See Example 4, using the same x-values.

35. $y = 7x$ **36.** $y = 9x$ **37.** $y = -5x$ **38.** $y = 8x$

39. $y = -3x + 1$ **40.** $2x + y = 3$ **41.** $4x - y = 2$ **42.** $3x - y = 9$

43. $5x - y = 10$ **44.** $2x - y = 8$

For Exercises 45–50, write a linear equation in two variables to express the relation between x and y.

45. y equals four times x.

46. y equals one less than twice x.

47. The sum of x and y equals 12.

48. The product $5x$ minus y equals 4.

49. One-third of x plus one-fifth of y equals 1.

50. x minus the product of 7 and y equals 3.

Solve Exercises 51–54.

51. Let x represent the number of tolls collected at a certain toll plaza on Green Parkway. Each toll is 25 cents. The amount of money, A, collected in tolls daily is given by the equation $A = .25x$. If on a given Sunday 4000 tolls are collected, what is the amount of money collected?

52. Suppose that x represents the number of minutes or part of a minute that the phone is being used during a call. The cost of a call C is given by $C = .20x$. What will it cost to make a 4.5-minute call?

53. Suppose that x represents the number of hand calculators manufactured by Digit, Inc. The daily profit P, for Digit, Inc., is given by the equation $P = 10x - 500$. What is the profit P if on a certain day Digit, Inc., produced and sold 90 calculators?

54. A person bought a 2-year-old car for $7500. Its value V, x years after purchase, is given by the equation $V = 7500 - 600x$. What is the value of the car 5 years after the person bought the car?

6.3 Graph of a Linear Equation

In Section 6.2 we saw that solutions to a linear equation, such as $x + y = 4$, are ordered pairs of numbers (a, b). In this section we graph the solutions to a linear equation in two variables on a rectangular coordinate system. As a result, we obtain a geometric figure or "picture" of *all* the solutions to a linear equation $Ax + By = C$.

Let us consider the equation $x + y = 4$. Recall from Section 6.2 that such an equation has an infinite number of solutions. It is easy to see that some of the solutions to $x + y = 4$ are $(-1, 5)$, $(0, 4)$, $(2, 2)$, $(4, 0)$, and $(5, -1)$. The graph of these ordered pairs are the points shown in Figure 6.7.

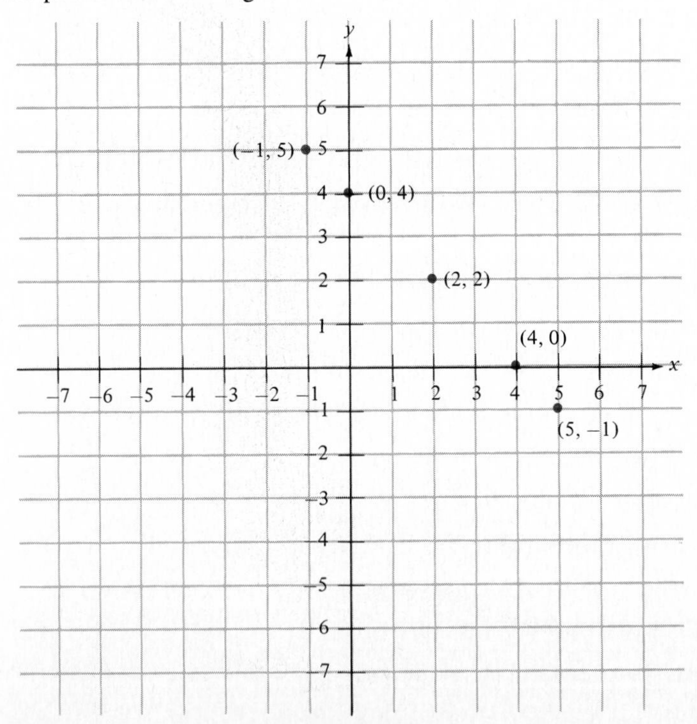

Figure 6.7

Notice that these points all lie on a straight line. In fact, the graphs (or points) of all solutions to $x + y = 4$ lie on the straight line shown in Figure 6.8.

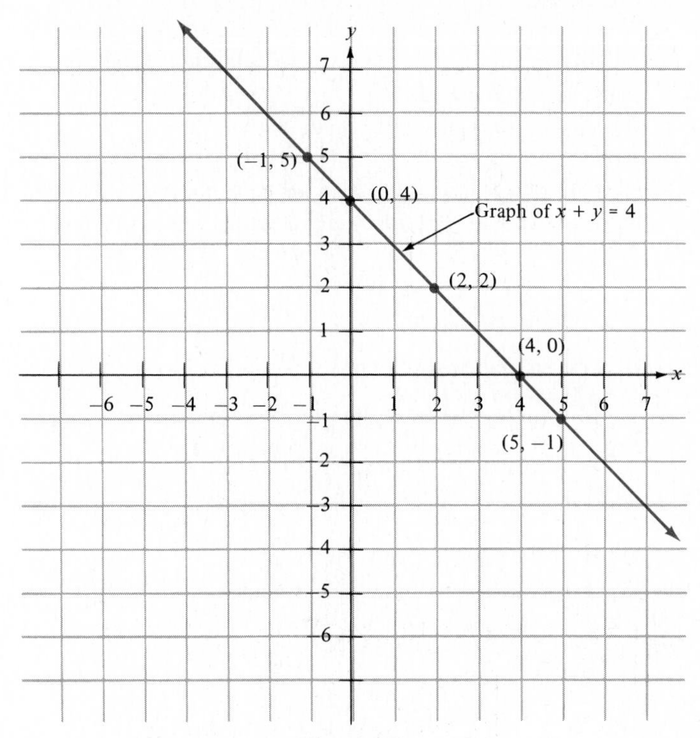

Figure 6.8

The arrows on the end of the line indicate that the line extends indefinitely in either direction. Moreover, any point on this line has coordinates, (a, b) that satisfy the equation $x + y = 4$. As a result, we say that the graph of $x + y = 4$ is a straight line. This line is the geometric picture of the set of *all* solutions to $x + y = 4$. Because of this, first degree equations in two variables, such as $x + y = 4$, are called *linear equations*.

> The graph of a linear equation, in two variables x and y, $Ax + By = C$, is a straight line.

Note that

1. Any point on the straight line has coordinates (a, b) that satisfy the equation $Ax + By = C$.
2. Any solution (a, b) to $Ax + By = C$ will have its graph on the straight line.

From elementary geometry we know that two different points determine a straight line. Therefore, to graph a linear equation on a rectangular coordinate system, we follow the steps outlined below.

1. Find two ordered pairs of numbers that are solutions to the equation.
2. Plot or graph these two solutions.
3. Draw a straight line through the two points.

In addition, we usually find a third solution to the equation as a checkpoint. The graph of the third solution should be on the straight line that we have drawn.

Example 1. Graph the equation $2x + y = 4$.

Solution: First we find two solutions to $2x + y = 4$. We proceed to find solutions as outlined in Section 6.2.

$$\text{Let } x = 0: \quad \begin{aligned} 2(0) + y &= 4 \\ 0 + y &= 4 \\ y &= 4. \end{aligned} \qquad \text{Let } y = 0: \quad \begin{aligned} 2x + 0 &= 4 \\ 2x &= 4 \\ x &= 2. \end{aligned}$$

Therefore, (0, 4) and (2, 0) are both solutions.

We then draw a straight line through the points (2, 0) and (0, 4) as shown in Figure 6.9.

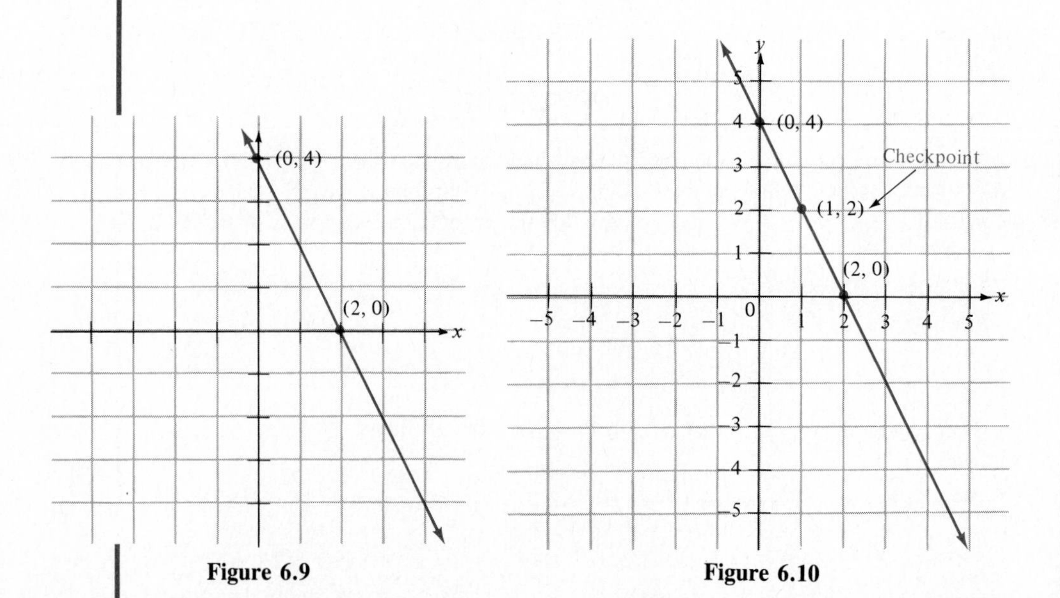

Figure 6.9

Figure 6.10

We now find a third solution in order to check that we have the correct line.

$$\begin{aligned} \text{Let } x = 1: \quad 2(1) + y &= 4 \\ 2 + y &= 4 \\ y &= 2. \end{aligned}$$

Therefore, (1, 2) is another solution. Plotting (1, 2), we see that it lies on the straight line drawn in Figure 6.10. The straight line shown is the graph of $2x + y = 4$.

The x-coordinate of the point at which a straight line crosses the x-axis is called the *x-intercept*. Similarly, the y-coordinate of the point where a line crosses the y-axis is called the *y-intercept*. Thus, in Example 1 the line crosses the x-axis at (2, 0), so the x-intercept is 2. It crosses the y-axis at (0, 4); therefore, the y-intercept is 4. Notice that for a given linear equation, $Ax + By = C$:

1. To find the x-intercept, let $y = 0$ and solve for x.
2. To find the y-intercept, let $x = 0$ and solve for y.

It is usually easier to plot the graph of a linear equation by finding the intercepts, provided that it has two. In some special cases, as we shall see later, there is only one intercept.

Example 2. Graph equation $2x - 3y = -6$ on a rectangular coordinate system.

Solution: First we find the x and y intercepts.

$$\begin{aligned} \text{Let } y = 0: \quad 2x - 3(0) &= -6 \\ 2x - 0 &= -6 \\ 2x &= -6 \\ x &= -3. \qquad \textit{x-intercept} \end{aligned}$$

Therefore, $(-3, 0)$ is a solution.

$$\begin{aligned} \textit{Let } x = 0: \quad 2(0) - 3y &= -6 \\ 0 - 3y &= -6 \\ y &= 2. \qquad \textit{y-intercept} \end{aligned}$$

Also, (0, 2) is a solution.

We now find a third solution to use as a checkpoint. If we let $x = 3$ in the equation, we would find that $y = 4$. Therefore, (3, 4) is another solution.

Sometimes we plot all three solutions at once. Plotting the three solutions $(-3, 0)$, (0, 2), and (3, 4), we see that all three points lie on a straight line. Drawing a straight line through the points, we obtain the graph of $2x - 3y = -6$ as shown in Figure 6.11.

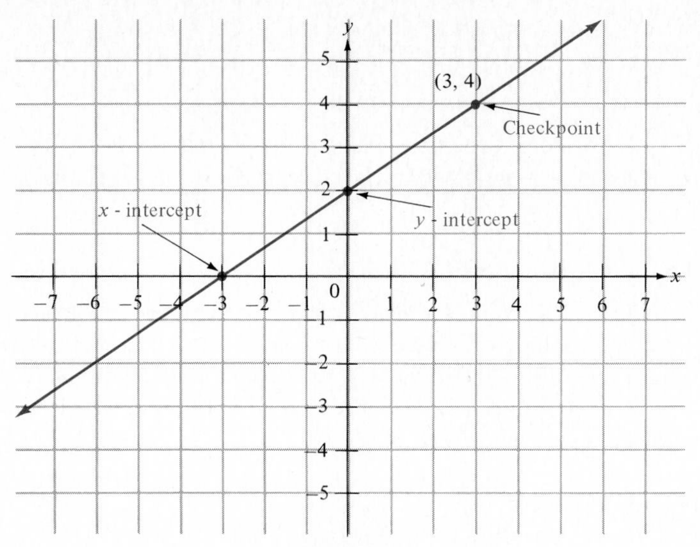

Figure 6.11

Example 3. Graph $y = -2x + 3$.

Solution: To find three solutions, we let $x = 0, 1, 2$. We then find the corresponding y-values. The result is the table of values on page 215. Plotting the points $(0, 3)$, $(1, 1)$, and $(2, -1)$ we obtain the graph shown in Figure 6.12.

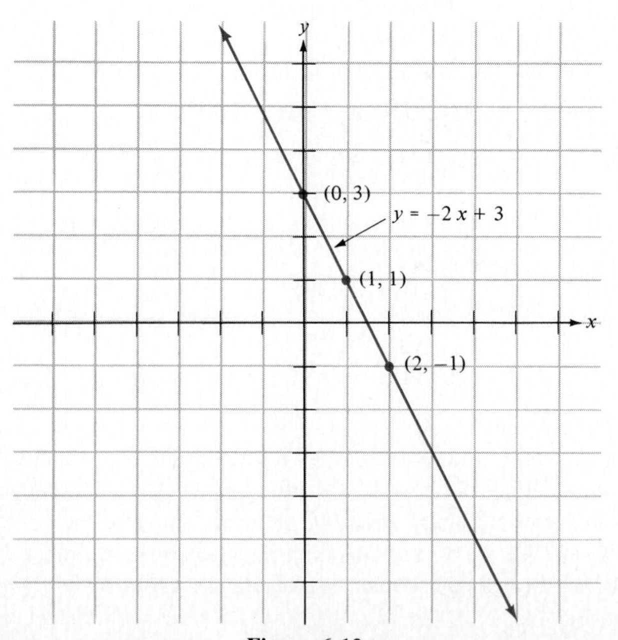

Figure 6.12

x	0	1	2
y	3	1	−1

Example 4. Graph each equation on a rectangular coordinate system.
(a) $x = 3$. (b) $y = 2$.

Solution

(a) The equation $x = 3$ does not appear to be a first-degree equation in two variables. However, if we use 0 to rewrite the equation $x = 3$ as $x + 0 \cdot y = 3$ we see that $x = 3$ represents a linear equation in *two variables*. Notice that whatever value we substitute for y, the result is always $x = 3$. Therefore, all solutions (x, y) have an x value of 3, no matter what value y has. Some solutions are (3, 0), (3, 1), and (3, 2). The graph of $x = 3$ is a vertical line as shown in Figure 6.13. Notice that it only has an x-intercept.

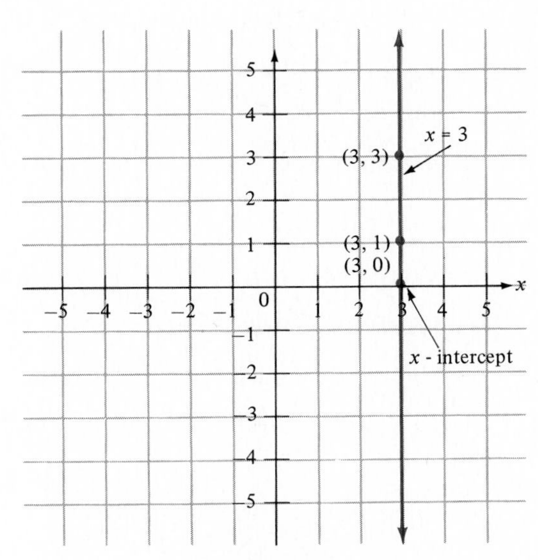

Figure 6.13

(b) The equation $y = 2$ can be rewritten as $0 \cdot x + y = 2$. Since $0 \cdot x = 0$, no matter what value we substitute for x in the equation $0 \cdot x + y = 2$, the resulting y-value will be 2. Some solutions are $(-3, 2)$, $(0, 2)$, and $(3, 2)$. The graph of $y = 2$ is the horizontal line shown in Figure 6.14. Notice that it only has a y-intercept.

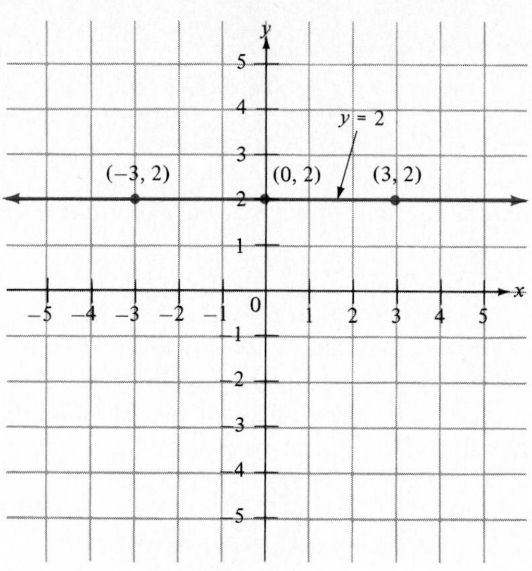

Figure 6.14

From Example 4 we see that any equation of the form

$x = a$ has as a graph a vertical line with x-intercept a

and any equation of the form

$y = b$ has as a graph a horizontal line with y-intercept b.

EXERCISES 6.3

For Exercises 1–8, find values of x or y so that the ordered pairs are solutions to the linear equation. Graph the equation on a rectangular coordinate system.

1. $3x - y = 6$; $(x, 0)$, $(0, y)$, $(1, y)$

2. $x - y = 2$; $(x, 0)$, $(0, y)$, $(1, y)$

3. $2x - 3y = 6$; $(x, 0)$, $(0, y)$, $(6, y)$

4. $3x + 4y = 12$; $(x, 0)$, $(0, y)$, $(x, 1)$

5. $3x + y = -6$; $(x, 0)$, $(0, y)$, $(x, -3)$

6. $x - 5y = 5$; $(x, 0)$, $(0, y)$, $(x, -2)$

7. $2x - y = 0$; $(x, 0)$, $(1, y)$, $(2, y)$

8. $x - 3y = 0$; $(x, 0)$, $(3, y)$, $(-3, y)$

For Exercises 9–30, graph the equation. See Examples 1–3.

9. $3x - y = 3$ **10.** $x - y = 4$ **11.** $2x + 5y = 10$ **12.** $2x + y = -2$

13. $x - y = 5$ **14.** $x + y = 1$ **15.** $y = 3x - 2$ **16.** $y = -3x - 2$

17. $y = x$ **18.** $y = -x$ **19.** $2x + y = -2$ **20.** $x - 3y = -6$

21. $-2x + y = 6$ **22.** $3x + 2y = -6$ **23.** $x - 2y = -4$ **24.** $3x - 4y = -12$

25. $x + y = \frac{7}{2}$ **26.** $x - 2y = -6$ **27.** $y = -5x + 2$ **28.** $y = -6x + 1$

29. $y = \frac{1}{3}x$ **30.** $y = -\frac{1}{2}x$

For Exercises 31–38, graph the equation. See Example 4.

31. $x = 4$ **32.** $x = -3$ **33.** $y = -5$ **34.** $y = 3$

35. $x = 0$ **36.** $y = 0$ **37.** $x + 2 = 0$ **38.** $y + 3 = 0$

Solve Exercises 39–42.

39. **(a)** Graph $y = 2x$, $y = 2x + 6$, and $y = 2x - 4$ on the same coordinate system.
(b) Will the graphs of these equations intersect?
(c) What are the x and y intercepts of each equation?

40. **(a)** Graph $y = -3x$, $y = -3x + 6$, and $y = -3x - 3$ on the same coordinate system.
(b) Will the graphs of these equations intersect?
(c) What are the x and y intercepts of each equation?

41. Suppose that x represents the number of inexpensive hand calculators sold per person, and y the price per calculator. Further, suppose that x and y are related by the linear equation $5x + 2y = 10$.
(a) Graph the equation.
(b) What is the price per calculator when 0 are sold per person (i.e., when $x = 0$)?
(c) If the calculators are given as free samples (i.e., the price $y = 0$), how many would a person "buy" or "take"?

42. Suppose that x represents the number of hours after 12 A.M. on a cold January morning, and y represents the temperature in degrees Fahrenheit. Further suppose that x and y are related by the linear equation $2x + y = 8$ for the next 6 hours.
(a) Graph the equation.
(b) What is the temperature at 12 A.M. when $x = 0$?
(c) At what time is the temperature $y = 0$?

Graphing Linear Inequalities 6.4

In this section we discuss the graph of the solution set of a first-degree or linear inequality in *two variables*. Because there are two variables involved, we graph the solution set on a rectangular coordinate system. A few examples of linear inequalities in two variables, x and y, are

$$x + y \leq 5, \quad y < 3x + 1, \quad 2x + y \geq 5, \quad y > 2x + 3.$$

A *linear inequality in two variables* x and y is an inequality that can be put in the form

$$Ax + By \leq C,$$

where A, B, and C are real numbers (A and B not both 0). In the definition above, the inequality sign $\leq$ may be replaced by any one of the inequality symbols, $<$, $\geq$, or $>$.

A *solution* to a linear inequality in two variables is an ordered pair of numbers (a, b) such that when x is replaced by a, and y is replaced by b, the resulting statement is true. The ordered pair (a, b) is said to *satisfy* the inequality.

Consider the inequality $2x + y \leq 6$. Which of the ordered pairs (2, 1), and (3, 2) are solutions to the inequality? We see that (2, 1) is a solution because if we replace x with 2 and y with 1 in $2x + y \leq 6$, we obtain

$$2(2) + 1 \leq 6 \quad \text{or} \quad 5 \leq 6, \quad \text{which is a } \textit{true statement}.$$

However, (3, 2) is *not* a solution because if we let $x = 3$ and $y = 2$ in $2x + y \leq 6$, we obtain

$$2(3) + 2 \leq 6 \quad \text{or} \quad 8 \leq 6, \quad \text{a } \textit{false statement}.$$

The *graph* of an inequality in two variables is the set of all points on the rectangular coordinate system whose coordinates (a, b) satisfy the inequality.

Let us consider the graph of $2x + y \leq 6$. As a first step, we graph the equation $2x + y = 6$ as shown in Figure 6.15. The graph is a straight line in the rectangular coordinate plane. Notice that the line divides the plane into two *regions* or *half-planes*. One half-plane lies above the line; the other half-plane lies below the line. The line is called the *boundary* of each of the half-planes. It is not part of either half-plane. This is illustrated in Figure 6.15.

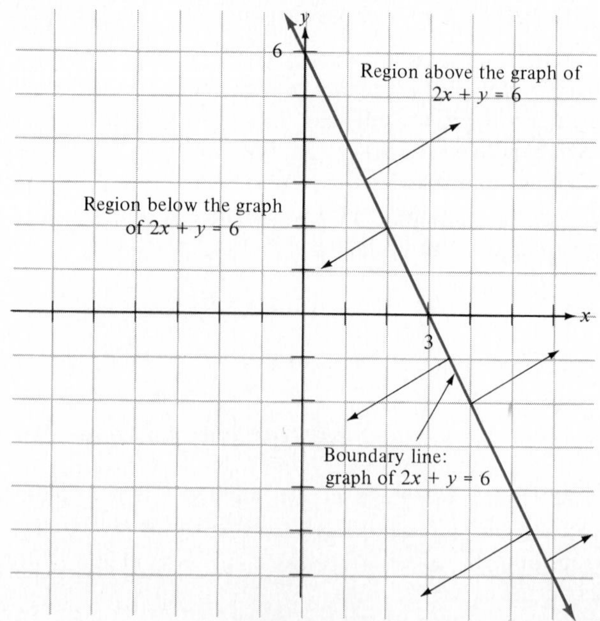

Figure 6.15

One of the half-planes represents the graph of the restricted inequality $2x + y < 6$. The other half-plane contains points whose coordinates do *not* satisfy the inequality. All points on the line have coordinates that satisfy $2x + y = 6$. Thus the boundary line together with a half-plane represents the graph of $2x + y \leq 6$. To determine which half-plane to choose, we need only select a test point in one of the half-planes. If the coordinates of the point satisfy the inequality, we have chosen the correct half-plane. If the coordinates of the point do *not* satisfy the inequality, we select the other half-plane. Thus, for $2x + y \leq 6$, we choose a convenient point in the half-plane below the line, such as the origin (0, 0). Usually, the origin is a good point to use for a test point. Replacing both x and y with 0 in $2x + y \leq 6$, we obtain $2 \cdot 0 + 0 \leq 6$ or $0 \leq 6$, a *true statement*. Thus the graph of $2x + y \leq 6$ consists of the half-plane below the line, together with the boundary line. We indicate this by shading the region below the line as shown in Figure 6.16.

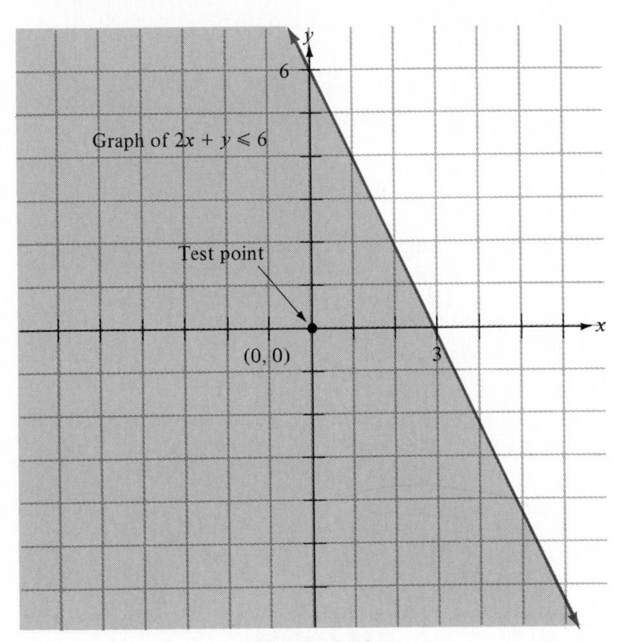

Figure 6.16

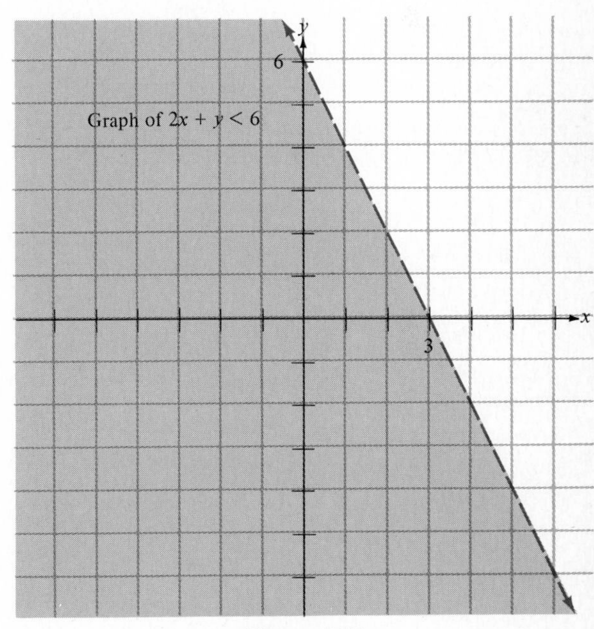

Figure 6.17

Note that if the original inequality does not include the symbol of equality, such as $2x + y < 6$, then the points on the boundary line do not satisfy the original inequality. Therefore, the boundary line is not part of the graph of the inequality. We indicate this in the graph by using a dashed line as shown in Figure 6.17.

In general, to graph the solution set of linear inequality in two variables, do the following:

1. Replace the symbol of inequality ($\leq, <, \geq, >$) with the equality symbol, $=$. Graph the resulting equation, which is a straight line. This is the boundary line between the two half-planes. Draw a solid line if the inequality symbol is $\leq$ or $\geq$. If the symbol is $<$ or $>$, draw a dashed line.

2. Choose a test point in one of the two half-planes. Generally, if the inequality symbol is $\leq$ or $<$, it is best to choose a test point below the boundary line. If the inequality is $\geq$ or $>$, choose a test point above the line. Do not choose a point on the boundary line.
3. If the coordinates of the test point satisfy the given inequality, "shade in" the half-plane in which the point lies. If the coordinates do not satisfy the inequality, shade in the other half-plane.

Example 1. Graph the inequality $x + y > 2$.

Solution: First we graph the equation $x + y = 2$. We draw a dashed line representing the graph because points on the line do not represent solutions to $x + y > 2$ (see Figure 6.18).

To decide which half-plane is the graph of $x + y > 2$, we choose a test point. Since the inequality symbol is $>$, we choose a point above the line such as (0, 3). Replacing x with 0 and y with 3 in

$$x + y > 2,$$

we obtain

$$0 + 3 > 2 \qquad \text{or} \qquad 3 > 2. \qquad \text{true}$$

Thus by the rules outlined above for graphing inequalities, the graph of $x + y > 2$ lies above the line as shown in Figure 6.18.

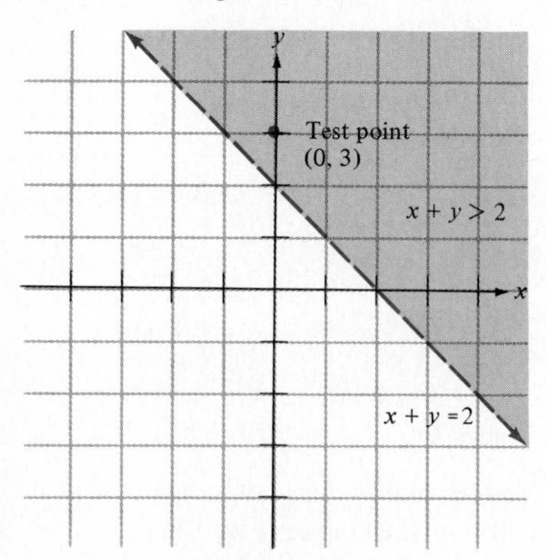

Figure 6.18

Example 2. Graph each inequality on a rectangular coordinate system. (a) $x \leq 4$. (b) $y > -3$.

Solution

(a) We graph the corresponding equation in two variables $x + 0 \cdot y = 4$ or $x = 4$, which is a vertical line. In this case the half-planes lie to the left and right of the line. Notice that points with x-coordinates less than 4 lie to the left of the vertical line. To check this, we use as a test point (0, 0), which lies in the left half-plane. Substituting in $x \leq 4$, we obtain $0 \leq 4$, a *true statement*. The graph of $x \leq 4$ is shown in Figure 6.19.

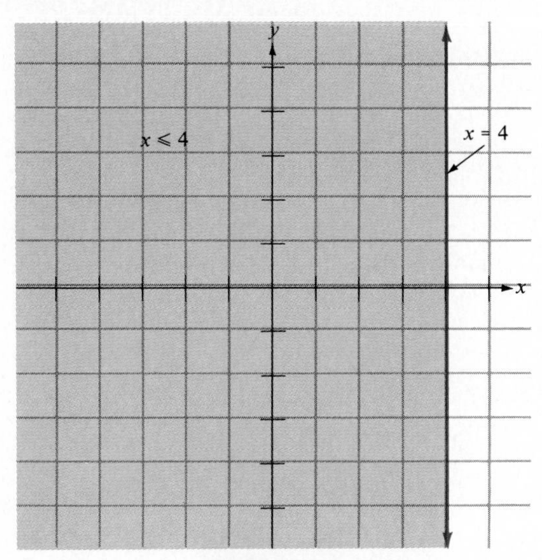

Figure 6.19

(b) To graph all the ordered pairs satisfying $y > -3$, we first graph $y = -3$. We represent the graph of $y = -3$ as a dashed horizontal line, since the boundary line is not part of the graph. Next we choose as a test point, the origin (0, 0),

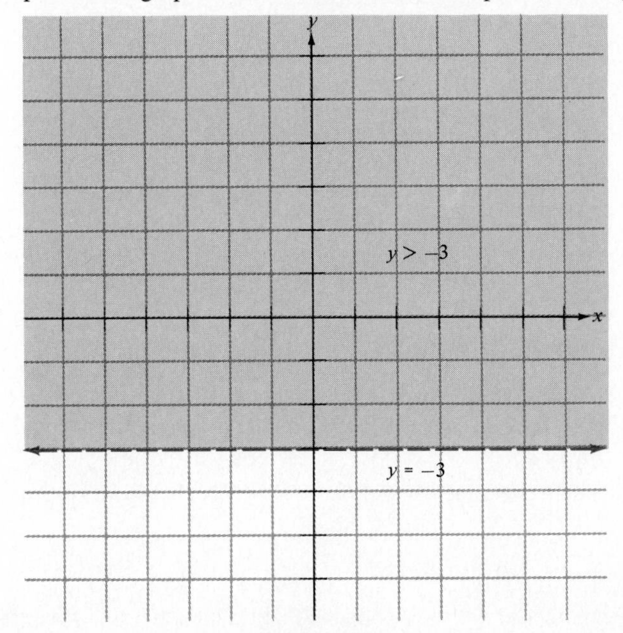

Figure 6.20

which lies in the half-plane above the line. Substituting 0 for y, we see that

$$0 > -3. \qquad \text{true}$$

Thus the graph of $y > -3$ is the half-plane above the line $y = -3$, as shown in Figure 6.20 at the bottom of page 221.

Example 3. Graph the inequality $y \geq 2x$.

Solution: First we graph $y = 2x$ which is our boundary line. We cannot choose the origin (0, 0) as a test point, since it is on the boundary line. Therefore, we choose another test point in the half-plane above the boundary line, such as (0, 3). Substituting these coordinates in our inequality, we obtain

$$3 \geq 2 \cdot 0 = 0 \qquad \text{true}$$

Thus the graph of $y \geq 2x$ is the boundary line $y = 2x$ and the half-plane above the line as shown in Figure 6.21.

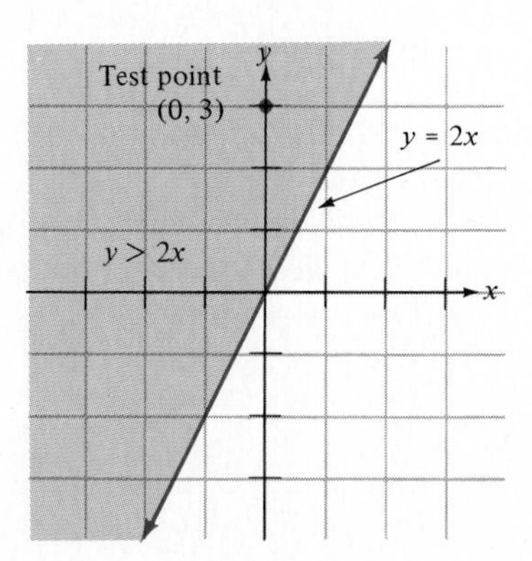

Figure 6.21

Example 4. Graph the inequality $3x - 2y \geq 6$.

Solution: We graph the line corresponding to $3x - 2y = 6$ as a solid line as shown in Figure 6.22 because it is part of the graph of the inequality. Since the inequality symbol is $\geq$, we choose a test point above the line to decide which half-plane is part of the graph. For convenience we choose (0, 0). Substituting 0 for both x and y in $3x - 2y \geq 6$, we obtain

$$3(0) - 2(0) \geq 6 \qquad \text{or} \qquad 0 \geq 6. \qquad \text{false}$$

Since the test point did *not* satisfy the inequality, the other half-plane below the line is the correct half-plane. The graph of $3x - 2y \geq 6$ is shown in Figure 6.22.

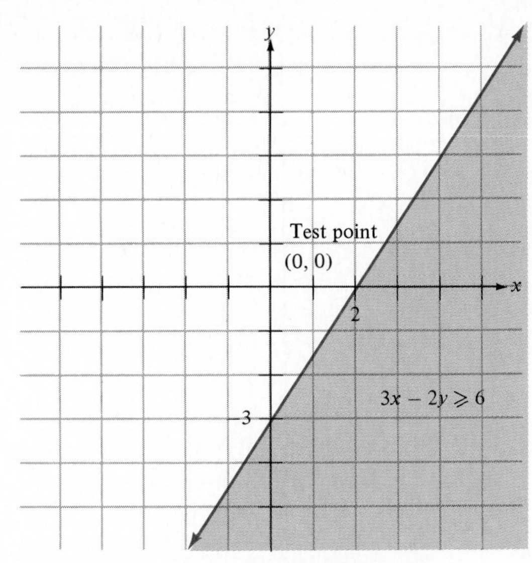

Figure 6.22

Most of the time for $\geq$ or $>$, a test point above the line satisfies the inequality. In Example 4 it did not. Thus we should always use a test point in the inequality to verify that we have chosen the correct half-plane.

EXERCISES 6.4

In Exercises 1–10, graph the inequality. The boundary line for each inequality is given. Shade in the correct half-plane to complete the graph. See Examples 1–4.

1. $x + y < 5$

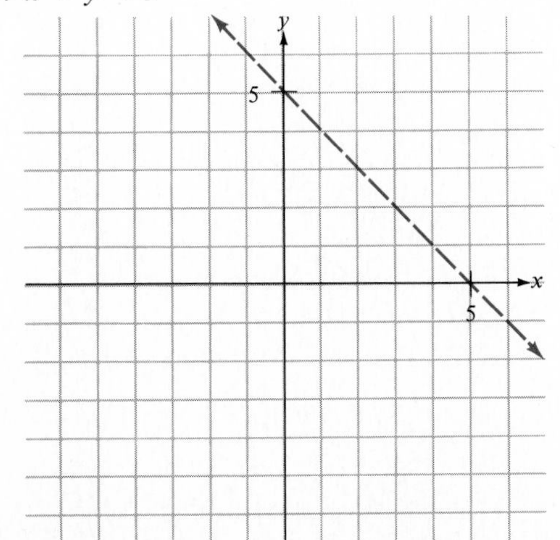

2. $x - y \geq 5$

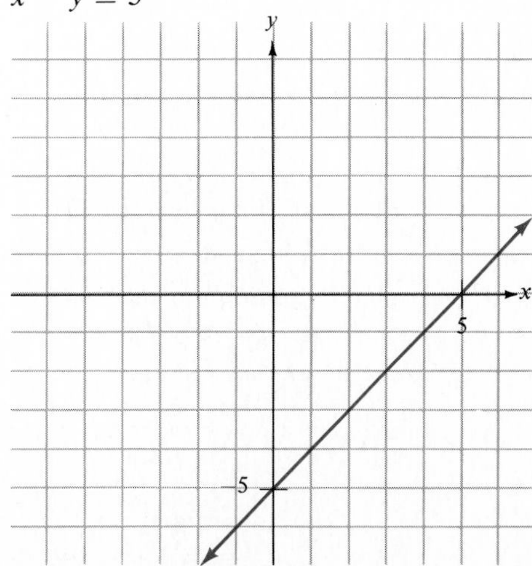

3. $-3x + 5y \geq 15$

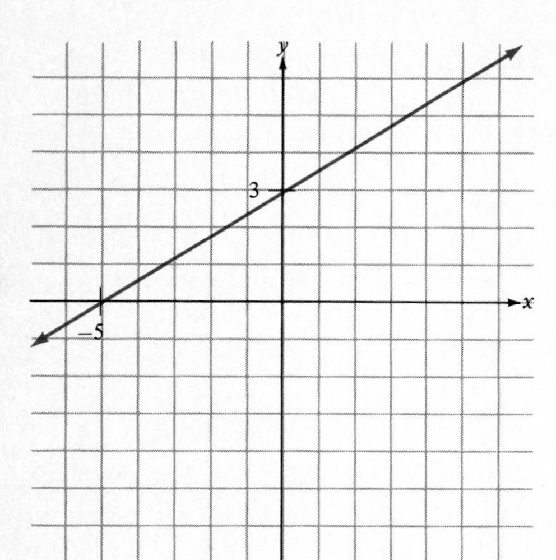

4. $-x - y > 3$

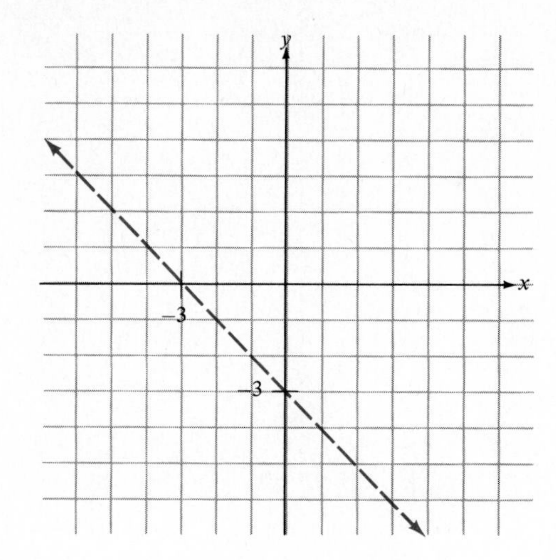

5. $y < 4x$

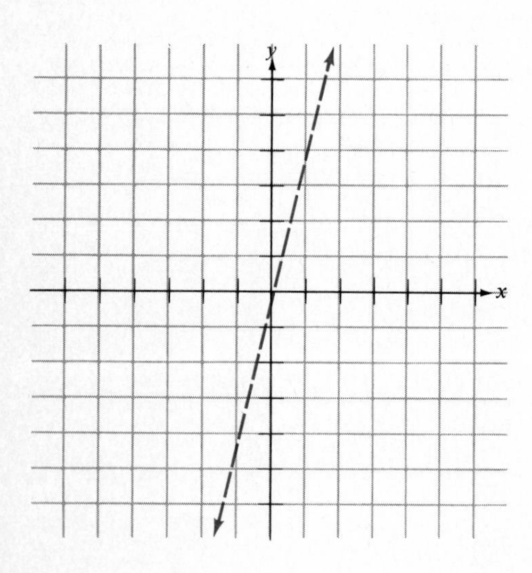

6. $y \geq -3x$

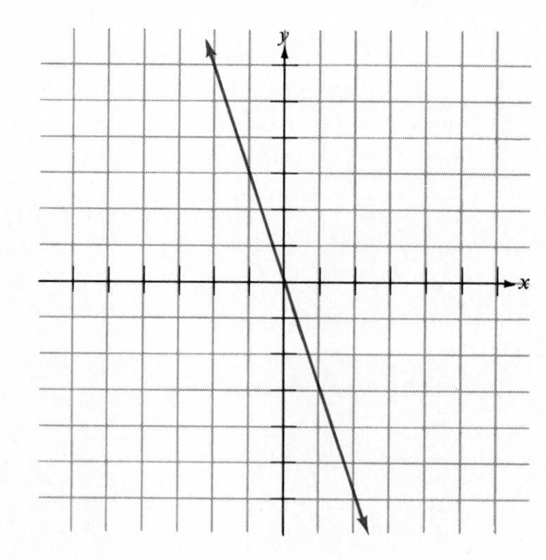

7. $3x + 2y \leq 9$

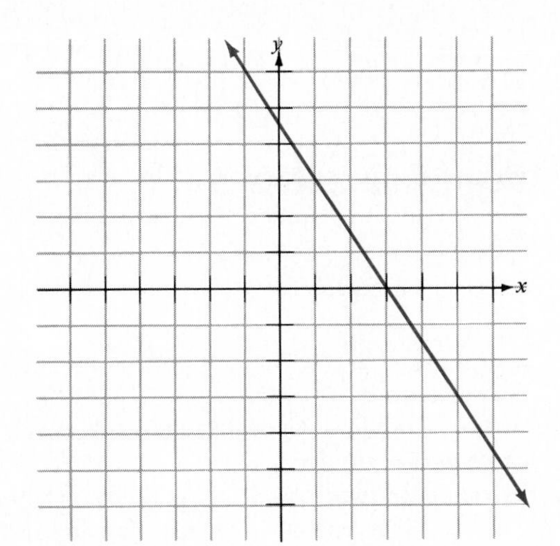

8. $4x + 2y < 10$

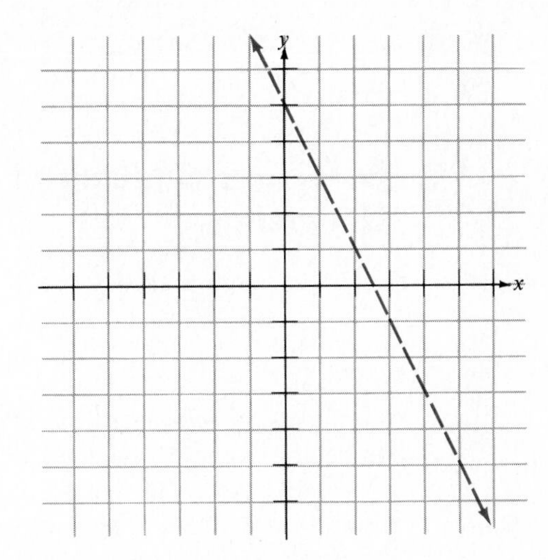

9. $y < 2$

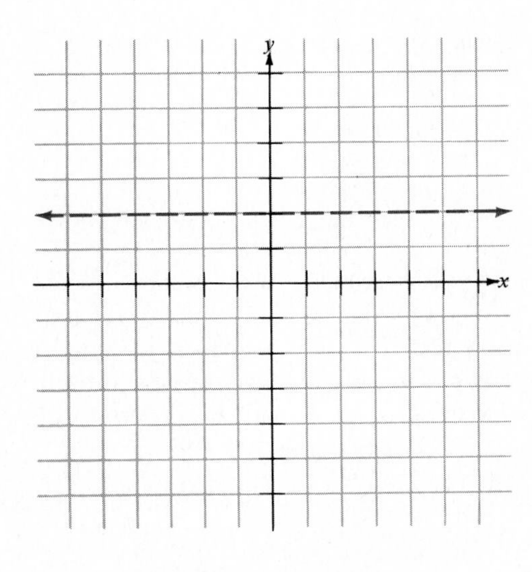

10. $x \geq 3$

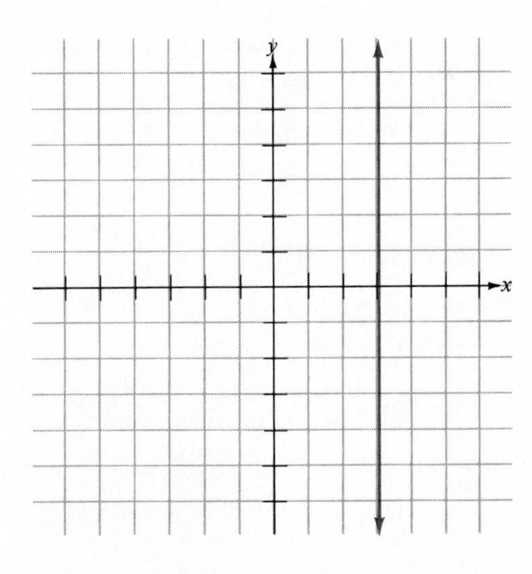

For Exercises 11–40, graph the linear inequality. See Examples 1–4.

11. $-3x + 2y \leq 6$ **12.** $x - 2y > 4$ **13.** $2x + 5y \leq 10$ **14.** $-3x + 4y \leq 12$

15. $4x - y < 4$ **16.** $x - 5y > 5$ **17.** $4x + 3y \geq 12$ **18.** $-3x + 5y \leq 15$

19. $x + y > -2$ **20.** $3x - 2y \geq -6$ **21.** $x < -2$ **22.** $x > 5$

23. $y < -2$ **24.** $y > 3$ **25.** $2x + y \geq -4$ **26.** $x + 2y \leq 6$

27. $y < x$ **28.** $y \geq x + 2$ **29.** $y < 4x + 4$ **30.** $y \geq 2x$

31. $y \geq -3x$ **32.** $5y \leq 10x$ **33.** $x - y < 0$ **34.** $4x - y > 0$

35. $x \geq 0$ **36.** $x \leq 0$ **37.** $y \leq 0$ **38.** $y \geq 0$

39. $y < \frac{1}{2}x - 4$ **40.** $y > -\frac{2}{5}x - 2$

6.5 Solving a System of Linear Equations by Graphing

A pair of linear equations in two variables x and y such as

$$\begin{aligned} 2x - y &= 1 \\ x + y &= 5 \end{aligned}$$

is called a *system of linear equations in two variables*. Sometimes the system is described as a pair of *simultaneous linear equations in two unknowns*.

A *solution* of a system of two linear equations in two variables is an ordered pair of numbers (a, b) that satisfies both equations. Thus the ordered pair $(2, 3)$ is a solution of the system of equations

$$\begin{aligned} 2x - y &= 1 \\ x + y &= 5 \end{aligned}$$

because when we substitute $x = 2$ and $y = 3$ in each equation of the system, we get a true statement. Thus

for $2x - y = 1$:

$$\begin{aligned} 2(2) - 3 &= 1 \\ 4 - 3 &= 1 \\ 1 &= 1. \quad \text{true} \end{aligned}$$

for $x + y = 5$:

$$\begin{aligned} 2 + 3 &= 5 \\ 5 &= 5. \quad \text{true} \end{aligned}$$

When we find the solutions to a system of equations, we say that we have *solved* the system.

One method of finding solutions to a system of linear equations in two variables is called the *graphical method*. This method of solving a system of linear equations will enable us to see visually that a system of linear equations in two variables may have

1. Exactly one solution.
2. An infinite number of solutions.
3. No solutions.

We illustrate the graphical method for solving a system of linear equations in two variables in the following examples.

Example 1. Solve the following system of linear equations by the graphical method.

$$\begin{aligned} 2x + y &= 6 \\ x - y &= -3. \end{aligned}$$

Solution: We graph both linear equations on the same coordinate system by finding the intercepts of each equation. To find the intercepts for the equation $2x + y = 6$, we proceed as in Section 6.3

Let $x = 0$: $2(0) + y = 6$
$y = 6.$

Let $y = 0$: $2x + 0 = 6$
$x = 3.$

Then we plot the points (0, 6) and (3, 0) and draw a straight line through them to obtain the graph of $2x + y = 6$. The graph of $x - y = -3$ is obtained using the same method. The two graphs intersect at the point (1, 4) as shown in Figure 6.23.

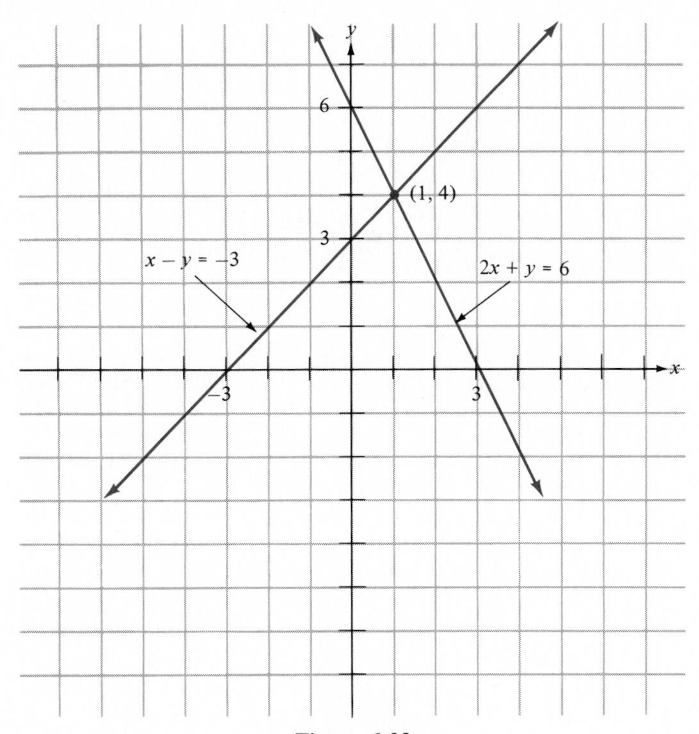

Figure 6.23

Since the point (1, 4) is on the graphs of both $2x + y = 6$ and $x - y = -3$, it must satisfy both equations. Thus (1, 4) is the solution to the given linear system.

CHECK: Substitute $x = 1$ and $y = 4$ in each equation.

For $2x + y = 6$:
$2(1) + 4 = 6$
$2 + 4 = 6$
$6 = 6.$ true

For $x - y = -3$:
$1 - 4 = -3$
$-3 = -3.$ true

When a system of linear equations in two variables has exactly one solution, as in Example 1, the system is said to be *consistent*.

Example 2. Solve the following system by graphing.

$$x + y = 3$$
$$x + y = 5.$$

Solution: First we find the intercepts of each equation and draw their graphs on the same coordinate system, as shown in Figure 6.24. The graphs of these equations are two parallel lines. The two graphs have *no* points in common. Therefore, there is no ordered pair that satisfies both equations simultaneously. Thus this system of linear equations has no solutions.

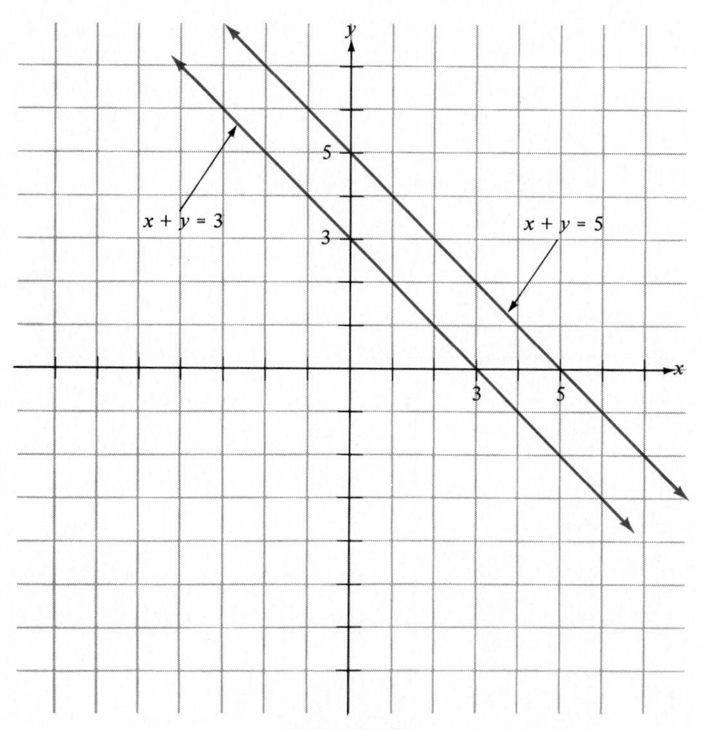

Figure 6.24

If a system of linear equations in two variables does *not* have a solution, we say that the system is *inconsistent*.

The system in Example 2 is inconsistent. This is because we are trying to find two numbers (x, y) whose sum is 3 $(x + y = 3)$ and at the same time has a sum of 5 $(x + y = 5)$. Clearly, there is no such pair of numbers.

Example 3. Solve the following system by graphing.

$$x + 2y = 2$$
$$2x + 4y = 4.$$

Solution: When we graph both equations on the same rectangular coordinate system, we find that the two graphs coincide (Figure 6.25).

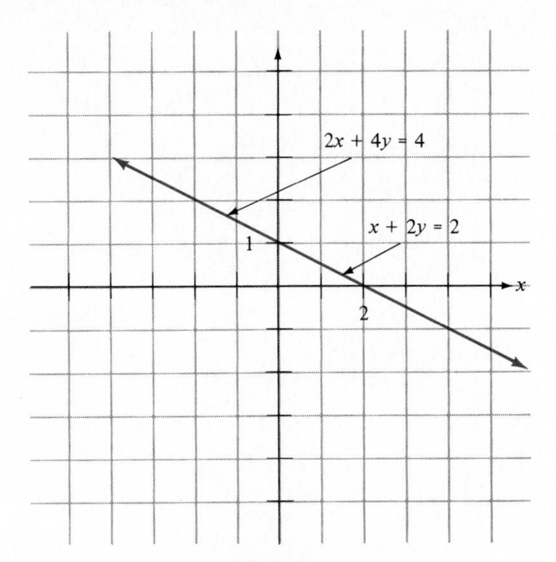

Figure 6.25

Thus, every point on the graph of $x + 2y = 2$ is on the graph of $2x + 4y = 4$, and vice versa. Therefore, every ordered pair of numbers (a, b) associated with a point on the line satisfies both equations of the system

$$\begin{aligned} x + 2y &= 2 \\ 2x + 4y &= 4. \end{aligned}$$

Thus there are an infinite number of solutions to this system of equations. Notice that the equation $2x + 4y = 4$ may be obtained by multiplying each side of the equation $x + 2y = 2$ by 2. Recall that multiplying each side of an equation by a nonzero constant does not affect the solution set of the equation. Therefore, the two equations, $x + 2y = 2$ and $2x + 4y = 4$ have exactly the same solution set.

Systems of linear equations in two variables with an infinite number of solutions are called *dependent* and *consistent* systems. They are dependent because one equation is a multiple of the other equation.

EXERCISES 6.5

For Exercises 1–28, solve the system of linear equations by the graphical method. See Example 1.

1. $x + y = 5$
$x - y = 3$

2. $x + y = 4$
$x - y = 2$

3. $-x + y = 6$
$x + y = 2$

4. $-x + y = 7$
$x + y = 3$

5. $2x + y = 8$, $x - y = 4$

6. $5x + 2y = 10$, $x - 2y = 2$

7. $2x - 3y = 6$, $x - 3y = 6$

8. $3x - 5y = 15$, $-2x + y = 4$

9. $2x + y = 7$, $x - y = 2$

10. $6x - 2y = 12$, $-x + 2y = 3$

11. $-2x + 3y = 12$, $x + 3y = 3$

12. $7x + y = 7$, $5x - y = 5$

13. $5x + 2y = -10$, $x + 2y = 6$

14. $2x + 3y = 12$, $x - 2y = -1$

15. $x + 2y = 0$, $x + y = 1$

16. $x + 3y = -6$, $x - y = 6$

17. $-3x + 2y = 12$, $3x + y = -3$

18. $2x + y = 4$, $x - 3y = 9$

19. $-3x + 4y = 12$, $-x + 2y = 8$

20. $4x - y = 8$, $2x + y = -2$

21. $5x - 3y = 15$, $-x + 2y = 4$

22. $-3x + y = 9$, $2x - y = -6$

23. $-3x + 4y = 24$, $x + 4y = 8$

24. $3x - y = 0$, $x + y = 4$

25. $-x + y = 3$, $y = 2$

26. $4x + y = 4$, $x = -1$

27. $3x + y = 3$, $x = 2$

28. $5x - y = 5$, $y = 5$

For Exercises 29–36, determine whether the system is dependent or inconsistent. See Examples 2 and 3.

29. $2x + 3y = 6$, $2x + 3y = -6$

30. $3x + 4y = 12$, $3x + 4y = -12$

31. $3x + 5y = 15$, $6x + 10y = 30$

32. $x + 2y = 4$, $2x + 4y = 8$

33. $2x + y = 2$, $4x + 2y = 8$

34. $3x + y = 6$, $9x + 3y = 9$

35. $4x + 5y = 20$, $8x + 10y = 40$

36. $7x - 2y = 14$, $14x - 4y = 28$

6.6 Solving Systems of Linear Equations by the Addition Method

The graphical method for solving a system of linear equations is instructive because it illustrates the possible solutions of the system. But it is not an accurate method because the solution we obtain

1. Depends on the accuracy with which the graphs are drawn.
2. It is often difficult to estimate solutions involving fractions such as $\left(-\frac{5}{6}, \frac{3}{8}\right)$.

However, there are two accurate methods for solving systems of equations that use algebraic methods. The two algebraic methods are called the *addition* method (or *elimination* method) and the *substitution* method. In this section we discuss the addition method. The substitution method is discussed in Section 6.7.

In Chapter 2 we stated the addition, subtraction, multiplication, and division properties of equality. To solve systems of linear equations, we use these properties and a more general form of the addition property of equality which we state as follows:

> If equal quantities are added to equal quantities, the resulting sums are equal.

This property is expressed symbolically as follows, where A, B, C, and D represent algebraic expressions:

Addition Property of Equality

If

$$A = B \quad \textit{and} \quad C = D,$$

then

$$A + C = B + D.$$

To solve systems of equations by the addition method, we use the properties of equality mentioned above as well as the following property:

If the ordered pair of real numbers (a, b) satisfies both equations of the system

$$\begin{aligned} &(1) \quad Ax + By = C \\ &(2) \quad Dx + Ey = F \end{aligned} \quad \text{(capital letters represent real numbers)}$$

then (a, b) satisfies the equation obtained by adding equations (1) and (2). We number each equation of the system for easy reference.

Now we show how to use this property to solve the system of equations

$$\begin{aligned} &(3) \quad x + y = 6 \\ &(4) \quad x - y = 2. \end{aligned}$$

Adding equations (3) and (4), we obtain the equation

$$(5) \quad (x + y) + (x - y) = 6 + 2.$$

After collecting like terms, equation (5) can be rewritten as

$$(5) \quad 2x + 0 \cdot y = 8.$$

Now any solution (x, y) to the original system of equations (3) and (4) must satisfy equation (5). But because of the form of equation (5), the x-values of all its solutions are the same. We now solve equation (5) for x.

$$\begin{aligned} 2x + 0 \cdot y &= 8 \\ 2x + 0 &= 8 \\ 2x &= 8 \\ x &= 4. \end{aligned}$$

Thus for any solution (x, y) to the equation

$$(5) \quad 2x + 0 \cdot y = 8$$

the x-value is 4. Now, as we have already noted, all solutions (x, y) to the original system of equations (3) and (4) must satisfy equation (5). Thus $x = 4$ is also the x-value of the solution to the original system of equations (3) and (4).

To find the y-value of the solution to the system, we substitute $x = 4$ in either one of the equations of the original system. Let us substitute $x = 4$ into equation (3):

(3) $$x + y = 6.$$

Let $x = 4$:

$$\begin{aligned} 4 + y &= 6 \\ y &= 2. \end{aligned}$$

Thus we claim that the solution to the system of equations is (4, 2).

CHECK: We substitute $x = 4$ and $y = 2$ into both equations of the system.

(3) $$\begin{aligned} x + y &= 6 \\ 4 + 2 &= 6 \\ 6 &= 6. \quad \text{true} \end{aligned}$$

(4) $$\begin{aligned} x - y &= 2 \\ 4 - 2 &= 2 \\ 2 &= 2. \quad \text{true} \end{aligned}$$

The method used above for solving a system of linear equations is called the *addition* or *elimination method.*

Example 1. Solve the system using the addition method.

(1) $$-2x + y = 6$$
(2) $$2x + 3y = 2.$$

Solution: First we add the two equations:

(1) $$-2x + y = 6$$
(2) $$2x + 3y = 2$$

$$0 \cdot x + 4y = 8$$
(3) $$4y = 8.$$

Next we find the y-value that satisfies equation (3).

$$\begin{aligned} 4y &= 8 \\ y &= 2. \end{aligned}$$

Thus the y-value of the solution to the system is 2. To find the x-value, we substitute in either equation (1) or (2) of the original system. Substituting $y = 2$ in equation (1) and solving for x, we get

$$\begin{aligned} -2x + 2 &= 6 \\ -2x &= 4 \\ x &= -2. \end{aligned}$$

The solution to the system is $(-2, 2)$. The reader should check that this is the correct solution.

Solving a system of equations by the addition method depends on obtaining a zero coefficient for one of the variables after adding the two equations. When we add the equations as in Example 1, we say that the variable with a zero coefficient has been ''eliminated'' or ''drops out.'' To eliminate a variable, we sometimes have to multiply *one* or *both* equations by some nonzero number. Multiplying either of the equations of a system by a nonzero number results in an equivalent system. *Equivalent systems* of equations have *exactly* the same solutions.

In the next two examples we discuss systems whose solution involves multiplying one or both equations by a nonzero number so that a variable can be eliminated.

Example 2. Solve the system

$$
\begin{aligned}
&(1) \qquad & 2x + 3y &= 9\\
&(2) \qquad & x - 2y &= 1.
\end{aligned}
$$

Solution: If we add equations (1) and (2) of the system, we obtain the equation

$$(3) \qquad 3x + y = 10.$$

But from equation (3), because we have not eliminated one of the variables, it is not possible to determine the x-value or y-value of the solution to the system. Notice that in order to eliminate a variable by the addition method, the coefficients of the variable must be opposites (or negatives of each other). Thus, if we choose to eliminate the x-variable, we can multiply both sides of equation (2) by -2. As a result, we obtain the equation

$$
(4) \qquad
\begin{aligned}
-2(x - 2y) &= -2(1)\\
-2x + 4y &= -2.
\end{aligned}
$$

We now replace the original system by the equivalent system consisting of equations (1) and (4) and add the equations. Notice that the x-variables now have coefficients that are opposites.

$$
\begin{aligned}
&(1) \qquad & 2x + 3y &= 9\\
&(4) \qquad & -2x + 4y &= -2\\
\hline
& & 0 \cdot x + 7y &= 7\\
&(5) \qquad & 7y &= 7.
\end{aligned}
$$

Therefore, $y = 1$. Substituting $y = 1$ into either equation of our original system yields $x = 3$. The solution to the system is $(3, 1)$. A check would reveal that this is the correct solution.

Example 3. Solve the system

$$
\begin{aligned}
&(1) \qquad & 3x + 5y &= 2\\
&(2) \qquad & 5x + 2y &= -3.
\end{aligned}
$$

Solution: We can eliminate either the x-variable or the y-variable. Let us eliminate the y-variable. To do this, we have to rewrite the equations so that the coefficients of the y-variable are opposites. Thus we multiply equation (1) by 2 and equation (2) by -5. As a result we obtain the equivalent system of equations (3) and (4), and then add.

$$\begin{array}{llll} (3) & 2(3x + 5y) = 2(2) & \longrightarrow & 6x + 10y = 4 \\ (4) & -5(5x + 2y) = -5(-3) & \longrightarrow & -25x - 10y = 15 \\ \hline & & \text{adding:} & -19x + 0 \cdot y = 19 \\ & & & -19x = 19 \\ & & & x = -1. \end{array}$$

We substitute $x = -1$ in equation (1) or (2) to determine the y-value of the solution. The reader should verify that $y = 1$. Thus the solution is $(-1, 1)$.

The systems of equations discussed so far had exactly one solution. In the next two examples we examine the solution of two special systems by the addition method. One system is an inconsistent system (no solutions); the other is a dependent system (an infinite number of solutions).

Example 4. Solve the system

$$\begin{array}{ll} (1) & x + y = 10 \\ (2) & x + y = 5. \end{array}$$

Solution: Multiply equation (2) by -1 to obtain

$$-1 \cdot (x + y) = -1 \cdot (5)$$

$$(3) \qquad -x - y = -5.$$

Next we add equations (1) and (3) in an equivalent system.

$$\begin{array}{ll} (1) & x + y = 10 \\ (3) & -x - y = -5 \\ \hline (4) & 0 \cdot x + 0 \cdot y = 5. \end{array}$$

Thus for any ordered pair (x, y),

(4) $\quad 0 \cdot x + 0 \cdot y = 5 \quad$ or $\quad 0 = 5$ is a *false statement*.

If we were to graph the equations of the system, the result would be two parallel lines. Thus they have no point in common, and as a result there is no solution to the given system. Recall from Example 2 in Section 6.5 that such systems are inconsistent.

Example 4 illustrated that when we attempt to solve a system of equations that is *inconsistent,* by eliminating a variable, the result is always a *false statement*.

We note that if we had studied the equations given in Example 4, we would have seen that the system was inconsistent because we could not possibly find two numbers x and y whose sum is *both* 5 and 10. Thus, before solving a system, we should inspect it to see if it is obviously inconsistent.

In the next example we show what happens when we try to solve a dependent system. Notice that the system is dependent because equation (2) can be obtained from equation (1) by multiplying both sides by $\frac{1}{2}$.

Example 5. Solve the system

$$\begin{aligned} &(1) \qquad 4x + 2y = 10 \\ &(2) \qquad 2x + y = 5. \end{aligned}$$

Solution: Multiply equation (2) by -2; the resulting equation is

$$(3) \qquad -4x - 2y = -10.$$

Next we add equations (1) and (3).

$$\begin{aligned} &(1) & 4x + 2y &= 10 \\ &(3) & -4x - 2y &= -10 \\ \hline & & 0 \cdot x + 0 \cdot y &= 0 \\ &(4) & 0 &= 0. \end{aligned}$$

Notice that any ordered pairs of numbers (x, y) satisfies equation (4) since $0 \cdot x + 0 \cdot y = 0$ or $0 = 0$ is a *true statement*. Thus from equation (4) it is not possible to determine the solution of the original system of equations. The graph of both equations is the same line. But, as we saw in Example 3 of Section 6.5, any solution of one equation is also a solution of the other equation. Therefore, the system has an infinite number of solutions and is a dependent system.

Example 5 illustrated that when we solve a *dependent* system of two equations with two variables by the addition method, the result is the *true statement* $0 = 0$.

We summarize this section with the following observations. To find the solution for a system of linear equations, use the following steps.

1. Put all equations of the system in standard form $Ax + By = C$.
2. If necessary, multiply one or both equations by numbers chosen so that the coefficients of one of the variables become opposites (or negatives) of each other.
3. Add the resulting equations, thus eliminating at least one of the variables.
4. If the system is not inconsistent or dependent, solve the resulting equation for the other variable.
5. Substitute the value found in step 4 in either one of the equations of the original system to determine the value of the other variable.
6. Check the solution by substituting in both equations of the original system.

EXERCISES 6.6

For Exercises 1–12, solve the system by the addition method. See Example 1.

1. $x + y = 7$, $x - y = 1$

2. $x + y = -3$, $x - y = 1$

3. $2x + y = -1$, $x - y = 10$

4. $3x - y = 8$, $2x + y = 2$

5. $-3x + 2y = 14$, $3x + 4y = -8$

6. $2x - 5y = -13$, $-3x + 5y = 12$

7. $-2x + 3y = 8$, $2x - y = 2$

8. $4x - 2y = 2$, $6x + 2y = 3$

9. $7x + 4y = 14$, $-3x - 4y = 10$

10. $-9x + 3y = 4$, $9x - y = 2$

11. $5x - y = -7$, $-5x + 2y = 15$

12. $11x + y = 10$, $7x - y = 8$

For Exercises 13–30, solve the system by the addition method. See Example 2.

13. $5x + y = 4$, $3x - 4y = 30$

14. $-x + 3y = -1$, $2x - y = 12$

15. $2x - y = 1$, $3x + 2y = 19$

16. $4x + y = -5$, $3x - 4y = -18$

17. $-9x + 4y = -1$, $3x + 2y = 2$

18. $3x + 2y = -1$, $6x - 4y = 10$

19. $2x - 3y = 1$, $3x + y = 7$

20. $4x - 3y = -9$, $3x + y = -10$

21. $3x + 11y = 12$, $x + 4y = 5$

22. $4x + 3y = 3$, $-x + 2y = 13$

23. $-x + 3y = 0$, $4x + 6y = 6$

24. $-x + 5y = 0$, $5x + 3y = 14$

25. $10x - 6y = 48$, $6x + 2y = -16$

26. $4x - 7y = 12$, $3x + y = 9$

27. $3x - 5y = 5$, $3x - y = 13$

28. $3x - y = 7$, $4x - 5y = -9$

29. $x - 6y = -15$, $7x - 5y = 6$

30. $2x - 5y = -11$, $x - 4y = -7$

For Exercises 31–42, solve the system by the addition method. See. Example 3.

31. $-3x + 4y = 1$, $2x - 3y = -3$

32. $-4x + 5y = -6$, $3x - 4y = 4$

33. $5x - 7y = 9$, $-2x + 9y = 15$

34. $7x - 3y = 1$, $2x - 5y = -8$

35. $6x - 4y = 1$, $-4x + 3y = 1$

36. $3x + 4y = 5$, $2x + 5y = -6$

37. $2x + 3y = 11$, $7x + 5y = -11$

38. $2x - 5y = 19$, $3x + 4y = -6$

39. $3x - 4y = 20$, $2x + 9y = -10$

40. $3x - 4y = -6$, $-2x + 3y = 5$

41. $2x + 3y = 11$, $5x - 2y = -20$

42. $7x - 5y = 3$, $-4x + 3y = 4$

For Exercises 43–50, determine whether the system is inconsistent or dependent. See Examples 4 and 5.

43. $4x - 3y = 1$, $-8x + 6y = -2$

44. $-5x + y = 3$, $10x - 2y = -6$

45. $3x + 4y = 1$, $-6x - 8y = 3$

46. $5x - y = 3$, $-15x + 3y = 2$

47. $2x + 2y = 4$, $3x + 3y = 1$

48. $7x - 2y = 1$, $-14x + 4y = 3$

49. $-6x + 5y = 4$, $18x - 15y = -12$

50. $3x - 4y = 7$, $-12x + 16y = -28$

Solving Linear Systems by the Substitution Method 6.7

The addition method for solving a system of two linear equations in two variables is based on eliminating one of the variables and then solving for the remaining variable. Another way of eliminating a variable when solving a system of equations is called the *substitution method*. This method is most useful when a variable in at least one equation of the system is expressed in terms of the other variable. An example of such a system is

(1) $$x + y = 10$$
(2) $$y = 4x.$$

Notice in equation (2) that y is expressed in terms of x. We apply the substitution method to solve this system.

Equation (2) states that y equals $4x$. Since we seek the y-value common to both equations, the y-value in equation (1) must also equal $4x$. Thus we replace the y-value in equation (1),

$$x + y = 10$$

with $4x$. The result of this substitution is an equation in one variable, x.

(2) $x + 4x = 10.$ y variable eliminated

We then solve this equation for x.

$$5x = 10$$
$$x = 2.$$

This is the x-value of the solution that is common to both equations. To find the y-value, we substitute 2 for x in either equation (1) or (2) of the original system. The easiest equation to use is equation (2), since y is given in terms of x. Thus, in equation (2), which is

$$y = 4x,$$

let $x = 2$; therefore, $y = 4(2) = 8$. The solution to the system is $(2, 8)$.

In the following examples we use the substitution method to solve different systems of equations.

Example 1. Solve the following system using the substitution method.

(1) $$5x - 3y = 9$$
(2) $$y = 3x + 1.$$

Solution: Equation (2) of the system tells us that y equals $3x + 1$. Thus we replace the y variable in equation (1) with $3x + 1$. The resulting equation in one variable is

(3) $5x - 3(3x + 1) = 9.$ y variable eliminated

Now we solve this last equation for x.

(3)
$$\begin{aligned} 5x - 3(3x + 1) &= 9 \\ 5x - 9x - 3 &= 9 \\ -4x - 3 &= 9 \\ -4x &= 12 \\ x &= -3. \end{aligned}$$

Thus the x-value of the solution to the system is -3. To find the y-value, we replace x with -3 in equation (2),

(2)
$$y = 3x + 1$$

because it expresses y in terms of x. As a result, we obtain

$$\begin{aligned} y &= 3(-3) + 1 \\ y &= -9 + 1 = -8. \end{aligned}$$

We see therefore that $(-3, -8)$ satisfies equation (2). A check would verify that $(-3, -8)$ also satisfies equation (1). Hence the solution to the system is $(-3, -8)$.

Example 2. Solve the following system using the substitution method.

(1)
$$6x + 4y = 6$$
(2)
$$3x - 2y = 6$$

Solution: To solve this system by the substitution method, we first rewrite one of the equations so that one of the variables is expressed in terms of the other variable. We use equation (2) to express x in terms of y. To do this, we solve equation (2) for x. Therefore, from

(2)
$$3x - 2y = 6$$

we see that

$$\begin{aligned} 3x &= 2y + 6 \\ x &= \frac{1}{3}(2y + 6) \end{aligned}$$

(3)
$$x = \frac{2}{3}y + 2.$$

Next, we substitute $\frac{2}{3}y + 2$ for x in equation (1).

(1)
$$6x + 4y = 6.$$

Thus

$$6\left(\frac{2}{3}y + 2\right) + 4y = 6$$

$$4y + 12 + 4y = 6$$
$$8y + 12 = 6$$
$$8y = -6$$
$$y = -\frac{6}{8} = -\frac{3}{4}.$$

To find the x-value of our solution, we will substitute $-\frac{3}{4}$ for y in equation (3), which is the rewritten form of equation (2). In equation

(3) $$x = \frac{2}{3}y + 2$$

let $y = -\frac{3}{4}$:

$$x = \frac{2}{3}\left(-\frac{3}{4}\right) + 2$$
$$x = -\frac{2}{3} \cdot \frac{3}{4} + 2$$
$$x = -\frac{1}{2} + 2 = \frac{3}{2}.$$

The solution to the system is $\left(\frac{3}{2}, -\frac{3}{4}\right)$. The reader should check that this solution satisfies the original equations.

The previous examples illustrated the use of the following steps for solving a system of two linear equations in two variables by the substitution method.

1. Rewrite one equation so that one of the variables is written in terms of the other variable (i.e., isolate one variable on one side of the equation).
2. Substitute the expression obtained in step 1 into the other equation of the system, thus obtaining one equation in the other variable only.
3. Solve the equation obtained in step 2. This is one value of the ordered-pair solution to the system.
4. Substitute the value found in step 3 into the expression obtained in step 1, and find the value of the remaining variable.
5. Check by substitution in both given equations of the original system.

Example 3. Solve by substitution:

$$(1) \qquad \frac{x}{6} + \frac{y}{3} = \frac{1}{2}$$

$$(2) \qquad \frac{3x}{10} + \frac{y}{2} = 1.$$

Solution: It will be easier to solve this system if we first clear both equations of fractions. Thus we multiply both sides of equation (1) by 6 and both sides of equation (2) by 10. As a result, we obtain the following equivalent system of equations:

$$(3) \qquad 6\left(\frac{x}{6} + \frac{y}{3}\right) = 6\left(\frac{1}{2}\right)$$

$$(4) \qquad 10\left(\frac{3x}{10} + \frac{y}{2}\right) = 10(1)$$

or

$$(3) \qquad x + 2y = 3$$
$$(4) \qquad 3x + 5y = 10.$$

The easiest way to proceed is to use equation (3) to write x in terms of y. Thus we have

$$(5) \qquad x = 3 - 2y.$$

Substituting $3 - 2y$ for x in equation (4),

$$(4) \qquad 3x + 5y = 10,$$

we obtain

$$3(3 - 2y) + 5y = 10$$
$$9 - 6y + 5y = 10.$$

We now solve for y

$$9 - 6y + 5y = 10$$
$$-6y + 5y = 10 - 9$$
$$-y = 1$$
$$y = -1.$$

This is the y-value of the solution to the system. To find the x-value of the solution we use equation (5), since x is expressed in terms of y.

$$(5) \qquad x = 3 - 2y$$

Let $y = -1$:

$$x = 3 - 2(-1)$$
$$x = 5.$$

Thus the solution to the system is $(5, -1)$. A check would confirm that this is the desired solution.

When use of the substitution method to solve a system of equations results in the elimination of both variables and a true statement, then the system is *dependent*.

If the use of the substitution method to solve a system of equations results in a false statement, then such a system is *inconsistent*.

EXERCISES 6.7

Solve the following systems by the substitution method.

For Exercises 1–10, see Example 1.

1. $x + y = 21$, $y = 2x$

2. $x + y = 13$, $y = 12x$

3. $x + y = 18$, $y = 4x - 7$

4. $x - y = 15$, $y = 5x + 9$

5. $2x - y = -12$, $x = 2y$

6. $2x - 3y = 9$, $x = 3y$

7. $4x - 9y = 23$, $x = 3y + 2$

8. $2x - 5y = 10$, $x = 2y + 3$

9. $7x - 6y = 5$, $y = 2x + 5$

10. $6x - 2y = 4$, $y = 6x + 13$

For Exercises 11–32, see Example 2.

11. $3x - 4y = -14$, $2y = 3x + 4$

12. $4x - 9y = -3$, $3y = 4x - 15$

13. $8x + 6y = 12$, $4x = 3y$

14. $5x + 2y = 30$, $5x = 2y + 10$

15. $4x - 7y = -20$, $4x = y + 4$

16. $3x + 5y = 12$, $5y = 10x - 1$

17. $3x + 5y = 5$, $x + 2y = 1$

18. $3x + 2y = 7$, $2x + y = 4$

19. $2x - 5y = 1$, $x - 4y = -1$

20. $x - 2y = 3$, $2x + y = 16$

21. $3x - 5y = 39$, $2x + y = 0$

22. $x + 8y = 46$, $3x + y = 0$

23. $4x + 7y = 45$, $x - 2y = 0$

24. $9x - 2y = 28$, $x - y = 0$

25. $6x - 8y = 32$, $y = 2$

26. $13x - 2y = 25$, $y = 7$

27. $7x - 2y = 21$, $x - 3 = 0$

28. $11x - 9y = 44$, $x - 4 = 0$

29. $5x + 4y = 5$, $3x + 2y = 1$

30. $2x - 6y = 20$, $-4x + 3y = 5$

31. $-4x + 10y = 8$, $8x - 5y = 14$

32. $6x - 14y = 14$, $-12x + 7y = -7$

For Exercises 33–36, see Example 3.

33. $\frac{x}{3} - \frac{y}{2} = \frac{7}{6}$
$\frac{x}{3} - \frac{y}{6} = \frac{1}{6}$

34. $\frac{x}{3} + \frac{y}{6} = 1$
$\frac{x}{2} + \frac{3y}{10} = 2$

35. $\frac{3x}{8} + \frac{y}{4} = 1$
$\frac{3x}{2} + \frac{y}{2} = -1$

36. $\frac{x}{2} + \frac{y}{8} = \frac{-3}{2}$
$\frac{2x}{3} - \frac{y}{3} = -2$

For Exercises 37–42, determine whether the system is inconsistent or dependent. See the discussion following Example 3.

37. $2x + 4y = 1$
$x = -2y + 5$

38. $9x + 3y = 8$
$y = 4 - 3x$

39. $4x - 20y = 8$
$x = 5y + 2$

40. $-12x + 2y = 14$
$y = 6x + 7$

41. $18x + 2y = 10$
$y = -9x + 5$

42. $2x - 6y = 16$
$x = 3y + 8$

6.8 Applied Problems

Many practical problems lead to linear equations or a system of linear equations. In this section we deal with practical problems whose solution involves a system of two linear equations in two variables. Although some of these problems can be solved using one variable and one linear equation, it is usually easier to use two equations and two variables.

Because these problems are stated in written form, the most difficult part of their solution is translating the written information into mathematical equations. However, with practice and care we will build confidence in solving "word" problems. The following guidelines should be used to solve such problems.

1. Read the problem carefully and determine what unknown quantities are to be found.
2. Let x and y or any two appropriate letters represent the unknown quantities.
3. Read the problem carefully again. Use the given information to write two equations, each equation relating the two unknown quantities.
4. Solve the system of equations obtained in step 3 by either the addition method or the method of substitution.
5. Check your answers against the information given in the problem, not the equations obtained in step 3.

The following examples illustrate the use of these steps to solve word problems involving a system of two linear equations in two variables.

Fxample 1. The sum of two numbers is 102. One number is 3 more than twice the other number. Find the two numbers.

Solution

The unknown quantities are two numbers.
Let x represent the smaller of the two numbers.
Let y represent the larger of the two numbers.
The statement that "the sum of two numbers is 102" translates into the equation

$$x + y = 102.$$

The other information given about the numbers "one number is 3 more than twice the other" translates into the equation

$$y = \underbrace{2x}_{\text{twice the other}} \quad \underbrace{+\ 3.}_{\text{3 more}}$$

To find the two numbers, we find the common solution to both equations. Thus we have to solve the system of equations

$$\begin{aligned} x + y &= 102 \\ y &= 2x + 3. \end{aligned}$$

We use the substitution method to solve the system. The second equation states that $y = 2x + 3$; therefore, we substitute the expression $2x + 3$ for y in the first equation. As a result we obtain the following equation, which we solve for x.

$$\begin{aligned} x + (2x + 3) &= 102 \\ 3x + 3 &= 102 \\ 3x &= 99 \\ x &= 33. \end{aligned}$$

Substituting 33 for x in the second equation, we see that

$$\begin{aligned} y &= 2(33) + 3 \\ &= 66 + 3 \\ &= 69. \end{aligned}$$

The two numbers are 69 and 33.

CHECK: The sum of 69 and 33 is 102, and 69 is 3 more than twice 33.

Example 2. A farmer charges people going to a nearby outdoor rock concert for parking on one of his empty fields. He charges \$3.00 for compact cars and \$5.50 for larger cars. After 328 cars have parked, the field is full. He made a total of \$1196.50. How many compact cars and how many large cars were able to park on the field?

Solution: Let b represent the number of large cars and c the number of compact cars that parked on the farmer's field. The total number of parked cars is 328. From this information we write the equation

$$b + c = 328.$$

The amount the farmer made from b large cars is $\$5.50b$ ($5.50 times b cars). Similarly, the amount he received from c small cars at $3.00 per car is $\$3.00c$. Since the total amount made by the farmer was $1196.50, we write the following equation.

$$5.50b \quad + \quad 3.00c \quad = \quad \$1196.50$$

$$\begin{pmatrix}\text{amount in dollars}\\\text{received from}\\\text{large cars}\end{pmatrix} + \begin{pmatrix}\text{amount in dollars}\\\text{received from}\\\text{small cars}\end{pmatrix} = \begin{pmatrix}\text{total amount in}\\\text{dollars received}\\\text{from all cars}\end{pmatrix}$$

To find b and c, we solve the following system by the addition method.

$$\begin{aligned} &(1) & b + c &= 328 \\ &(2) & 5.50b + 3.00c &= 1196.50. \end{aligned}$$

It will be easier to solve the system if we first multiply both sides of the second equation by 10 in order to clear decimals.

$$\begin{aligned} &(1) & b + c &= 328 \\ &(3) & 55b + 30c &= 11{,}965. \end{aligned}$$

Next we multiply both sides of equation (1) by -30, and add the equations to eliminate the variable c. We then solve the equation obtained for b.

$$\begin{aligned} -30b - 30c &= -9{,}840 \\ 55b + 30c &= 11{,}965 \\ \hline 25b &= 2{,}125 \\ b &= 85. \end{aligned}$$

Now we substitute $b = 85$ into equation (1) of the original system of equations and solve for c. Thus we obtain

$$\begin{aligned} &(1) & 85 + c &= 328 \\ & & c &= 243. \end{aligned}$$

We see that the farmer's field was able to accommodate 243 compact cars and 85 big cars. As a check we observe that the sum of 85 and 243 is 328. Eighty-five big cars at $5.50 a car is $467.50. Similarly, 243 small cars at $3.00 a car is $729.00. The sum of $467.50 and $729.00 is $1196.50, the amount that the farmer received for allowing the cars to park on his field.

Example 3. Silica is a hard, glassy mineral found in sand and other substances. To make window glass a manufacturer needs a 900-kilogram (kg) batch of sand that contains 80% silica. The company has some sand with 60% silica and another batch of sand with 90% silica. How many kilograms of each type will be needed to make a mixture of 900 kg of sand containing 80% silica?

Solution

Let x represent the required amount of sand with 60% silica content.

Let y represent the required amount of sand with 90% silica content. The total amount of sand with 80% silica content needed is 900 kg. On the basis of this information we obtain the equation

$$x + y = 900.$$

$$\begin{pmatrix}\text{kg of sand}\\ \text{with 60\%}\\ \text{silica}\end{pmatrix} + \begin{pmatrix}\text{kg of sand}\\ \text{with 90\%}\\ \text{silica}\end{pmatrix} = \begin{pmatrix}\text{total kg of}\\ \text{sand with 80\%}\\ \text{silica}\end{pmatrix}$$

Note that 60% of x kg and 90% of y kg is pure silica. That is, there is

1. $.60x$ kg of silica in x kg of the 60% batch.
2. $.90y$ kg of silica in y kg of the 90% batch.
3. $.80(900) = 720$ kg of silica in 900 kg of sand with 80% silica content.

Translating the information above into an equation, we get

$$.60x + .90y = 720.$$

$$\begin{pmatrix}\text{kg of silica}\\ \text{in } x \text{ kg of}\\ \text{60\% batch}\end{pmatrix} + \begin{pmatrix}\text{kg of silica}\\ \text{in } y \text{ kg of}\\ \text{90\% batch}\end{pmatrix} = \begin{pmatrix}\text{kg of silica}\\ \text{in 900 kg of sand}\\ \text{with 80\% silica}\end{pmatrix}$$

We multiply both sides of the last equation by 10 to clear decimals.

$$6x + 9y = 7200.$$

Thus we need to solve the following system to find the solution to the problem.

$$\begin{aligned} &(1) \qquad & x + y &= 900\\ &(2) \qquad & 6x + 9y &= 7200. \end{aligned}$$

If we multiply both sides of equation (1) by -6 and add the two equations, we eliminate the x-variable.

$$\begin{aligned} -6x - 6y &= -5400\\ 6x + 9y &= 7200\\ \hline 3y &= 1800. \end{aligned}$$

Now solve for y.

$$\begin{aligned} 3y &= 1800\\ y &= 600. \end{aligned}$$

Substituting $y = 600$ in equation (1) of the original system, we obtain

$$\begin{aligned} x + 600 &= 900\\ x &= 300. \end{aligned}$$

Thus the manufacturer needs 300 kilograms of sand with 60% silica content and 600 kilograms of sand with 90% silica content in order to make a batch of 900 kilograms of sand with 80% silica content. As a check we note that $300 + 600 = 900$. Also, 60% of 300 is 180 and 90% of 600 is 540. Thus the amount of silica in the 900 kilograms is $180 + 540 = 720$. This is the correct amount of silica needed in 900 kilograms of sand because 80% of 900 is 720.

Example 4. A small airplane flying directly against the wind (headwind) takes 8 hours to cover a distance of 720 miles. On the return trip, flying with the wind directly behind it, it takes 5 hours to cover the same distance. Find the velocity of the wind and the speed of the airplane in still air.

Solution

Let a represent the speed of the airplane in still air.
Let w represent the velocity of the wind.
We need to recall that $rt = d$ (rate $\times$ time = distance).
The speed of the airplane flying against the wind is

$$a \quad - \quad w.$$

$$\binom{\text{airplane speed}}{\text{in still air}} - \binom{\text{velocity of}}{\text{the wind}}$$

Therefore, the statement ''airplane flying against the wind takes 8 hours to cover a distance of 720 miles'' translates into the equation

$$\begin{aligned}(a - w)\ \cdot \quad 8 \quad &= \quad 720\\ \text{(rate)} \ \cdot \text{(time)} &= \text{distance}\end{aligned}$$

or

$$8a - 8w = 720.$$

On the return flight with the aid of the wind the airplane speed is equal to

$$a \quad + \quad w.$$

$$\binom{\text{airplane speed}}{\text{in still air}} + \binom{\text{velocity of}}{\text{the wind}}$$

The return trip covers 720 miles in 5 hours. This information give us the second equation,

$$\begin{aligned}(a + w) \cdot 5 &= 720\\ 5a + 5w &= 720.\end{aligned}$$

We must now solve the system

(1) $$8a - 8w = 720$$
(2) $$5a + 5w = 720.$$

We multiply both sides of equation (1) by $\frac{1}{8}$ and both sides of equation (2) by $\frac{1}{5}$ and then add the new equations in order to eliminate the w variable. We then solve the resulting equation for a.

$$\begin{aligned} (3) \qquad a - w &= 90 \\ (4) \qquad a + w &= 144 \\ \hline 2a \qquad &= 234 \\ a &= 117. \end{aligned}$$

Now we solve for w by substituting $a = 117$ into equation (4).

$$\begin{aligned} 117 + w &= 144 \\ w &= 27. \end{aligned}$$

Thus the speed of the plane in still air was 117 miles per hour. The wind velocity was 27 miles per hour. The reader should check that this is the solution to the problem.

EXERCISES 6.8

Solve the following exercises.

For Exercises 1–10, see Examples 1–4.

1. The sum of two numbers is 214. Their difference is 40. Find the two numbers.
2. The sum of two numbers is 360. Their difference is 70. Find the two numbers.
3. One number is 8 more than three times another number. Their sum is 112. Find the two numbers.
4. One number exceeds five times another number by 2. Their sum is 224. Find the two numbers.
5. A mother and daughter receive an inheritance of $8000. The will states that the mother shall receive $600 more than three times the daughter's share. How much of the inheritance should each receive?
6. A maintenance man has 100 feet of fence. He wants to use the fence to enclose a rectangular region against a factory wall. The length of the enclosed region must be twice the width. Find the length and width of the enclosed region.
7. A building that is under construction will need 450 electrical fuse boxes. Some of the fuse boxes will be of the circuit-breaker type. The remaining boxes will contain plug fuses. The plans for the building call for 250 more of the circuit-breaker boxes than the plug-fuse boxes. How many of each type must be ordered?
8. A plumber installing copper tubing in an office building needs to make 275 connections. Some of the connections are made by soldering. The remaining connections are made using pressure joints. The number of soldered connections that he needs will be 20 more than four times the number of pressure connections. How many of each should he plan on doing?
9. The toll at a toll plaza before a heavily traveled section of a highway is $2.00 per car for those who drive alone and $1.20 for passenger cars with two or more occupants. On a certain day 1400 cars paid a total of $2000 in tolls. How many cars had only the driver and how many cars had two or more occupants?

10. On a certain night a municipal court judge fined 43 people for two types of traffic violations, passing through a red light and speeding. No person was guilty of both violations. Speeders were fined \$25 and those passing through a red light \$15. The court clerk collected \$935 from the 43 offenders. How many traffic violations of each type were there?

11. A roadside ice cream stand sells only cones and sundaes. Cones cost \$.60 and sundaes \$1.20. At the end of a summer day the owner counted receipts totaling \$180.00. The owner knew from the cash register that he sold a total of 220 cones and sundaes, but not how many of each kind. Find how many cones and how many sundaes he sold that day.

12. On a 20-question true-or-false test the score is based on 7 points for each correct answer and minus 2 points for each incorrect answer. A student answers all questions on the test and receives a score of 77. How many correct answers and how many incorrect answers did the student have on the test?

13. An actor receives \$70 for a shaving cream commercial and \$55 for a hair shampoo commercial each time they are shown on a certain TV station. The two commercials were shown a total of 30 times during one week. For that week the actor received a check for \$1800. How many times was each commercial shown during that week?

14. A farmer sets aside 110 acres for planting potatoes and onions. At the end of the year the profit per acre for potatoes was \$80 and for onions \$90. His total profit for the 110 acres was \$9500. How many acres of potatoes and how many acres of onions did he plant?

15. A copy machine makes regular-size copies ($8\frac{1}{2} \times 11$) for \$.15 each and legal-size copies ($8\frac{1}{2} \times 14$) for \$.25 each. A serviceman collects \$51.00 from the machine. A counter tells him that a total of 300 copies were made. The counter that tells him how many of each type was made was broken. How many regular-size and how many legal-size copies did the machine make that week?

16. A man wishes to invest \$6000 in bonds and savings certificates. The bonds will yield 9% annually on the amount invested and the savings certificate 12% annually. How much should he invest at each interest rate if he wishes to earn a total of \$660 in interest annually?

17. A toy manufacturer makes a large teddy bear and a small teddy bear. When producing the teddy bears, 3% of the large ones will be defective and 4% of the small ones will be defective. During one week the manufacturer produced a total of 1000 teddy bears. Thirty-three of them were defective. How many of each size of teddy bear was produced?

18. At a school fundraiser, raffle tickets are sold for a prize. A \$.50 raffle gives the buyer a chance to win a prize worth less than \$50. A \$1.00 raffle ticket gives the buyer a chance to win a prize greater than \$50. The total receipts from the raffle tickets was \$242.00. Twice as many \$.50 tickets were sold as \$1.00 tickets. How many \$.50 raffle tickets were sold and how many \$1.00 raffle tickets were sold?

19. A picture of each class of children is taken at an elementary school. Each child also poses for an individual photo. It is possible to purchase the class picture alone for \$3.50, or a photo package consisting of the class picture and individual pictures for \$7.50. Three times as many parents purchased the entire photo package as those purchasing the class picture alone. The total amount received by the school for the photos was \$1040.00. How many of each kind were purchased?

20. A state has two types of passenger motor vehicle registration fees: \$35 for compact cars and \$55 for standard-size cars. On a given day one agency office issued 100 passenger car registration certificates for a total amount of \$4800. How many of each type of registration certificate was issued?

21. A chemist needs 20 liters of solution with 50% (by volume) alcohol. He has solutions of 30% alcohol and 80% alcohol. How many liters of each must be mixed to make the 20 liters of 50% solution?

22. How many gallons of oil that is 90% olive oil must be added to oil that is 60% olive oil in order to produce 45 gallons of oil that is 80% olive oil?

23. A jewelry manufacturer needs 60 grams of 35% platinum alloy. He has in stock only 15% platinum alloy and 65% platinum alloy. How many grams of each must he use to make the 60 grams of 35% platinum alloy?

24. A 50% solution of antifreeze in a radiator with a 20-quart capacity will give freezing protection to −34°F. The full radiator now contains a 20% antifreeze solution. How many quarts must be replaced by pure antifreeze to obtain a 50% antifreeze solution?

25. A candy store owner wishes to mix chocolate containing nuts worth $3.00 a pound with plain chocolates worth $2.00 a pound to make a 50-pound assortment of chocolates worth $2.40 a pound. How many pounds of each kind of chocolate should he use to make an assortment worth $2.40 a pound?

26. A student going to college has two part-time jobs that she wishes to keep. She can fix the number of hours that she works on each job. One job pays $3.50 per hour, the other job pays $5.00 per hour. She wants to work exactly 25 hours a week. She also needs to earn at least $113.00 a week to meet expenses. How many hours should she work at each part-time job to earn exactly $113.00 a week?

27. The jet stream is wind that blows across the country from west to east. A jet airliner makes an 800-mile nonstop trip in 2 hours from Newark International Airport to Chicago's O'Hare Airport flying directly against the jet stream. The jet liner makes the return trip with the jet stream directly behind it in 1.6 hours. Find the speed of the jet liner in still air and the speed of the jet stream.

28. A boat travels 16 miles upstream in 4 hours. On the return trip with the aid of the current, it takes 2 hours to cover the same distance. Find the speed of the boat in still water and the speed of the stream's current.

29. A freight train and an express passenger train pass each other going in opposite directions. After 3 hours they are 270 miles apart. The passenger train travels 40 miles per hour faster than the freight train. Find the speed of each train.

30. The radar of a jet fighter on a defensive patrol indicates that there is an unidentified flying object (UFO) 600 miles away. The fighter pursues the UFO on the same straight-line path for 3 hours before finally catching up with it. The sum of the speeds of the fighter and the UFO was 1000 miles per hour. What was the average speed of the jet fighter and UFO during the chase?

31. A limousine and bus service shuttle passengers over a 100-mile expressway connecting an airport and a large city. The limousines and buses are allowed to travel at fixed, but different, speeds on the expressway. A bus leaving the airport at the same time a limousine leaves the city will meet the limousine in exactly 1 hour. Another limousine that is 15 miles behind a bus overtakes the bus in $1\frac{1}{2}$ hours. At what speeds are the buses and limousines allowed to travel back and forth between the city and airport?

32. A boat travels for 3 hours with a current of 3 miles per hour and returns the same distance against the current in 4 hours. How far did it travel each way, and what is its speed in still water?

Key Words

Consistent system
Dependent system
Graph
Inconsistent system
Independent system
Linear equation
Linear inequality
Ordered pair
Origin
Quadrant
Rectangular (Cartesian) coordinate system
Solving linear systems by:
  Addition
  Graphing
  Substitution
System of linear equations
x-axis
y-axis
x-coordinate (abscissa)
y-coordinate (ordinate)
x-intercept
y-intercept

CHAPTER 6 TEST

1. Find the coordinates of the points labeled A to D in the accompanying figure.

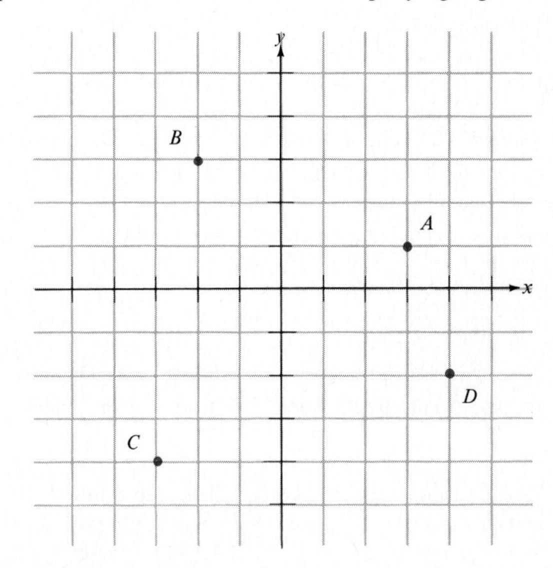

2. Which of the ordered pairs are solutions to the equation $5x - 2y = 6$?

$$(2, 2); \quad (-4, 13); \quad (6, 12); \quad (2, 8)$$

3. Find the values of x or y so that the ordered pairs are solutions to the equation $4x - y = 12$.

$$(x, 0); \quad (0, y); \quad (1, y); \quad (x, 4)$$

4. Complete the table for the equation $y = 2x - 4$.

x	0		-1	2
y		2		

5. Graph the equation $2x - y = 4$.

6. Graph the linear inequality $y > -2x + 6$.

7. A medical student sets aside a total of at most 8 hours a day for homework, exercise, and relaxation. Let x represent the hours spent on homework and y the hours spent on exercise and relaxation. Write an inequality expressing the relation between x and y.

For Problems 8 and 9, solve the system of linear equations by the graphing method.

8. $\begin{aligned} x + y &= 2 \\ x - y &= 4 \end{aligned}$

9. $\begin{aligned} 2x + y &= 4 \\ x - y &= -1 \end{aligned}$

10. Complete the following statements.

(a) If the graphs of a system of linear equations are parallel lines, the system is called an ______________ system.

(b) If the graphs of a system of linear equations are lines that coincide, the system is called an ______________ system.

For Problems 11–14, solve the system by the addition method.

11. $\begin{aligned} x + y &= 10 \\ x - y &= 2 \end{aligned}$

12. $\begin{aligned} 2x + y &= 6 \\ x + y &= 8 \end{aligned}$

13. $\begin{aligned} 3x - y &= 7 \\ x + 2y &= 7 \end{aligned}$

14. $\begin{aligned} 4x + 3y &= 20 \\ 5x - 2y &= 2 \end{aligned}$

For Problems 15–17, solve the system by the substitution method.

15. $\begin{aligned} 7x - 2y &= 5 \\ y &= 6x \end{aligned}$

16. $\begin{aligned} 3x - 5y &= 5 \\ 5y &= 2x - 10 \end{aligned}$

17. $\begin{aligned} 5x + 4y &= 10 \\ 3x + y &= 6 \end{aligned}$

Solve Problems 18–20.

18. The sum of two numbers is 155. One number is 5 less than three times the other number. Find the two numbers.

19. How many liters each of a 50% acid solution and 10% acid solution must be mixed to obtain 2000 liters of a 20% acid solution?

20. An airplane carrier is traveling east. A helicopter takes off from the carrier flying east. After $2\frac{1}{2}$ hours it is 300 miles from the carrier. It then turns around and returns to the carrier in 2 hours. What are the speeds of the aircraft carrier and helicopter?

CHAPTER 7
Roots and Radicals

Roots and radicals appear extensively in mathematics. They also have practical applications for traffic accidents. Police use the formula $s = \sqrt{30\ fd}$ to estimate the speed s(mph) that an automobile was traveling at when it skidded d feet. The letter f depends on whether the road was wet or dry, and what kind of road it was (asphalt or concrete).

CHAPTER 7
CHAPTER 7
CHAPTER 7
CHAPTER 7
CHAPTER 7
CHAPTER 7
CHAPTER 7
CHAPTER 7
CHAPTER 7
CHAPTER 7
CHAPTER 7
CHAPTER 7
CHAPTER 7
CHAPTER 7
CHAPTER 7
CHAPTER 7
CHAPTER 7
CHAPTER 7
CHAPTER 7
CHAPTER 7
CHAPTER 7
CHAPTER 7
CHAPTER 7
CHAPTER 7
CHAPTER 7
CHAPTER 7
CHAPTER 7
CHAPTER 7
CHAPTER 7
CHAPTER 7
CHAPTER 7
CHAPTER 7
CHAPTER 7
CHAPTER 7

In this chapter we examine what it means to take the root of a real number. Particular attention is given to the square root of a number. Taking the root of a number, as we shall see, is the opposite of raising a number to a power. The root of a number is represented by a radical. We discuss how to evaluate and simplify radicals. We also perform the standard operations (addition, subtraction, multiplication, division) with radicals.

7.1 Roots of Real Numbers and Radicals

A square root of the nonnegative number b is a number a such that a *squared* equals b. More precisely,

> If a and b are real numbers such that
>
> $$a^2 = b,$$
>
> then a is called a *square root* of b.

The square roots of 9 are $+3$ and -3 because

$$(+3)^2 = 9 \qquad \text{and} \qquad (-3)^2 = 9.$$

In fact, every positive real number has two square roots, one positive, the other negative. The positive root of a number, such as 9, is represented by $\sqrt{9}$. Thus we write

$$\sqrt{9} = 3.$$

The symbol $\sqrt{\ }$ is called a *radical sign;* the number under the radical sign, in this case 9, is called the *radicand*. The complete expression, $\sqrt{9}$, is called a *radical*. To represent the negative square root of 9, we write

$$-\sqrt{9} = -3.$$

The number 0 has only one square root, 0. Thus $\sqrt{0} = 0$.

Note that the square root of a negative number cannot be a real number. For example, if there were a real number a such that $a = \sqrt{-4}$, then by the definition of a square root, we would have

$$a^2 = -4.$$

But the square of a real number a cannot be negative because:

If a is positive, then

$$a^2 = a \cdot a = (\text{positive}) \cdot (\text{positive}) \text{ is } \textit{positive}$$

or if a is negative, then

$$a^2 = a \cdot a = (\text{negative}) \cdot (\text{negative}) \text{ is } \textit{positive}.$$

Thus we see that $a^2 \neq -4$, because a^2 is never negative. Therefore, $a = \sqrt{-4}$ cannot represent a real number. Negative numbers do have square roots, but they are not real numbers. We discuss square roots of negative numbers later in this book.

In general:

> If b is a positive real number,
>
> $$a = \sqrt{b}$$
>
> represents the *positive square root* of b and means
>
> $$a^2 = (\sqrt{b})^2 = b.$$

The positive square root of b is also called the *principal square root*. The negative square root of b is denoted by $-\sqrt{b}$. To represent both square roots of b compactly, we write $\pm\sqrt{b}$. For example, to represent both square roots of 9, +3, and −3, we write $\pm\sqrt{9} = \pm 3$.

Example 1. Evaluate each radical.

(a) $\sqrt{81}$. (b) $-\sqrt{100}$. (c) $\sqrt{\dfrac{9}{4}}$. (d) $\sqrt{625}$.

Solution

(a) The symbol $\sqrt{81}$ indicates a positive number whose square is 81. Thus $\sqrt{81} = 9$ because $9^2 = 81$.

(b) Remember that $-\sqrt{100}$ represents the negative square root of 100. Thus $-\sqrt{100} = -10$ because $(-10)^2 = 100$.

(c) Note that $\sqrt{\dfrac{9}{4}} = \dfrac{3}{2}$ because $\left(\dfrac{3}{2}\right)^2 = \dfrac{9}{4}$.

(d) $\sqrt{625} = 25$ because $(25)^2 = 625$.

A rational number is called a *perfect square* if it is the *square* of a rational number. Examples of perfect squares are

$$\frac{4}{9} = \left(\frac{2}{3}\right)^2, \qquad 9 = 3^2, \qquad 25 = 5^2, \qquad 49 = 7^2.$$

It can be shown that the square root of a number that is *not* a perfect square is an irrational number. The number 2 is not the square of a rational number; therefore, $\sqrt{2}$ is an irrational number. Recall from Chapter 1 that irrational numbers are real numbers that have decimal representations that are nonterminating and nonrepeating. Thus the statement that

$$\sqrt{2} = 1.4142136 \ldots$$

expresses the fact that we cannot indicate the exact value of $\sqrt{2}$ using decimal notation. However, we can approximate every irrational number by a rational number. We use the symbol $\simeq$ to mean "approximately equal to." With this understanding we write

$$\sqrt{2} \simeq 1.414.$$

The rational number 1.414 is a four-digit approximation of $\sqrt{2}$. At the end of this book there is a table of positive square roots with four-to-five-digit accuracy for the whole numbers from 1 to 200. Here we list the square roots of the first 10 whole numbers in Table 7.1. Note that $\sqrt{1}$, $\sqrt{4}$, and $\sqrt{9}$ are rational numbers, since 1, 4, and 9 are perfect squares.

Table 7.1

n	$\sqrt{n}$	n	$\sqrt{n}$
1	1.000	6	2.449
2	1.414	7	2.646
3	1.732	8	2.828
4	2.000	9	3.000
5	2.236	10	3.162

Most hand calculators with a $\sqrt{\ }$ key will enable one to find an eight-digit approximation of square roots of rational numbers. Some have only four-digit approximations. We can use the table at the back of the book or a hand calculator. If we use a hand calculator or parts of the table that give more than a four-digit answer, we will round off to obtain four-digit accuracy.

Example 2. Approximate each radical to four-digit accuracy.
(a) $\sqrt{19}$. (b) $\sqrt{87}$. (c) $-\sqrt{90}$.

Solution: Using the square root table or a hand calculator, we find that
(a) $\sqrt{19} \simeq 4.359$. (b) $\sqrt{87} \simeq 9.327$. (c) $-\sqrt{90} \simeq -9.487$.

In the following discussion we define third or cube roots and fourth roots. It will then be clear how to evaluate fifth, sixth, and higher roots of real numbers.

If a and b are real numbers such that

$$a^3 = b,$$

then a is called the *cube root* or *third root* of b.

We denote the cube root of b by $\sqrt[3]{b}$. The number 3 is called the *index* or *order* of the *radical* $\sqrt[3]{b}$. (In the square root symbol, $\sqrt{\ }$, the index is understood to be 2.) Thus, by definition,

$$\sqrt[3]{8} = 2 \qquad \text{because} \qquad 2^3 = 8.$$

Notice that 8 has only one real number that is its cube root, since $(-2)^3 = -8$. Unlike square roots, negative numbers have real cube roots. Let us consider the cube root of -8, denoted by $\sqrt[3]{-8}$. We see that

$$\sqrt[3]{-8} = -2 \qquad \text{because} \qquad (-2)^3 = -8.$$

The previous discussion illustrates the fact that a real number has exactly one real cube root. Note that the cube root of a positive number is positive and that the cube root of a negative number is always negative. Also, we see that $\sqrt[3]{0} = 0$ because $0^3 = 0$.

We now define the fourth root of a real number.

> If a and b are real numbers such that
>
> $$a^4 = b,$$
>
> then a is called a *fourth root* of b.

Every positive real number has two real fourth roots, one positive, the other negative. The *radical* $\sqrt[4]{b}$ represents the positive fourth root of b, and $-\sqrt[4]{b}$ represents the negative fourth root. The index of the radical $\sqrt[4]{b}$ is 4. Observe that the number 16 has two fourth roots, 2 and -2, since $(2)^4 = 16$ and $(-2)^4 = 16$. Thus we write

$$\sqrt[4]{16} = 2 \qquad \text{and} \qquad -\sqrt[4]{16} = -2.$$

The fourth root of a negative real number is not a real number. The reason for this is that if a is any real number, a^4 is a positive number or zero. Thus a^4 is never negative. This is similar to the situation for square roots.

The previous discussion illustrates the following properties of the roots of a real number.

1. **Even roots:**
   a. Even roots are square roots, fourth roots, sixth roots, and so on.
   b. *Every* positive real number has two even real roots, one positive, the other negative. The *positive root* is called the *principal root*. The number 0 has only one even root, itself.
   c. Even roots of negative real numbers are *not* real numbers. They belong to a set of numbers not yet discussed in this book.
2. **Odd roots:**
   a. Odd roots are cube or third roots, fifth roots, seventh roots, and so on.
   b. All real numbers have exactly one odd real root.
   c. Odd real roots of positive numbers are positive. Odd real roots of negative numbers are negative. Any odd root of 0 is 0.

Whenever we consider roots of real numbers, we should be aware of the properties listed above, and the differences between odd and even roots.

Example 3. Evaluate.
(a) $\sqrt[3]{125}$. (b) $\sqrt[4]{81}$. (c) $\sqrt[3]{-64}$. (d) $-\sqrt[4]{1296}$. (e) $\sqrt[5]{32}$.

Solution

(a) $\sqrt[3]{125} = 5$ because $5^3 = 125$.
(b) $\sqrt[4]{81} = 3$ because $3^4 = 81$.
(c) $\sqrt[3]{-64} = -4$ because $(-4)^3 = -64$.
(d) $-\sqrt[4]{1296} = -6$ because $(-6)^4 = 1296$.
(e) $\sqrt[5]{32}$ indicates the fifth root of 32. Following the pattern of definitions for cube roots and fourth roots in parts (a) to (d), we have that $\sqrt[5]{32} = 2$ because $2^5 = 32$.

Example 4. Determine which of the following expressions are not real numbers. Evaluate those that represent real numbers.
(a) $\sqrt[4]{-16}$. (b) $\sqrt[5]{-32}$. (c) $\sqrt[6]{-32}$. (d) $\sqrt[6]{64}$.

Solution

(a) $\sqrt[4]{-16}$ is not a real number, because even roots of negative numbers are not real numbers.
(b) $\sqrt[5]{-32} = -2$ because $(-2)^5 = -32$.
(c) $\sqrt[6]{-32}$ is also not a real number for the same reason given in part (a).
(d) $\sqrt[6]{64} = 2$ because $2^6 = 64$.

We now discuss radicals with variables under the radical sign. In the following example all variables appearing under the square root sign or under radical signs with an even index are assumed to have only nonnegative values.

Example 5. Evaluate each expression.
(a) $\sqrt{25x^2}$. (b) $\sqrt[3]{8y^3}$. (c) $\sqrt[4]{625z^4}$. (d) $\sqrt[3]{-27s^3}$. (e) $\sqrt{x^4}$.

Solution

(a) To evaluate $\sqrt{25x^2}$, we need to find an expression whose square is $25x^2$. Since $(5x)^2 = (5x)(5x) = 25x^2$, we have $\sqrt{25x^2} = 5x$.
(b) The expression $\sqrt[3]{8y^3}$ represents an expression whose cube is $8y^3$. Note that $(2y)^3 = 2^3 \cdot y^3 = 8y^3$. Therefore, $\sqrt[3]{8y^3} = 2y$.
(c) $\sqrt[4]{625z^4} = 5z$ because $(5z)^4 = 5^4 \cdot z^4 = 625z^4$.
(d) Notice that $\sqrt[3]{-27s^3} = -3s$, since $(-3s)^3 = (-3)^3s^3 = -27s^3$.
(e) $\sqrt{x^4} = x^2$ because $(x^2)^2 = x^4$.

EXERCISES 7.1

For Exercises 1–20, find the square root. See Example 1.

1. $\sqrt{36}$
2. $\sqrt{49}$
3. $-\sqrt{36}$
4. $\sqrt{64}$
5. $-\sqrt{121}$
6. $\sqrt{100}$
7. $\sqrt{169}$
8. $-\sqrt{144}$
9. $\sqrt{256}$
10. $-\sqrt{256}$
11. $\sqrt{\frac{1}{4}}$
12. $\sqrt{\frac{1}{9}}$
13. $\sqrt{\frac{4}{9}}$
14. $-\sqrt{\frac{9}{16}}$
15. $\sqrt{1}$
16. $-\sqrt{1}$
17. $\sqrt{400}$
18. $\sqrt{289}$
19. $\sqrt{1024}$
20. $\sqrt{961}$

For Exercises 21–28, approximate the square root to four-digit accuracy. See Example 2.

21. $\sqrt{30}$
22. $\sqrt{37}$
23. $-\sqrt{68}$
24. $-\sqrt{15}$
25. $\sqrt{70}$
26. $\sqrt{77}$
27. $-\sqrt{65}$
28. $\sqrt{55}$

For Exercises 29–40, evaluate the root. See Example 3.

29. $\sqrt[3]{-1}$
30. $\sqrt[3]{27}$
31. $\sqrt[3]{64}$
32. $\sqrt[3]{216}$
33. $\sqrt[4]{625}$
34. $\sqrt[4]{1296}$
35. $\sqrt[3]{\frac{8}{27}}$
36. $\sqrt[4]{\frac{81}{625}}$
37. $\sqrt[4]{\frac{16}{81}}$
38. $\sqrt[5]{-1}$
39. $\sqrt[5]{243}$
40. $\sqrt[5]{\frac{32}{243}}$

For Exercises 41–52, determine which expressions are not real numbers. Evaluate those expressions that represent real numbers. See Example 4.

41. $-\sqrt{-1}$
42. $-\sqrt{-36}$
43. $\sqrt[3]{-27}$
44. $\sqrt[3]{-125}$
45. $-\sqrt[4]{81}$
46. $\sqrt[3]{-216}$
47. $\sqrt[4]{-81}$
48. $-\sqrt[6]{-1}$
49. $\sqrt[5]{-243}$
50. $\sqrt[3]{343}$
51. $\sqrt[6]{-729}$
52. $\sqrt[6]{0}$

For Exercises 53–68, evaluate the radical. All variables under the square root sign represent positive real numbers. See Example 5.

53. $\sqrt{y^2}$
54. $\sqrt{x^2}$
55. $\sqrt{36y^2}$
56. $\sqrt{49x^2}$
57. $\sqrt{x^2y^2}$
58. $\sqrt{a^2b^2}$
59. $\sqrt{4x^2y^2}$
60. $\sqrt{9a^2b^2}$
61. $\sqrt{(x + y)^2}$
62. $\sqrt{(a + b)^2}$
63. $\sqrt[3]{8x^3}$
64. $\sqrt[3]{27z^3}$
65. $\sqrt[3]{27^3y^3}$
66. $\sqrt[3]{64a^3b^3}$
67. $\sqrt{x^4}$
68. $\sqrt{x^6}$

For Exercises 69–76, express the answer to each part as integers and then compare. What conclusion can you make?

69. $\sqrt{(16 + 9)}$; $\sqrt{16} + \sqrt{9}$
70. $\sqrt{(25 + 144)}$; $\sqrt{25} + \sqrt{144}$
71. $\sqrt{(25 - 9)}$; $\sqrt{25} - \sqrt{9}$
72. $\sqrt{169 - 144}$; $\sqrt{169} - \sqrt{144}$
73. $\sqrt{(225 + 64)}$; $\sqrt{225} + \sqrt{64}$
74. $\sqrt{(576 + 49)}$; $\sqrt{576} + \sqrt{49}$
75. $\sqrt{(289 - 225)}$; $\sqrt{289} - \sqrt{225}$
76. $\sqrt{(625 - 576)}$; $\sqrt{625} - \sqrt{576}$

7.2 First Law of Radicals

There are two basic properties of radicals that are used for multiplying and dividing algebraic expressions that contain radicals. These properties are also used to express radicals in simpler forms. In this section we state and apply the basic property of radicals related to multiplication. The other property dealing with division of radicals is discussed in the next section. These properties are given in terms of square roots but also apply to all radicals of higher order (cube roots, fourth roots, etc.).

The following equations will suggest a rule for finding the product of two radicals. Note that

$$\sqrt{4} \cdot \sqrt{9} = 2 \cdot 3 = 6 \qquad \sqrt{4} = 2;\ \sqrt{9} = 3$$

and

$$\sqrt{4 \cdot 9} = \sqrt{36} = 6.$$

Both expressions $\sqrt{4} \cdot \sqrt{9}$ and $\sqrt{4 \cdot 9}$ equal 6; therefore, they are equal to each other. As a result, the following equality is true:

$$\sqrt{4 \cdot 9} = \sqrt{4} \cdot \sqrt{9}.$$

The last equation illustrates the following property involving the multiplication of radicals.

First Law of Radicals

> If a and b are nonnegative real numbers (positive or 0), then
>
> $$\sqrt{a \cdot b} = \sqrt{a}\,\sqrt{b}.$$

Note that $\sqrt{a}\sqrt{b}$ is the same as $\sqrt{a} \cdot \sqrt{b}$. The following examples illustrate some applications of this property.

Example 1. Find each product.
(a) $\sqrt{5}\sqrt{20}$. (b) $\sqrt{3}\sqrt{7}$.

Solution: For this problem, we use the first law of radicals, reading from right to left. Thus

(a) $\sqrt{5}\sqrt{20} = \sqrt{5 \cdot 20}$ first law of radicals
$= \sqrt{100}$
$= 10.$

(b) $\sqrt{3}\sqrt{7} = \sqrt{3 \cdot 7}$
$= \sqrt{21}.$

Example 2. Use the First Law of Radicals to approximate $\sqrt{600}$.

Solution: Notice that 600 is too large for our table of square roots. However, we can rewrite $\sqrt{600}$ as follows:

$$\begin{aligned} \sqrt{600} &= \sqrt{100 \cdot 6} & \\ &= \sqrt{100}\sqrt{6} & \text{first law of radicals} \\ &= 10\sqrt{6} & \sqrt{100} = 10 \end{aligned}$$

From the table of square roots we find that $\sqrt{6} \simeq 2.449$.

As a result, we obtain the following approximation:

$$\sqrt{600} = 10\sqrt{6} \simeq 10(2.449) = 24.49.$$

Thus

$$\sqrt{600} \simeq 24.49.$$

We may also use the first law of radicals to simplify a radical representing the square root of a positive integer.

> A radical $\sqrt{n}$ is in *simplified form* if the radicand n (where n is a positive integer) does not have a perfect square factor other than 1.

Let us consider the two radicals $\sqrt{10}$ and $\sqrt{72}$. Notice that $\sqrt{10}$ is in simplified form because none of the factors of 10—2, 5, and 10 (other than 1)—are perfect squares. However, $\sqrt{72}$ is not in simplified form because 72 has perfect square factors of 9 and 36. Let us use the first law of radicals to express $\sqrt{72}$ in simplified form. The most efficient way to proceed is to find the largest perfect square factor of 72, which is 36. Thus we write

$$72 = 36 \cdot 2.$$

Next we use the first law of radicals to rewrite $\sqrt{72}$ in an equivalent form. Thus

$$\begin{aligned} \sqrt{72} &= \sqrt{36 \cdot 2} & \\ &= \sqrt{36}\sqrt{2} & \text{first law of radicals} \\ &= 6\sqrt{2}. & \sqrt{36} = 6 \end{aligned}$$

Notice that 2 has no perfect square factor other than 1. As a result, $\sqrt{72}$ can be written in simplified form as $6\sqrt{2}$.

Example 3. Simplify $\sqrt{48}$.

Solution: We see that 16 is the largest perfect square factor of 48. That is, $48 = 16 \cdot 3$. Thus

$$\begin{aligned} \sqrt{48} &= \sqrt{16 \cdot 3} & \\ &= \sqrt{16}\sqrt{3} & \text{first law of radicals} \\ &= 4\sqrt{3}. & \sqrt{16} = 4 \end{aligned}$$

The number 3 under the radical sign has no perfect factors other than 1; therefore, $4\sqrt{3}$ is the simplified form of $\sqrt{48}$.

Simplifying a square root radical by determining the largest perfect square factor, or any perfect square factor, of a number may prove difficult. Therefore, it is sometimes more efficient to factor a number or numbers under the radical sign into its prime factors. We illustrate both methods of simplifying a radical in the next example.

Example 4. Find the product $\sqrt{15}\sqrt{10}$ and simplify.

Solution: We will solve this problem in two ways. One method using prime factors and the other not using the prime factors. First we use prime factors. Thus

$$
\begin{aligned}
\sqrt{15}\sqrt{10} &= \sqrt{15 \cdot 10} && \text{first law of radicals} \\
&= \sqrt{(3 \cdot 5)(5 \cdot 2)} && \text{factor 15 and 10 into prime factors; } 15 = 5 \cdot 3,\ 10 = 5 \cdot 2 \\
&= \sqrt{5^2 \cdot 3 \cdot 2}. && \\
&= \sqrt{5^2}\sqrt{3 \cdot 2} && \text{first law of radicals} \\
&= 5\sqrt{6} && \sqrt{5^2} = 5
\end{aligned}
$$

Therefore, the product $\sqrt{15}\sqrt{10}$ equals $5\sqrt{6}$, which is in simplified form. We now do this problem a little differently without using prime factors. We see that

$$
\begin{aligned}
\sqrt{15 \cdot 10} &= \sqrt{150} && \text{first law of radicals} \\
&= \sqrt{25 \cdot 6} && \text{25 is the largest perfect square factor of 150} \\
&= \sqrt{25}\sqrt{6} && \text{first law of radicals} \\
&= 5\sqrt{6}. && \sqrt{25} = 5
\end{aligned}
$$

Example 4 shows two methods for simplifying a radical. The most efficient way depends on the particular problem. When simplifying a radical, the reader should use the method that he or she finds most comfortable.

The first law of radicals also applies to square root radicals with variables under the radical sign. Remember that the variables must represent positive numbers. A radical representing a square root with a radicand that contains variables raised to positive integer powers is in *simplified form* if it has no perfect square factors other than 1.

Example 5. Simplify $\sqrt{75x^3}$.

Solution: First we find the largest perfect square factor of $75x^3$. This factor is $25x^2$. Therefore,

$$75x^3 = 25x^2(3x).$$

Thus

$$\begin{aligned}\sqrt{75x^3} &= \sqrt{25x^2(3x)} \\ &= \sqrt{25x^2}\sqrt{3x} && \text{first law of radicals} \\ &= 5x\sqrt{3x}. && \sqrt{25x^2} = 5x\end{aligned}$$

The radicand $3x$ appearing in the last radical has no perfect square factor other than 1. Consequently, $5x\sqrt{3x}$ is the simplified form of $\sqrt{75x^3}$.

As mentioned at the beginning of this section, the first law of radicals is also true for higher-order radicals. Accordingly, we have

$$\begin{aligned}\sqrt[3]{ab} &= \sqrt[3]{a}\sqrt[3]{b} \\ \sqrt[4]{ab} &= \sqrt[4]{a}\sqrt[4]{b} && (a,\ b \text{ positive real numbers})\end{aligned}$$

and so on. Also, the definition of the simplified form of higher-order radicals is essentially the same as that given for square roots. Thus the cube root of an expression must not contain a factor that is a perfect cube, the fourth root of an expression must not contain a factor that is a perfect fourth power, and so on. The following example illustrates how to simplify radicals involving cube roots.

Example 6. Simplify.

(a) $\sqrt[3]{24}$. (b) $\sqrt[3]{54x^5}$.

Solution

(a) We wish to express $\sqrt[3]{24}$ so that the radicand does not have a perfect cube as a factor. We observe that $24 = 8 \cdot 3$ and that 8 is a perfect cube. That is, $8 = 2^3$. Thus

$$\begin{aligned}\sqrt[3]{24} &= \sqrt[3]{8 \cdot 3} \\ &= \sqrt[3]{8}\sqrt[3]{3} && \text{first law of radicals for higher-order radicals} \\ &= 2\sqrt[3]{3}. && \sqrt[3]{8} = 2\end{aligned}$$

Notice that 3 has no factors that are perfect cubes. Therefore, $2\sqrt[3]{3}$ is the simplified form of $\sqrt[3]{24}$.

(b) To simplify $\sqrt[3]{54x^5}$, we look for the largest perfect cube that is a factor of $54x^5$. This factor is $27x^3$. Thus

$$54x^5 = (27x^3)2x^2.$$

As a result, we may write

$$\sqrt[3]{54x^5} = \sqrt[3]{(27x^3)2x^2}$$
$$= \sqrt[3]{27x^3}\sqrt[3]{2x^2} \qquad \text{first law of radicals}$$
$$= 3x\sqrt[3]{2x^2}. \qquad \sqrt[3]{27x^3} = 3x$$

There are no perfect *cube* factors under the last radical sign; therefore, $3x\sqrt[3]{2x^2}$ is the simplified form of $\sqrt[3]{54x^5}$.

EXERCISES 7.2

For Exercises 1–18, use the first law of radicals to express the given radical expression in simplified form. See Examples 1, 3, and 4.

1. $\sqrt{12}\sqrt{3}$ **2.** $\sqrt{3}\sqrt{27}$ **3.** $\sqrt{7}\sqrt{7}$ **4.** $\sqrt{13}\sqrt{13}$ **5.** $\sqrt{11}\sqrt{11}$
6. $\sqrt{19}\sqrt{19}$ **7.** $\sqrt{18}$ **8.** $\sqrt{20}$ **9.** $\sqrt{27}$ **10.** $\sqrt{45}$
11. $\sqrt{98}$ **12.** $\sqrt{75}$ **13.** $\sqrt{96}$ **14.** $\sqrt{54}$ **15.** $\sqrt{128}$
16. $\sqrt{162}$ **17.** $\sqrt{200}$ **18.** $\sqrt{242}$

For Exercises 19–36, simplify the radical. Assume that all variables represent positive real numbers. See Examples 4 and 5.

19. $\sqrt{49y^2}$ **20.** $\sqrt{36x^2}$ **21.** $\sqrt{81x^2y^2}$ **22.** $\sqrt{100y^2z^2}$
23. $\sqrt{24x^3}$ **24.** $\sqrt{28y^3}$ **25.** $\sqrt{175x^3y^3}$ **26.** $\sqrt{300a^3b^2}$
27. $\sqrt{160x^4}$ **28.** $\sqrt{90a^4}$ **29.** $\sqrt{360b^5}$ **30.** $\sqrt{180a^7}$
31. $\sqrt{20}\sqrt{6}$ **32.** $\sqrt{15}\sqrt{6}$ **33.** $\sqrt{21}\sqrt{35}$ **34.** $\sqrt{22}\sqrt{33}$
35. $\sqrt{65}\sqrt{39}$ **36.** $\sqrt{45}\sqrt{7 \cdot 4}$

For Exercises 37–48, simplify the radical. See Example 6.

37. $\sqrt[3]{56}$ **38.** $\sqrt[3]{40}$ **39.** $\sqrt[3]{250}$ **40.** $\sqrt[3]{-128}$ **41.** $\sqrt[3]{81x^4}$
42. $\sqrt[3]{48y^5}$ **43.** $\sqrt[3]{32x^6}$ **44.** $\sqrt[3]{x^3y^4}$ **45.** $\sqrt[3]{x^4y^5}$ **46.** $\sqrt[4]{32}$
47. $\sqrt[4]{162x^5}$ **48.** $\sqrt[4]{x^5y^6}$

For Exercises 49–56 find a decimal approximation. See Example 2.

49. $\sqrt{800}$ **50.** $\sqrt{1100}$ **51.** $\sqrt{125}$ **52.** $\sqrt{360}$
53. $\sqrt{270}$ **54.** $\sqrt{240}$ **55.** $\sqrt{6500}$ **56.** $\sqrt{5500}$

Second Law of Radicals 7.3

The following two equations suggest a second property of square root radicals related to division. Note that

$$\sqrt{\frac{36}{9}} = \sqrt{4} = 2 \qquad \text{and that} \qquad \frac{\sqrt{36}}{\sqrt{9}} = \frac{6}{3} = 2.$$

Since the expressions on the left-hand side of these equations are both equal to 2, we assert the following equality:

$$\sqrt{\frac{36}{9}} = \frac{\sqrt{36}}{\sqrt{9}}.$$

This equation illustrates the following property involving square roots of *quotients*.

Second Law of Radicals

> If a and b are positive real numbers, then
>
> $$\sqrt{\frac{a}{b}} = \frac{\sqrt{a}}{\sqrt{b}}.$$

Usually, the second law of radicals is stated as follows. The *square root of a fraction* is the *square root of the numerator* over the *square root of the denominator*.

The property stated above is also true for radicands involving variables provided that the variables represent positive numbers.

The following examples illustrate its application to quotients involving radicals.

Example 1. Simplify $\sqrt{\frac{121}{169}}$.

Solution

$$\sqrt{\frac{121}{169}} = \frac{\sqrt{121}}{\sqrt{169}} \qquad \text{second law of radicals}$$

$$= \frac{11}{13}. \qquad \sqrt{121} = 11,\ \sqrt{169} = 13$$

Example 2. Simplify $\frac{\sqrt{125}}{\sqrt{5}}$.

Solution: To simplify the given expression, we use the second law of radicals reading from right to left. Thus we have

$$\frac{\sqrt{125}}{\sqrt{5}} = \sqrt{\frac{125}{5}} \qquad \text{second law of radicals}$$

$$= \sqrt{25}$$

$$= 5.$$

Example 3. Simplify $\sqrt{.09}$.

Solution: We know that $.09 = \dfrac{9}{100}$. Thus

$$\sqrt{.09} = \sqrt{\frac{9}{100}}$$

$$= \frac{\sqrt{9}}{\sqrt{100}} \qquad \text{second law of radicals}$$

$$= \frac{3}{10}$$

$$= .3 .$$

Example 4. Simplify $\sqrt{\dfrac{5}{9}}$.

Solution

$$\sqrt{\frac{5}{9}} = \frac{\sqrt{5}}{\sqrt{9}} \qquad \text{second law of radicals}$$

$$= \frac{\sqrt{5}}{3}. \qquad \sqrt{9} = 3$$

Recall that in Section 7.2 the first law of radicals applied not only to square roots but to higher-order radicals as well. Similarly, the second law of radicals is valid for higher-order radicals. In the next example we apply the second law to cube roots.

Example 5. Simplify.

(a) $\sqrt[3]{\dfrac{27}{256}}$. (b) $\dfrac{\sqrt[3]{250}}{\sqrt[3]{2}}$.

Solution

(a) $\sqrt[3]{\frac{27}{256}} = \frac{\sqrt[3]{27}}{\sqrt[3]{256}}$ second law of radicals

$= \frac{3}{6}$ $3^3 = 27;\ 6^3 = 256$

$= \frac{1}{2}.$

(b) $\frac{\sqrt[3]{250}}{\sqrt[3]{2}} = \sqrt[3]{\frac{250}{2}}$ second law of radicals

$= \sqrt[3]{125}$

$= 5.$

In the previous examples the radicals appearing in the denominator of a fraction were either square roots of perfect squares or cube roots of perfect cubes. As a result we can express these denominators as rational numbers without radicals. This often proves useful when working with radicals in algebraic expressions. In situations where the radical in the *denominator* of a fraction represents an irrational number, such as

$$\frac{1}{\sqrt{3}}$$

it sometimes proves convenient to also rewrite it as a fraction with a rational number in the denominator. To accomplish this for the fraction $\frac{1}{\sqrt{3}}$ we simply multiply the numerator and denominator of $\frac{1}{\sqrt{3}}$ by $\sqrt{3}$. This procedure eliminates the radical from the denominator as follows:

$\frac{1}{\sqrt{3}} = \frac{1}{\sqrt{3}} \cdot \frac{\sqrt{3}}{\sqrt{3}}$ $\frac{\sqrt{3}}{\sqrt{3}} = 1$

$= \frac{\sqrt{3}}{\sqrt{9}}$ $\sqrt{3}\sqrt{3} = \sqrt{9}$ by the first law of radicals

$= \frac{\sqrt{3}}{3}.$

Thus we were able to convert $\frac{1}{\sqrt{3}}$ to an equivalent fraction $\frac{\sqrt{3}}{3}$, with a rational number as a denominator. In general, the process of converting a fraction with a radical in the denominator representing an irrational number to an equivalent fraction with a rational number in the denominator is called *rationalizing the denominator*.

Note that it is not incorrect to write $\frac{1}{\sqrt{3}}$ rather than its equivalent form $\frac{\sqrt{3}}{3}$. But often $\frac{\sqrt{3}}{3}$ is easier to work with. This is one reason why $\frac{\sqrt{3}}{3}$ is called a *simplified form* of $\frac{1}{\sqrt{3}}$.

Example 6. Simplify $\frac{3}{\sqrt{5}}$ by rationalizing the denominator.

Solution: To rationalize $\frac{3}{\sqrt{5}}$, we multiply numerator and denominator by $\sqrt{5}$ as shown in the following equation. Note that multiplying the numerator and denominator of a fraction by the same nonzero value does not change the value of the fraction. Thus

$$\frac{3}{\sqrt{5}} = \frac{3}{\sqrt{5}} \cdot \frac{\sqrt{5}}{\sqrt{5}} \qquad \frac{\sqrt{5}}{\sqrt{5}} = 1$$

$$= \frac{3\sqrt{5}}{\sqrt{5^2}} \qquad \sqrt{5} \cdot \sqrt{5} = \sqrt{5^2}$$

$$= \frac{3\sqrt{5}}{5}. \qquad \sqrt{5^2} = 5$$

Thus the simplified form of $\frac{3}{\sqrt{5}}$ is $\frac{3\sqrt{5}}{5}$.

Example 7. Approximate $\frac{1}{\sqrt{2}}$ by rationalizing the denominator.

Solution

$$\frac{1}{\sqrt{2}} = \frac{1}{\sqrt{2}} \cdot \frac{\sqrt{2}}{\sqrt{2}} \qquad \frac{\sqrt{2}}{\sqrt{2}} = 1$$

$$= \frac{\sqrt{2}}{\sqrt{2 \cdot 2}} \qquad \text{first law of radicals}$$

$$= \frac{\sqrt{2}}{\sqrt{4}}$$

$$= \frac{\sqrt{2}}{2}.$$

From the table of square roots we see that $\sqrt{2} \simeq 1.414$ (many people memorize this fact). Therefore,

$$\frac{\sqrt{2}}{2} \simeq \frac{1.414}{2} = .707.$$

Thus .707 is an approximate value for $\frac{1}{\sqrt{2}}$.

Recall that in Section 7.2 we simplified the products of radicals by using the First Law of Radicals. In this section we used the Second Law of Radicals to simplify quotients involving radicals. As a rule, radicals in simplified form are easier to manipulate. However, it is not always clear what the simplified form of a radical expression looks like. A *radical expression* is an expression that involves radicals such as square roots, cube roots, or higher-order roots. The following definition provides a guide to recognizing simplified forms of radical expressions.

An expression containing radicals is in *simplified form* if the following three conditions are satisfied:

1. The radicand (expression under the radical sign) of a square root radical does not have a perfect square factor. Cube roots do not have radicands with perfect cube factors, and so forth.
2. The denominator of a fraction does not contain a radical.
3. A radical does not have a fraction under the radical sign.

The process of simplifying a radical or an expression containing radicals often requires the use of both the first and second law of radicals. We illustrate this in the examples that follow.

Example 8. Simplify $\sqrt{\frac{50}{121}}$.

Solution: Notice that this radical is not in simplified form because the radicand is a fraction. Applying the second law of radicals, we obtain

$$\sqrt{\frac{50}{121}} = \frac{\sqrt{50}}{\sqrt{121}}$$

$$= \frac{\sqrt{50}}{11}. \qquad \sqrt{121} = 11$$

The expression $\frac{\sqrt{50}}{11}$ is not yet in simplified form because 50 has a perfect square factor 25. Therefore, we continue the process of simplification

$$\frac{\sqrt{50}}{11} = \frac{\sqrt{25 \cdot 2}}{11}$$

$$= \frac{\sqrt{25}\sqrt{2}}{11} \qquad \text{first law of radicals}$$

$$= \frac{5\sqrt{2}}{11}.$$

Thus $\frac{5\sqrt{2}}{11}$ is the simplified form of $\sqrt{\frac{50}{121}}$.

Example 9. Simplify $\frac{3}{\sqrt{32}}$.

Solution: The given expression is not in simplified form because the denominator of the fraction contains a radical. To put $\frac{3}{\sqrt{32}}$ in simplified form, we must rationalize the denominator. We could do this by multiplying numerator and denominator by the radical appearing in the denominator. However, this is not always the most efficient way to rationalize the denominator of a fraction. In this example we notice that $32 \cdot 2 = 64$ is a perfect square. Therefore, we may obtain a rational number in the denominator of $\frac{3}{\sqrt{32}}$ by multiplying $\sqrt{32}$ by $\sqrt{2}$ because $\sqrt{32} \cdot \sqrt{2} = \sqrt{32 \cdot 2} = \sqrt{64} = 8$ is a rational number. Thus we simplify $\frac{3}{\sqrt{32}}$ as follows:

$$\frac{3}{\sqrt{32}} = \frac{3}{\sqrt{32}} \cdot \frac{\sqrt{2}}{\sqrt{2}} = \frac{3\sqrt{2}}{\sqrt{32 \cdot 2}} \qquad \text{first law of radicals}$$

$$= \frac{3\sqrt{2}}{\sqrt{64}}$$

$$= \frac{3\sqrt{2}}{8}. \qquad \sqrt{64} = 8$$

Thus $\frac{3\sqrt{2}}{8}$ is the simplified form of $\frac{3}{\sqrt{32}}$.

Example 10. Simplify $\frac{\sqrt{27x^3y}}{\sqrt{3x}}$. Assume that x and y represent positive numbers.

Solution: To simplify the fraction, we have to eliminate the radical in the denominator. For this example it is easier to rewrite the expression as the square root of a quotient because $3x$ divides evenly into $27x^3y$. Thus we have

$$\frac{\sqrt{27x^3y}}{\sqrt{3x}} = \sqrt{\frac{27x^3y}{3x}} \qquad \text{second law of radicals}$$

$$= \sqrt{9x^2y} \qquad \frac{27x^3y}{3x} = 9x^2y$$

$$= \sqrt{9x^2}\sqrt{y} \qquad \text{first law of radicals}$$

$$= 3x\sqrt{y}.$$

In the following example, we show how to simplify radical expressions involving cube roots. The techniques illustrated can be extended to the simplifying of expressions involving higher-order radicals.

Example 11. Simplify $\sqrt[3]{\frac{16}{27}}$.

Solution: To simplify the radical, we apply the second law of radicals. Thus

$$\sqrt[3]{\frac{16}{27}} = \frac{\sqrt[3]{16}}{\sqrt[3]{27}} \qquad \text{second law of radicals}$$

$$= \frac{\sqrt[3]{16}}{3}.$$

Note that $\sqrt[3]{16}$ is not in simplified form because 16 has a perfect cube as a factor (a perfect cube is the cube of a rational number). We observe that $16 = 8 \cdot 2$, and that $8 = 2^3$, a perfect cube. Continuing the process of simplification, we write

$$\frac{\sqrt[3]{16}}{3} = \frac{\sqrt[3]{8 \cdot 2}}{3}$$

$$= \frac{\sqrt[3]{8}\sqrt[3]{2}}{3} \qquad \text{first law of radicals}$$

$$= \frac{2\sqrt[3]{2}}{3}. \qquad \sqrt[3]{8} = 2$$

Thus $\frac{2\sqrt[3]{2}}{3}$ is the simplified form of $\sqrt[3]{\frac{16}{27}}$.

Example 12. Simplify each radical.

(a) $\frac{5}{\sqrt[3]{2}}$. (b) $\sqrt[3]{\frac{1}{2}}$.

Solution

(a) To rationalize the denominator (eliminate the cube root radical from the denominator), we must multiply $\sqrt[3]{2}$ by a radical so that we obtain a perfect cube under the radical sign. Note that $2 \cdot 4 = 8$, and that 8 is a perfect cube. Thus we multiply numerator and denominator of $\frac{1}{\sqrt[3]{2}}$ by $\sqrt[3]{4}$. As a result, we have

$$\frac{5}{\sqrt[3]{2}} = \frac{5}{\sqrt[3]{2}} \cdot \frac{\sqrt[3]{4}}{\sqrt[3]{4}} \qquad \frac{\sqrt[3]{4}}{\sqrt[3]{4}} = 1$$

$$= \frac{5\sqrt[3]{4}}{\sqrt[3]{2}\sqrt[3]{4}}$$

$$= \frac{5\sqrt[3]{4}}{\sqrt[3]{2 \cdot 4}} \qquad \text{first law of radicals}$$

$$= \frac{5\sqrt[3]{4}}{\sqrt[3]{8}}$$

$$= \frac{5\sqrt[3]{4}}{2}. \qquad \sqrt[3]{8} = 2$$

Therefore, $\frac{5\sqrt[3]{4}}{2}$ is the simplified form of $\frac{5}{\sqrt[3]{2}}$.

(b) To simplify the radical $\sqrt[3]{\frac{1}{2}}$, we observe that multiplying numerator and denominator of the fraction under the radical sign by 4 gives us an equivalent fraction with a denominator of 8 (a perfect cube).

$$\sqrt[3]{\frac{1}{2}} = \sqrt[3]{\frac{1}{2} \cdot \frac{4}{4}} = \sqrt[3]{\frac{4}{8}}.$$

Then we apply the second law of radicals to obtain

$$\sqrt[3]{\frac{4}{8}} = \frac{\sqrt[3]{4}}{\sqrt[3]{8}} = \frac{\sqrt[3]{4}}{2}.$$

Therefore, $\frac{\sqrt[3]{4}}{2}$ is the simplified form of $\sqrt[3]{\frac{1}{2}}$.

EXERCISES 7.3

For Exercises 1–24, use the second law of radicals to express the given radical in simplified form. See Example 1–5.

1. $\sqrt{\frac{36}{49}}$ **2.** $\sqrt{\frac{64}{81}}$ **3.** $\sqrt{\frac{25}{16}}$ **4.** $\sqrt{\frac{100}{9}}$ **5.** $\frac{\sqrt{27}}{\sqrt{3}}$

6. $\dfrac{\sqrt{8}}{\sqrt{2}}$ 7. $\dfrac{\sqrt{216}}{\sqrt{6}}$ 8. $\dfrac{\sqrt{343}}{\sqrt{7}}$ 9. $\sqrt{.36}$ 10. $\sqrt{.25}$

11. $\sqrt{.49}$ 12. $\sqrt{1.21}$ 13. $\sqrt{1.44}$ 14. $\sqrt{1.69}$ 15. $\sqrt{.0081}$

16. $\sqrt{.0064}$ 17. $\sqrt{\dfrac{3}{4}}$ 18. $\sqrt{\dfrac{7}{36}}$ 19. $\sqrt{\dfrac{11}{144}}$ 20. $\sqrt{\dfrac{17}{169}}$

21. $\sqrt[3]{\dfrac{8}{125}}$ 22. $\sqrt[3]{\dfrac{1}{64}}$ 23. $\dfrac{\sqrt[3]{81}}{\sqrt[3]{3}}$ 24. $\dfrac{\sqrt[3]{432}}{\sqrt[3]{2}}$

For Exercises 21–52, simplify the radical. Use both the first and second law of radicals if necessary. See Examples 6, 8, and 9.

25. $\dfrac{6}{\sqrt{6}}$ 26. $\dfrac{2}{\sqrt{2}}$ 27. $\dfrac{7}{\sqrt{7}}$ 28. $\dfrac{11}{\sqrt{11}}$

29. $\dfrac{9}{\sqrt{8}}$ 30. $\dfrac{2}{\sqrt{10}}$ 31. $\dfrac{3}{\sqrt{11}}$ 32. $\dfrac{4}{\sqrt{5}}$

33. $\dfrac{7}{\sqrt{6}}$ 34. $\dfrac{5}{\sqrt{7}}$ 35. $\dfrac{5}{\sqrt{18}}$ 36. $\dfrac{3}{\sqrt{50}}$

37. $\sqrt{\dfrac{27}{64}}$ 38. $\sqrt{\dfrac{75}{100}}$ 39. $\sqrt{\dfrac{98}{144}}$ 40. $\sqrt{\dfrac{125}{49}}$

41. $\dfrac{5}{\sqrt{75}}$ 42. $\dfrac{3}{\sqrt{48}}$ 43. $\dfrac{11}{\sqrt{12}}$ 44. $\dfrac{1}{\sqrt{72}}$

45. $\dfrac{3}{\sqrt{27}}$ 46. $\dfrac{2}{\sqrt{13}}$ 47. $\dfrac{10}{\sqrt{17}}$ 48. $\dfrac{1}{\sqrt{98}}$

49. $\sqrt{\dfrac{4}{3}}\sqrt{\dfrac{27}{64}}$ 50. $\sqrt{\dfrac{2}{5}}\sqrt{\dfrac{125}{8}}$ 51. $\sqrt{\dfrac{98}{216}}\sqrt{\dfrac{6}{2}}$ 52. $\sqrt{\dfrac{50}{3}}\sqrt{\dfrac{30}{5}}$

For Exercises 53–64, simplify the radical. Assume that all variables represent positive numbers. Use both laws of radicals if necessary. See Example 10.

53. $\sqrt{\dfrac{25}{x^2}}$ 54. $\sqrt{\dfrac{49}{y^2}}$ 55. $\sqrt{\dfrac{16x^2}{z^2}}$ 56. $\sqrt{\dfrac{x^2y^2}{z^2}}$

57. $\dfrac{\sqrt{a^3}}{\sqrt{a}}$ 58. $\sqrt{\dfrac{a^4}{b^2}}$ 59. $\sqrt{\dfrac{a^6}{b^8}}$ 60. $\sqrt{\dfrac{x}{y}}\sqrt{\dfrac{y^3}{x^5}}$

61. $\sqrt{\dfrac{a^5}{b}}\sqrt{\dfrac{b^3}{a^7}}$ 62. $\sqrt{\dfrac{32x^2y^2}{2y}}$ 63. $\sqrt{\dfrac{75x^3y^4}{3x^2}}$ 64. $\dfrac{\sqrt{2a^2b}}{\sqrt{128ab^3}}$

For Exercises 65–76, simplify the radical. See Examples 11 and 12.

65. $\sqrt[3]{\dfrac{40}{27}}$ 66. $\sqrt[3]{\dfrac{48}{1000}}$ 67. $\sqrt[3]{\dfrac{1}{3}}$ 68. $\sqrt[3]{\dfrac{1}{4}}$

69. $\sqrt[3]{1/x^2}$ **70.** $\sqrt[3]{1/y^5}$ **71.** $\dfrac{\sqrt[3]{a^8}}{\sqrt[3]{a^2}}$ **72.** $\sqrt[3]{\dfrac{x^9}{y^6}}$

73. $\dfrac{2}{\sqrt[3]{3}}$ **74.** $\dfrac{5}{\sqrt[3]{16}}$ **75.** $\dfrac{x}{\sqrt[3]{y^2}}$ **76.** $\dfrac{z}{\sqrt[3]{9y}}$

For Exercises 77–84, find a decimal approximation by rationalizing the denominator. See Example 7.

77. $\dfrac{5}{\sqrt{7}}$ **78.** $\dfrac{3}{\sqrt{7}}$ **79.** $\dfrac{10}{\sqrt{5}}$ **80.** $\dfrac{\sqrt{3}}{\sqrt{5}}$

81. $\sqrt{\dfrac{16}{.10}}$ **82.** $\dfrac{6}{\sqrt{3}}$ **83.** $\sqrt{\dfrac{4}{3}}$ **84.** $\sqrt{\dfrac{25}{10}}$

7.4 Addition and Subtraction of Radicals

Like radicals are radicals that have the same index and radicand (the expression under the radical sign). Thus $\sqrt[3]{7}$ and $\sqrt{7}$ are unlike radicals because they have different indices. The square root radicals $\sqrt{5}$ and $\sqrt{7}$ are unlike radicals because their radicands are different. Expressions containing terms with *like* (or identical) radicals may be added or subtracted by using the distributive property. For the remainder of this chapter we concern ourselves only with square root radicals. Consider the expression

$$3\sqrt{7} + 2\sqrt{7}.$$

The terms of this expression, $3\sqrt{7}$ and $2\sqrt{7}$, are *like terms* because each term contains the same radical, $\sqrt{7}$. We can combine these terms just as we combined like terms of polynomials by applying the distributive property. Thus

$$\begin{aligned} 3\sqrt{7} + 2\sqrt{7} &= (3 + 2)\sqrt{7} && \text{distributive property} \\ &= 5\sqrt{7} && \text{addition} \end{aligned}$$

When we combine the terms of a radical expression and obtain an expression with fewer terms, we say that the original expression has been *simplified.*

Example 1. Simplify each radical expression by performing the operation indicated.
(a) $4\sqrt{3} + 3\sqrt{3}$. (b) $3\sqrt{2} - 5\sqrt{2}$. (c) $5\sqrt{5} - 2\sqrt{5} + 8\sqrt{5}$.

Solution

(a) $$\begin{aligned} 4\sqrt{3} + 3\sqrt{3} &= (4 + 3)\sqrt{3} && \text{distributive property} \\ &= 7\sqrt{3}. && \text{addition} \end{aligned}$$

(b) $$\begin{aligned} 3\sqrt{2} - 5\sqrt{2} &= (3 - 5)\sqrt{2} && \text{distributive property} \\ &= -2\sqrt{2}. && \text{subtraction} \end{aligned}$$

(c) $$\begin{aligned} 5\sqrt{5} - 2\sqrt{5} + 8\sqrt{5} &= (5 - 2 + 8)\sqrt{5} && \text{distributive property} \\ &= 11\sqrt{5}. && \text{subtraction, addition} \end{aligned}$$

Sometimes it is possible to combine terms in a radical expression that appear to be unlike terms by first simplifying the radicals. Other times, even after simplification of the radicals, we cannot combine the radicals. The following example illustrates the previous remarks.

Example 2. If possible, simplify each radical expression by combining terms. (a) $\sqrt{27} + \sqrt{12}$. (b) $\sqrt{45} + \sqrt{24}$.

Solution

(a) The radicals are not like radicals. Therefore, we simplify each radical appearing in the expression.

$$
\begin{aligned}
\sqrt{27} + \sqrt{12} &= \sqrt{9 \cdot 3} + \sqrt{4 \cdot 3} \\
&= \sqrt{9}\sqrt{3} + \sqrt{4}\sqrt{3} && \text{first law of radicals} \\
&= 3\sqrt{3} + 2\sqrt{3} && \sqrt{9} = 3,\ \sqrt{4} = 2 \\
&= (3 + 2)\sqrt{3} && \text{distributive property} \\
&= 5\sqrt{3}.
\end{aligned}
$$

Thus $\sqrt{27} + \sqrt{12} = 5\sqrt{3}$.

(b) We simplify each radical of the expression as in part (a).

$$
\begin{aligned}
\sqrt{45} + \sqrt{24} &= \sqrt{9 \cdot 5} + \sqrt{4 \cdot 6} \\
&= \sqrt{9}\sqrt{5} + \sqrt{4}\sqrt{6} && \text{first law of radicals} \\
&= 3\sqrt{5} + 2\sqrt{6}. && \sqrt{9} = 3,\ \sqrt{4} = 2
\end{aligned}
$$

The unlike radicals $\sqrt{5}$ and $\sqrt{6}$ cannot be simplified further. Thus we cannot combine the two radicals. Therefore, the terms of the original expression

$$\sqrt{45} + \sqrt{24}$$

cannot be combined. However, $3\sqrt{5} + 2\sqrt{6}$ is the simplified form of $\sqrt{45} + \sqrt{24}$.

Example 3. Simplify, by combining terms,

$$2\sqrt{18} + \sqrt{32}.$$

Solution

$$
\begin{aligned}
2\sqrt{18} + \sqrt{32} &= 2\sqrt{9 \cdot 2} + \sqrt{16 \cdot 2} \\
&= 2\sqrt{9}\sqrt{2} + \sqrt{16}\sqrt{2} && \text{first law of radicals} \\
&= 2(3)\sqrt{2} + 4\sqrt{2} \\
&= 6\sqrt{2} + 4\sqrt{2} \\
&= (6 + 4)\sqrt{2} && \text{distributive property} \\
&= 10\sqrt{2}.
\end{aligned}
$$

Example 4. Simplify, by combining as many terms as possible,

$$3\sqrt{24} - \sqrt{54} + 7\sqrt{5}.$$

Solution

$$3\sqrt{24} - \sqrt{54} + 7\sqrt{5} = 3\sqrt{4}\sqrt{6} - \sqrt{9}\sqrt{6} + 7\sqrt{5} \qquad \text{first law of radicals}$$
$$= 3(2)\sqrt{6} - 3\sqrt{6} + 7\sqrt{5}$$
$$= 6\sqrt{6} - 3\sqrt{6} + 7\sqrt{5}$$
$$= (6 - 3)\sqrt{6} + 7\sqrt{5} \qquad \text{distributive property}$$
$$= 3\sqrt{6} + 7\sqrt{5}.$$

The last two terms cannot be combined, since $\sqrt{6}$ and $\sqrt{5}$ are unlike radicals in simplified form.

Example 5. Simplify, by combining terms,

$$3\sqrt{28} + 5\sqrt{63} - \sqrt{7}.$$

Solution

$$3\sqrt{28} + 5\sqrt{63} - \sqrt{7} = 3\sqrt{4}\sqrt{7} + 5\sqrt{9}\sqrt{7} - \sqrt{7} \qquad \text{first law of radicals}$$
$$= 3(2)\sqrt{7} + 5(3)\sqrt{7} - \sqrt{7}$$
$$= 6\sqrt{7} + 15\sqrt{7} - 1\sqrt{7}$$
$$= (6 + 15 - 1)\sqrt{7} \qquad \text{distributive property}$$
$$= 20\sqrt{7}.$$

Example 6. Simplify the following expression. Assume that x and y represent positive numbers.

$$\sqrt{4x^2y} + x\sqrt{9y}.$$

Solution: We proceed by simplifying each radical term of the expression. Therefore, we find the largest perfect square factor of each radicand.

$$\sqrt{4x^2y} + x\sqrt{9y} = \sqrt{(4x^2)y} + x\sqrt{9 \cdot y}$$
$$= \sqrt{4x^2}\sqrt{y} + x\sqrt{9}\sqrt{y} \qquad \text{first law of radicals}$$
$$= 2x\sqrt{y} + 3x\sqrt{y}. \qquad \sqrt{4x^2} = 2x;\ \sqrt{9} = 3$$
$$= (2x + 3x)\sqrt{y} \qquad \text{distributive property}$$
$$= 5x\sqrt{y}. \qquad 2x + 3x = 5x$$

Thus

$$\sqrt{4x^2y} + x\sqrt{9y} = 5x\sqrt{y}.$$

In the expression $2\sqrt{3}$, 2 is called the coefficient of $\sqrt{3}$. Similarly, in the expression $4x\sqrt{y}$, $4x$ is the coefficient of $\sqrt{y}$. Note in the previous examples that like radicals may be combined by merely adding or subtracting the *coefficients of the like radicals*. The reason for this, as we have seen, is the distributive property. In general, when simplifying radical expressions, do the following:

1. Simplify each radical in the expression.
2. Combine the terms with like radicals by adding or subtracting the coefficients of the like radicals.

Example 7. Simplify the expression

$$3\sqrt{75} + \sqrt{300} - 6\sqrt{3}.$$

Solution: First, if possible, simplify each radical.

$$\begin{aligned} 3\sqrt{75} + \sqrt{300} - 6\sqrt{3} &= 3\sqrt{25}\sqrt{3} + \sqrt{100}\sqrt{3} - 6\sqrt{3} \\ &= 3(5)\sqrt{3} + 10\sqrt{3} - 6\sqrt{3} \\ &= 15\sqrt{3} + 10\sqrt{3} - 6\sqrt{3}. \end{aligned}$$

first law of radicals

Adding the numerical coefficients of the like radicals we obtain $19\sqrt{3}$. Thus

$$3\sqrt{75} + \sqrt{300} - 6\sqrt{3} = 19\sqrt{3}.$$

Example 8. Simplify the following expression. Assume that x and y represent positive real numbers.

$$x\sqrt{50xy} - \sqrt{2x^3y}.$$

Solution: We begin by simplifying each radical appearing in the expression.

$$\begin{aligned} x\sqrt{50xy} - \sqrt{2x^3y} &= x\sqrt{25(2xy)} - \sqrt{x^2(2xy)} \\ &= x\sqrt{25}\sqrt{2xy} - \sqrt{x^2}\sqrt{2xy} \\ &= 5x\sqrt{2xy} - x\sqrt{2xy} \end{aligned}$$

Subtracting coefficients of like radicals, we see that

$$5x\sqrt{2xy} - x\sqrt{2xy} = 4x\sqrt{2xy}.$$

Thus

$$x\sqrt{50xy} - \sqrt{2x^3y} = 4x\sqrt{2xy}.$$

EXERCISES 7.4

For Exercises 1–34, simplify the radical expression by combining terms, as far as possible. See Examples 1–5, and 7.

1. $3\sqrt{2} + 5\sqrt{2}$

2. $4\sqrt{3} + 6\sqrt{3}$

3. $8\sqrt{5} - 6\sqrt{5}$

4. $9\sqrt{7} - 12\sqrt{7}$

5. $5\sqrt{6} + \sqrt{6}$

6. $8\sqrt{6} - \sqrt{6}$

7. $2\sqrt{13} - 5\sqrt{13} + 6\sqrt{13}$

8. $5\sqrt{11} - 9\sqrt{11} + 2\sqrt{11}$

9. $-3\sqrt{2x} - 5\sqrt{2x}$

10. $-12\sqrt{3y} - 4\sqrt{3y}$

11. $5\sqrt{7} - 2\sqrt{7} + 3\sqrt{5}$

12. $7\sqrt{15} + 4\sqrt{15} - 11\sqrt{5}$

13. $3\sqrt{5} + \sqrt{20}$

14. $2\sqrt{7} + \sqrt{28}$

15. $6\sqrt{45} - 2\sqrt{5}$

16. $6\sqrt{50} - 3\sqrt{2}$

17. $\sqrt{48} + \sqrt{75}$

18. $\sqrt{63} + \sqrt{28}$

19. $\sqrt{20} - \sqrt{80}$

20. $\sqrt{45} - \sqrt{125}$

21. $2\sqrt{75} + 3\sqrt{12}$

22. $5\sqrt{72} + 2\sqrt{98}$

23. $7\sqrt{90} + 3\sqrt{40}$

24. $4\sqrt{150} + 2\sqrt{96}$

25. $\sqrt{27} + \sqrt{12} + \sqrt{48}$

26. $\sqrt{20} + \sqrt{45} + \sqrt{180}$

27. $\sqrt{32} + \sqrt{50} - \sqrt{98}$

28. $\sqrt{44} + \sqrt{99} - \sqrt{176}$

29. $4\sqrt{54} - 10\sqrt{24} - 2\sqrt{96}$

30. $3\sqrt{28} - 8\sqrt{63} - 2\sqrt{112}$

31. $2\sqrt{90} - 6\sqrt{160} + 3\sqrt{250}$

32. $5\sqrt{52} - 3\sqrt{13} + 4\sqrt{117}$

33. $4\sqrt{48} + 5\sqrt{75} - \sqrt{12}$

34. $3\sqrt{250} - 5\sqrt{360} + 2\sqrt{40}$

For Exercises 35–50 simplify the radical expression. Assume that all variables represent positive real numbers. See Examples 6 and 8.

35. $\sqrt{9y} + \sqrt{49y}$

36. $\sqrt{64y} + \sqrt{81y}$

37. $y\sqrt{50} + \sqrt{2y^2}$

38. $x\sqrt{48} + \sqrt{75x^2}$

39. $\sqrt{98z} + 3\sqrt{2z} - \sqrt{162z}$

40. $\sqrt{300r} + \sqrt{27r} - \sqrt{12r}$

41. $x\sqrt{72x} + \sqrt{8x^3}$

42. $m\sqrt{48m} + \sqrt{27m^3}$

43. $\sqrt{12n^3} + n\sqrt{27n}$

44. $\sqrt{20r^3} + 4r\sqrt{80r}$

45. $\sqrt{25x^2y^3} + \sqrt{36x^2y^3}$

46. $\sqrt{200r^2s} + \sqrt{50r^2s}$

47. $\sqrt{x^2y^2z} + 2x\sqrt{y^2z}$

48. $3\sqrt{r^4s^3t} + rs\sqrt{r^2st}$

49. $\sqrt{y^3} + \sqrt{49y^3} + y\sqrt{25y}$

50. $z\sqrt{z^3} + 2\sqrt{z^5} - 3z^2\sqrt{9z}$

7.5 Multiplication and Division with Radical Expressions

In this section we discuss how to multiply and divide radical expressions when at least one of the expressions contains two terms. Examples of such expressions are

$$3 + \sqrt{2}, \qquad \sqrt{3} + \sqrt{2}, \qquad 2\sqrt{5} - \sqrt{7}.$$

We call a radical expression with two terms a *binomial radical expression*.

We consider multiplication first. The following examples show how to multiply and simplify products involving binomial radical expressions.

Example 1. Multiply and then simplify if necessary.
(a) $\sqrt{3}(2 + \sqrt{3})$. (b) $\sqrt{3}(\sqrt{18} + \sqrt{50})$.

Solution

(a) We perform the indicated multiplication by using the distributive property. Thus

$$\begin{aligned} \sqrt{3}(2+\sqrt{3}) &= \sqrt{3}\cdot 2+\sqrt{3}\sqrt{3} && \text{distributive property} \\ &= 2\sqrt{3}+3. && \sqrt{3}\sqrt{3}=\sqrt{9}=3 \end{aligned}$$

Notice in the final expression that we write $\sqrt{3}\cdot 2$ as $2\sqrt{3}$ to make it clear that 2 is not under the radical sign.

(b) Using the distributive property, we see that

$$\begin{aligned} \sqrt{3}(\sqrt{18}+\sqrt{50}) &= \sqrt{3}\sqrt{18}+\sqrt{3}\sqrt{50} \\ &= \sqrt{54}+\sqrt{150}. \end{aligned}$$

We simplify the last radical expression as follows:

$$\begin{aligned} \sqrt{54}+\sqrt{150} &= \sqrt{9\cdot 6}+\sqrt{25\cdot 6} \\ &= \sqrt{9}\sqrt{6}+\sqrt{25}\sqrt{6} && \text{first law of radicals} \\ &= 3\sqrt{6}+5\sqrt{6} \\ &= 8\sqrt{6}. && \text{addition of like radicals} \end{aligned}$$

Thus

$$\sqrt{3}(\sqrt{18}+\sqrt{50}) = 8\sqrt{6}.$$

Next we discuss how to multiply two binomial radical expressions such as $(\sqrt{2}+5)(\sqrt{2}+4)$. We could use the distributive property to multiply such expressions. However, in Section 3.5 we saw that the distributive property was the basis for the FOIL method of multiplying two binomials quickly. We can also apply the FOIL method to multiply two binomial expressions, whose terms involve radicals. The following examples show how to multiply two binomial radical expressions using the FOIL method.

Example 2. Multiply the binomial radical expressions and then simplify if necessary.

$$(\sqrt{2}+5)(\sqrt{2}+4).$$

Solution: Apply the FOIL method. Thus

$$\begin{aligned} & \quad\;\; \text{F} \quad + \quad \text{O} \quad + \quad \text{I} \quad + \quad \text{L} \\ (\sqrt{2}+5)(\sqrt{2}+4) &= \sqrt{2}\sqrt{2}+4\sqrt{2}+5\sqrt{2}+5(4) \\ &= 2+9\sqrt{2}+20 \\ &= 22+9\sqrt{2}. \end{aligned}$$

The last radical expression is in simplified form.

Example 3. Multiply and then simplify if necessary.

$$(2\sqrt{10} + 3)(5\sqrt{5} + \sqrt{2}).$$

Solution: We follow the pattern for the FOIL method.

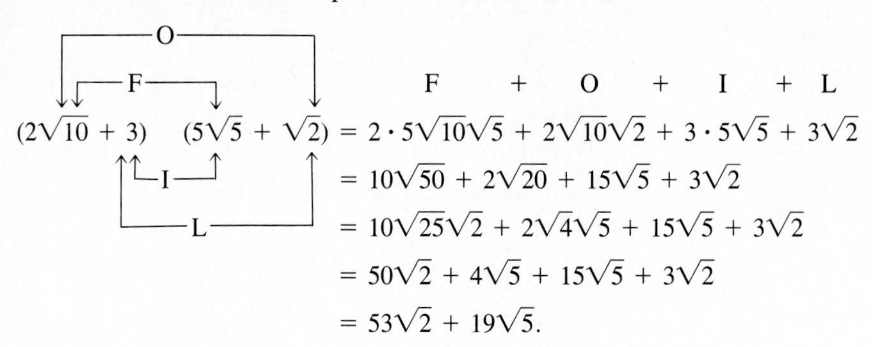

$$\begin{aligned}(2\sqrt{10} + 3)(5\sqrt{5} + \sqrt{2}) &= 2 \cdot 5\sqrt{10}\sqrt{5} + 2\sqrt{10}\sqrt{2} + 3 \cdot 5\sqrt{5} + 3\sqrt{2} \\ &= 10\sqrt{50} + 2\sqrt{20} + 15\sqrt{5} + 3\sqrt{2} \\ &= 10\sqrt{25}\sqrt{2} + 2\sqrt{4}\sqrt{5} + 15\sqrt{5} + 3\sqrt{2} \\ &= 50\sqrt{2} + 4\sqrt{5} + 15\sqrt{5} + 3\sqrt{2} \\ &= 53\sqrt{2} + 19\sqrt{5}.\end{aligned}$$

In the remaining examples of this section we shall need to multiply radical expressions of the form $(a + b)(a - b)$. In Section 3.5 we saw that this special product is the difference of two squares. That is,

$$(a + b)(a - b) = a^2 - b^2.$$

We use this formula in the examples that follow.

Example 4. Multiply and then simplify if necessary.
(a) $(3 + \sqrt{7})(3 - \sqrt{7})$. (b) $(\sqrt{10} + \sqrt{2})(\sqrt{10} - \sqrt{2})$.

Solution

(a) We use the formula given above because the given expression is of the form $(a + b)(a - b)$. Here $a = 3$ and $b = \sqrt{7}$; thus

$$\begin{aligned}(3 + \sqrt{7})(3 - \sqrt{7}) &= 3^2 - (\sqrt{7})^2 \\ &= 9 - 7 \qquad\qquad (\sqrt{7})^2 = 7 \\ &= 2.\end{aligned}$$

Therefore, $(3 + \sqrt{7})(3 - \sqrt{7}) = 2$.

(b) In this problem $a = \sqrt{10}$ and $b = \sqrt{2}$, and the formula gives us

$$\begin{aligned}(\sqrt{10} + \sqrt{2})(\sqrt{10} - \sqrt{2}) &= (\sqrt{10})^2 - (\sqrt{2})^2 \\ &= 10 - 2 \\ &= 8.\end{aligned}$$

The radical expressions $3 + \sqrt{7}$ and $3 - \sqrt{7}$ are called *conjugates* of each other. Similarly, $\sqrt{10} + \sqrt{2}$ and $\sqrt{10} - \sqrt{2}$ are conjugates. Notice in the last example that the product of conjugates is an expression without radicals. We use this property of conjugates to divide with binomial radical expressions as shown in the following examples.

Example 5. Perform the division and simplify if necessary.

$$\frac{12}{3-\sqrt{5}}.$$

Solution: We do the indicated division by rationalizing the denominator. Recall from Section 7.3 that rationalizing a denominator with a radical meant "eliminating" the radical from the denominator. So far, we have only rationalized the denominators of quotients with only one term in the denominator, such as $\frac{1}{\sqrt{2}}$. In order to rationalize a binomial denominator such as $3-\sqrt{5}$ in the given expression, we multiply numerator and denominator by its conjugate $3+\sqrt{5}$. We saw in Example 4 that the product of conjugates does not contain radicals. Thus

$$\begin{aligned}\frac{12}{(3-\sqrt{5})}\cdot\frac{(3+\sqrt{5})}{(3+\sqrt{5})} &= \frac{12(3+\sqrt{5})}{3^2-(\sqrt{5})^2}\\ &= \frac{12(3+\sqrt{5})}{9-5}\\ &= \frac{12(3+\sqrt{5})}{4}\\ &= 3(3+\sqrt{5}) = 9+3\sqrt{5}.\end{aligned}$$

Therefore,

$$\frac{12}{3-\sqrt{5}} = 9+3\sqrt{5}.$$

Example 6. Perform the division and simplify.

$$\frac{\sqrt{6}+9}{2\sqrt{2}+\sqrt{3}}.$$

Solution: To divide, we rationalize the denominator, $2\sqrt{2}+\sqrt{3}$, by multiplying numerator and denominator of the given expression by its conjugate $2\sqrt{2}-\sqrt{3}$.

$$\begin{aligned}\frac{(\sqrt{6}+9)(2\sqrt{2}-\sqrt{3})}{(2\sqrt{2}+\sqrt{3})(2\sqrt{2}-\sqrt{3})} &= \frac{2\sqrt{6}\sqrt{2}-\sqrt{6}\sqrt{3}+18\sqrt{2}-9\sqrt{3}}{(2\sqrt{2})^2-(\sqrt{3})^2}\\ &= \frac{2\sqrt{12}-\sqrt{18}+18\sqrt{2}-9\sqrt{3}}{2^2(\sqrt{2})^2-(\sqrt{3})^2}\\ &= \frac{2\sqrt{4}\sqrt{3}-\sqrt{9}\sqrt{2}+18\sqrt{2}-9\sqrt{3}}{4\cdot 2-3}\\ &= \frac{4\sqrt{3}-3\sqrt{2}+18\sqrt{2}-9\sqrt{3}}{8-3}\\ &= \frac{15\sqrt{2}-5\sqrt{3}}{5}\\ &= 3\sqrt{2}-\sqrt{3}.\end{aligned}$$

Thus

$$\frac{\sqrt{6} + 9}{2\sqrt{2} + \sqrt{3}} = 3\sqrt{2} - \sqrt{3}.$$

EXERCISES 7.5

For Exercises 1–16, multiply and then simplify if necessary. See Example 1.

1. $\sqrt{5}(4 + \sqrt{5})$
2. $\sqrt{2}(3 + \sqrt{2})$
3. $\sqrt{5}(\sqrt{5} + \sqrt{20})$
4. $\sqrt{8}(\sqrt{8} - \sqrt{2})$
5. $\sqrt{3}(\sqrt{15} - \sqrt{3})$
6. $\sqrt{2}(\sqrt{22} - \sqrt{2})$
7. $3\sqrt{6}(\sqrt{6} + \sqrt{24})$
8. $5\sqrt{12}(\sqrt{3} + \sqrt{12})$
9. $\sqrt{3}(2\sqrt{8} + \sqrt{18})$
10. $\sqrt{2}(5\sqrt{45} + \sqrt{80})$
11. $2\sqrt{7}(\sqrt{14} + 3\sqrt{7})$
12. $6\sqrt{5}(\sqrt{30} + 2\sqrt{5})$
13. $3\sqrt{5}(2\sqrt{5} + \sqrt{10})$
14. $2\sqrt{3}(5\sqrt{3} + \sqrt{21})$
15. $2\sqrt{2}(3\sqrt{6} - 5\sqrt{24})$
16. $3\sqrt{5}(4\sqrt{10} - 3\sqrt{40})$

For Exercises 17–38, multiply and then simplify if necessary. See Examples 2–4.

17. $(\sqrt{3} + 2)(\sqrt{3} + 4)$
18. $(\sqrt{5} + 1)(\sqrt{5} + 3)$
19. $(\sqrt{7} - 3)(\sqrt{7} - 5)$
20. $(\sqrt{11} - 2)(\sqrt{11} - 7)$
21. $(\sqrt{13} + 6)(\sqrt{13} - 2)$
22. $(\sqrt{6} + 8)(\sqrt{6} - 3)$
23. $(3\sqrt{6} + 1)(2\sqrt{2} + \sqrt{3})$
24. $(2\sqrt{5} + \sqrt{2})(3\sqrt{10} + 4)$
25. $(3\sqrt{5} + 2\sqrt{3})(4\sqrt{5} - 6\sqrt{3})$
26. $(3\sqrt{3} + \sqrt{7})(3\sqrt{3} - 2\sqrt{7})$
27. $(\sqrt{5} + \sqrt{3})^2$
28. $(\sqrt{11} - \sqrt{2})^2$
29. $(2\sqrt{21} + 3)(\sqrt{3} - 2\sqrt{7})$
30. $(2\sqrt{6} + 3)(3\sqrt{3} - \sqrt{2})$
31. $(4 + \sqrt{2})(4 - \sqrt{2})$
32. $(8 - \sqrt{14})(8 + \sqrt{14})$
33. $(5 + \sqrt{5})(5 - \sqrt{5})$
34. $(7 + \sqrt{7})(7 - \sqrt{7})$
35. $(\sqrt{7} + \sqrt{5})(\sqrt{7} - \sqrt{5})$
36. $(\sqrt{6} + \sqrt{3})(\sqrt{6} - \sqrt{3})$
37. $(\sqrt{11} + \sqrt{13})(\sqrt{11} - \sqrt{13})$
38. $(\sqrt{15} + \sqrt{19})(\sqrt{15} - \sqrt{19})$

For Exercises 39–54, divide and then simplify if necessary. See Examples 5 and 6.

39. $\dfrac{7}{3 + \sqrt{2}}$
40. $\dfrac{5}{2 + \sqrt{3}}$
41. $\dfrac{6}{2 - \sqrt{5}}$
42. $\dfrac{2}{4 - \sqrt{15}}$

43. $\dfrac{\sqrt{2}}{3 + \sqrt{8}}$
44. $\dfrac{\sqrt{3}}{5 + \sqrt{24}}$
45. $\dfrac{2}{\sqrt{6} - \sqrt{5}}$
46. $\dfrac{3}{\sqrt{8} - \sqrt{7}}$

47. $\dfrac{4}{\sqrt{5} + \sqrt{3}}$ 48. $\dfrac{12}{\sqrt{10} + \sqrt{7}}$ 49. $\dfrac{\sqrt{10} - 2}{\sqrt{5} + \sqrt{2}}$ 50. $\dfrac{\sqrt{6} - 2}{\sqrt{3} - \sqrt{2}}$

51. $\dfrac{\sqrt{15} - 1}{2\sqrt{3} + \sqrt{5}}$ 52. $\dfrac{3\sqrt{21} + 3}{3\sqrt{7} + \sqrt{3}}$ 53. $\dfrac{\sqrt{3} - 3\sqrt{5}}{4\sqrt{3} + 3\sqrt{5}}$ 54. $\dfrac{\sqrt{6} - 6}{2\sqrt{2} + \sqrt{3}}$

Equations Involving Square Root Radicals 7.6

In this section we are interested in solving equations such as

$$\sqrt{x + 1} = 4, \qquad \sqrt{2x - 1} + 2 = 5, \qquad \sqrt{3x + 7} - x = 1.$$

Notice in these equations that the variable x appears under the radical sign. We call such equations *radical equations*. We will confine our discussion to radical equations that involve only square roots.

To solve radical equations, we begin by eliminating the radicals from the equation by using the following property of equality.

> If two quantities are equal, then their squares are equal.

The statement is called the *squaring property of equality* and is written symbolically as follows.

Squaring Property of Equality

> If $a = b$, then $a^2 = b^2$.

Squaring both sides of an equation may introduce extra "solutions," which are sometimes referred to as "extraneous roots." For example, let us consider an equation that does not involve radicals, such as

$$x = 5.$$

Squaring both sides, we obtain

$$x^2 = 5^2 \qquad \text{or} \qquad x^2 = 25.$$

The last equation has two solutions $+5$ and -5. However, only 5 is the solution to the original equation, $x = 5$. Thus squaring both sides of the equation $x = 5$ introduced the extraneous root -5. Therefore, whenever we use the squaring property of equality to solve an equation, we must check all of the apparent solutions in the original equation. In the following examples we demonstrate the procedure for solving radical equations.

Example 1. Solve the radical equation

$$\sqrt{x+1} = 4.$$

Solution: To solve this equation, we square both sides of the equation in order to eliminate the radical.

$$(\sqrt{x+1})^2 = 4^2 \qquad \text{squaring property of equality}$$
$$x + 1 = 16$$
$$x = 15.$$

The solution of the last equation is 15. We check to see if 15 is the solution of the original equation.

$$\sqrt{x+1} = 4$$
$$\sqrt{15+1} = 4 \qquad \text{replace } x \text{ with } 15$$
$$\sqrt{16} = 4$$
$$4 = 4. \qquad \text{true}$$

Therefore, 15 is the desired solution.

Example 2. Solve the radical equation

$$\sqrt{2y-1} + 2 = 5.$$

Solution: First we must isolate the radical on one side of the equation in order to eliminate the radical by the squaring method. Therefore, we add -2 to both sides of the equation. Thus

$$\sqrt{2y-1} + 2 + (-2) = 5 + (-2)$$
$$\sqrt{2y-1} = 3.$$

Next we square both sides of the last equation.

$$(\sqrt{2y-1})^2 = 3^2 \qquad \text{squaring property of equality}$$
$$2y - 1 = 9$$
$$2y = 10$$
$$y = 5.$$

A check would reveal that 5 is the solution to the given equation.

Example 3. Solve the radical equation

$$\sqrt{t+8} = -3.$$

Solution: The equation

$$\sqrt{t+8} = -3$$

has no solution because the left side represents a nonnegative number. (Remember that the radical sign $\sqrt{\ }$ represents a nonnegative number.) However, suppose that we did not recognize this and attempted to solve the equation? What would result? Squaring both sides of the given equation, we have

$$\begin{aligned}(\sqrt{t + 8})^2 &= (-3)^2\\ t + 8 &= 9\\ t &= 1.\end{aligned}$$

The number 1 appears to be the desired solution. However, if we replace t with 1 in the original equation $\sqrt{t + 8} = -3$, we see that

$$\begin{aligned}\sqrt{1 + 8} &= -3\\ \sqrt{9} &= -3\\ 3 &= -3. \qquad \text{false}\end{aligned}$$

Therefore, 1 is an extraneous root. Thus we have confirmed that the original equation has no solution.

Example 4. Solve the equation

$$\sqrt{3x + 7} - x = 1.$$

Solution: We proceed as before by isolating $\sqrt{3x + 7}$ on the left side of the equation.

$$\sqrt{3x + 7} = x + 1. \qquad \text{add } x \text{ to both sides}$$

Next we square both sides.

$$\begin{aligned}(\sqrt{3x + 7})^2 &= (x + 1)^2\\ 3x + 7 &= x^2 + 2x + 1.\end{aligned}$$

The last equation is a quadratic equation. We rewrite this equation in standard form by adding $-3x - 7$ to both sides.

$$\begin{aligned}0 &= x^2 + 2x - 3x + 1 + (-7)\\ 0 &= x^2 - x - 6. \qquad \text{combine terms}\end{aligned}$$

We solve this quadratic equation by the factoring method.

$$0 = (x + 2)(x - 3)$$

or

$$(x + 2)(x - 3) = 0.$$

Setting both factors equal to 0, we have

$$\begin{aligned}x + 2 &= 0 \quad \text{or} \quad x - 3 = 0\\ x &= -2 \quad \text{or} \quad x = 3.\end{aligned}$$

Now we check these solutions in the original equation

$$\sqrt{3x + 7} - x = 1.$$

For -2:

$$\sqrt{3(-2)+7}-(-2)=1$$
$$\sqrt{-6+7}+2=1$$
$$\sqrt{1}+2=1$$
$$3=1. \quad \text{false}$$

For 3:

$$\sqrt{3(3)+7}-3=1$$
$$\sqrt{16}-3=1$$
$$4-3=1$$
$$1=1. \quad \text{true}$$

We conclude that the number 3 is the only solution to the original equation, and -2 is an extraneous root.

Example 5. Solve the equation

$$\sqrt{x^2+9}=5.$$

Solution: Square both sides of the given equation to remove the radical.

$$\left(\sqrt{x^2+9}\right)^2=5^2$$
$$x^2+9=25$$
$$x^2-16=0.$$

The last equation is a quadratic equation, which we solve by factoring.

$$x^2-16=0$$
$$(x+4)(x-4)=0$$
$$x=-4 \quad \text{or} \quad x=4.$$

A check would confirm that *both* -4 and 4 are solutions to the original equation.

$$\sqrt{x^2+9}=5.$$

Example 5 illustrates that the squaring method does not necessarily give us extraneous roots, when we obtain more than one solution.

The previous examples suggest the following procedures for solving radical equations:

1. Isolate the radical on one side of the equation.
2. Square both sides of the equation.
3. Solve the equation obtained in step 2 for the unknown.
4. Check the solutions obtained in step 3 in the original equation.

EXERCISES 7.6

For Exercises 1–42, solve the radical equation. Check all solutions in the original equation. See Examples 1–5.

1. $\sqrt{x}=2$ **2.** $\sqrt{y}=3$ **3.** $\sqrt{x+1}=3$

4. $\sqrt{y+6}=6$ **5.** $\sqrt{y-4}=5$ **6.** $\sqrt{y-3}=4$

7. $\sqrt{z - 10} = -2$
8. $\sqrt{z - 1} = -1$
9. $\sqrt{s + 5} = 4$
10. $\sqrt{s + 40} = 8$
11. $\sqrt{t - 5} = 0$
12. $\sqrt{t - 12} = 0$
13. $\sqrt{3x + 4} = 4$
14. $\sqrt{2x + 5} = 7$
15. $\sqrt{6x + 1} = 5$
16. $\sqrt{5x + 11} = 6$
17. $\sqrt{y} + 2 = 8$
18. $\sqrt{y} - 3 = -1$
19. $2\sqrt{x} - 1 = 5$
20. $3\sqrt{x} - 2 = 4$
21. $\sqrt{2x + 3} + 5 = 12$
22. $\sqrt{4x + 1} + 5 = 10$
23. $\sqrt{3x + 1} + 8 = 12$
24. $\sqrt{5x - 4} - 4 = 0$
25. $\sqrt{2x + 9} = \sqrt{3x + 1}$
26. $\sqrt{5x - 1} = \sqrt{2x + 8}$
27. $\sqrt{9x + 4} = \sqrt{7x + 14}$
28. $\sqrt{7x + 2} = \sqrt{3x + 6}$
29. $\sqrt{x + 3} = x + 1$
30. $\sqrt{4x + 12} = x + 3$
31. $\sqrt{8x + 4} = x - 2$
32. $\sqrt{2x + 1} = x - 7$
33. $\sqrt{3x + 6} + 4 = x$
34. $\sqrt{x - 1} + x = 3$
35. $\sqrt{4x + 1} - x = 1$
36. $\sqrt{2x + 4} + 2 = x$
37. $\sqrt{4x - 5} = 2x - 2$
38. $\sqrt{13x - 3} = 2x$
39. $\sqrt{x^2 + 144} = 13$
40. $\sqrt{x^2 + 441} = 29$
41. $\sqrt{x^2 + 225} = 17$
42. $\sqrt{x^2 + 900} = 50$

Key Words

Conjugates
Extraneous solutions
Index
Like radicals
Principal square root
Radical
Radical equations
Radical expressions
  Simplified form
Radical sign
Radicand
Rationalizing the denominator
Roots
  Square
  Cube
  Fourth
  Higher

CHAPTER 7 TEST

For Problems 1–6, find the roots.

1. $\sqrt{81}$
2. $-\sqrt{144}$
3. $\sqrt[3]{-27}$
4. $\sqrt[4]{81}$
5. $\sqrt[5]{-243}$
6. $\pm\sqrt{100}$
7. Which of the following radicals represent real numbers? Evaluate the radicals that represent real numbers.
   (a) $\sqrt{-16}$ (b) $-\sqrt{16}$ (c) $\sqrt[4]{16}$ (d) $\sqrt[4]{-16}$

For Problems 8–13, write the radical in simplified form.

8. $\sqrt{300}$
9. $\sqrt{50x^3}$
10. $\dfrac{\sqrt{250}}{\sqrt{10}}$
11. $\dfrac{\sqrt{50x^5y^3}}{\sqrt{2x^3}}$
12. $\sqrt{.64}$
13. $\dfrac{\sqrt[3]{250}}{\sqrt[3]{2}}$

For Problems 14 and 15, simplify the expression by rationalizing the denominator.

14. $\frac{12}{\sqrt{3}}$ **15.** $\frac{6}{\sqrt[3]{9}}$

For Problems 16–18, simplify the radical expression by combining terms.

16. $3\sqrt{11} + 5\sqrt{11} - 2\sqrt{11}$ **17.** $2\sqrt{27} + 3\sqrt{48}$ **18.** $\sqrt{20x^3} + x\sqrt{80x}$

For Problems 19–23, perform the operations indicated and simplify if necessary.

19. $4\sqrt{3}(\sqrt{6} + \sqrt{3})$ **20.** $(\sqrt{11} + 3)(\sqrt{11} - 3)$ **21.** $(\sqrt{8} - 3)(\sqrt{2} + 5)$

22. $(6\sqrt{2} + 5\sqrt{8})(3\sqrt{2} - \sqrt{8})$ **23.** $\frac{13}{4 - \sqrt{3}}$

For Problems 24–27, solve the radical equation.

24. $\sqrt{5x - 9} = 4$ **25.** $\sqrt{t - 1} + 5 = 7$ **26.** $\sqrt{4x + 1} = -9$ **27.** $\sqrt{2x + 1} = x - 7$

CHAPTER 8
Quadratic Equations

A landscaper is designing a 40 foot by 60 foot garden. She wants to design it with flower beds and paths as shown in the photograph. If the paths are to take up 16% of the area of the garden, then the solution to the quadratic equation $x^2 - 50x + 96 = 0$ will tell her how wide to make the paths.

PAR–NYC. © 1983

CHAPTER 8
CHAPTER 8
CHAPTER 8
CHAPTER 8
CHAPTER 8
CHAPTER 8
CHAPTER 8
CHAPTER 8
CHAPTER 8
CHAPTER 8
CHAPTER 8
CHAPTER 8
CHAPTER 8
CHAPTER 8
CHAPTER 8
CHAPTER 8
CHAPTER 8
CHAPTER 8
CHAPTER 8
CHAPTER 8
CHAPTER 8
CHAPTER 8
CHAPTER 8
CHAPTER 8
CHAPTER 8
CHAPTER 8
CHAPTER 8
CHAPTER 8
CHAPTER 8
CHAPTER 8
CHAPTER 8
CHAPTER 8
CHAPTER 8

In this chapter we study the general quadratic equation $ax^2 + bx + c = 0$ with $a \neq 0$. We use our knowledge of radicals to develop techniques for solving quadratic equations in one variable. In order to do this for any quadratic equation, we introduce complex numbers. We also discuss the graph of $y = ax^2 + bx + c$ and some applications that involve quadratic equations.

8.1 Extractions of Roots

Before considering the general quadratic equation, we solve some quadratic equations that have a special form. The equation

$$x^2 = 25$$

may be read as: "What numbers x, when squared, will equal 25?" From Chapter 7 we know that the solutions are

$$x = \sqrt{25} \quad \text{or} \quad x = -\sqrt{25}$$
$$= 5 \qquad\qquad = -5.$$

We also say that 5 and -5 are the *roots* of the quadratic equation $x^2 = 25$.

In general, when we solve an equation of the type $x^2 = a$ $(a \geq 0)$ by finding the square roots of a, we say that we are *extracting the roots* of the equation. For convenience, the solutions $x = \sqrt{a}$ or $x = -\sqrt{a}$ are often written as $x = \pm\sqrt{a}$. This is read as "x equals plus or minus the square root of a."

Example 1. Solve each equation.
(a) $x^2 = 9$. (b) $y^2 = 11$.

Solution

(a) Extracting the square roots of 9, we get

$$x = \sqrt{9} \quad \text{or} \quad x = -\sqrt{9}$$
$$= 3 \qquad\qquad x = -3.$$

Thus the solution set is $\{-3, 3\}$.

(b) Extracting the square roots of 11, we see that

$$y = \sqrt{11} \quad \text{or} \quad y = -\sqrt{11}.$$

Thus the solution set is $\{\pm\sqrt{11}\}$.

Sometimes, one side of a quadratic equation is a perfect square trinomial and the other side is a constant. When this occurs, the perfect square trinomial may be factored as the square of a binomial and the equation solved by extracting the square roots of the constant. We illustrate this approach in the following examples.

Example 2. Solve $(2x - 5)^2 = 49$.

Solution: The left-hand side of the equation is the square of a binomial. This equation can be solved by extracting the square roots of 49. This results in

$$2x - 5 = 7 \qquad \text{or} \qquad 2x - 5 = -7.$$

Thus

$$2x = 12 \qquad \text{or} \qquad 2x = -2.$$

Therefore,

$$x = 6 \qquad \text{or} \qquad x = -1.$$

CHECK

Let $x = 6$:

$$\begin{aligned} (2x - 5)^2 &= 49 \\ (2 \cdot 6 - 5)^2 &= 49 \\ (12 - 5)^2 &= 49 \\ 7^2 &= 49. \quad \text{true} \end{aligned}$$

Let $x = -1$:

$$\begin{aligned} (2x - 5)^2 &= 49 \\ [2(-1) - 5]^2 &= 49 \\ (-2 - 5)^2 &= 49 \\ (-7)^2 &= 49. \quad \text{true} \end{aligned}$$

We conclude that the solution set is $\{-1, 6\}$.

Example 3. Solve $y^2 - 6y + 9 = 10$.

Solution: The left-hand side, $y^2 - 6y + 9$, is a perfect square trinomial, which may be factored as

$$y^2 - 6y + 9 = (y - 3)^2.$$

Thus the original equation may be rewritten as

$$(y - 3)^2 = 10.$$

From this last equation we know that

$$y - 3 = \sqrt{10} \qquad \text{or} \qquad y - 3 = -\sqrt{10}.$$

Thus

$$y = 3 + \sqrt{10} \qquad \text{or} \qquad y = 3 - \sqrt{10}.$$

We now check the solution $3 - \sqrt{10}$ by substituting it in the original equation.

CHECK

$$\begin{aligned} (y - 3)^2 &= 10 \\ (3 - \sqrt{10} - 3)^2 &= 10 \\ (-\sqrt{10})^2 &= 10 \\ 10 &= 10. \quad \text{true} \end{aligned}$$

The reader should check that $3 + \sqrt{10}$ is the other solution.

Example 4. Solve $(6x - 3)^2 = 18$.

Solution: Finding the square roots of 18, we get

$$6x - 3 = \sqrt{18} = 3\sqrt{2} \quad \text{or} \quad 6x - 3 = -\sqrt{18} = -3\sqrt{2}.$$

Thus $6x = 3 + 3\sqrt{2}$ or $6x = 3 - 3\sqrt{2}$. Therefore,

$$x = \frac{3 + 3\sqrt{2}}{6} = \frac{1 + \sqrt{2}}{2} \quad \text{or} \quad x = \frac{3 - 3\sqrt{2}}{6} = \frac{1 - \sqrt{2}}{2}.$$

We check the solution $\frac{1 - \sqrt{2}}{2}$ in the given equation.

$$\left[6\left(\frac{1 - \sqrt{2}}{2}\right) - 3\right]^2 = 18$$

$$[3(1 - \sqrt{2}) - 3]^2 = 18$$

$$(-3\sqrt{2})^2 = 18$$

$$18 = 18. \qquad \text{true}$$

The reader should check that $\frac{1 + \sqrt{2}}{2}$ is also a solution. The solution set is

$$\left\{\frac{1 + \sqrt{2}}{2}, \frac{1 - \sqrt{2}}{2}\right\}$$

EXERCISES 8.1

Solve each equation by finding the square roots of both sides of the equation. Simplify all radicals and check your answers.

For Exercises 1–20, see Example 1.

1. $z^2 = 16$ **2.** $y^2 = 36$ **3.** $x^2 = 81$ **4.** $w^2 = 64$ **5.** $x^2 = 144$

6. $y^2 = 400$ **7.** $y^2 = 3$ **8.** $z^2 = 5$ **9.** $w^2 = 17$ **10.** $x^2 = 23$

11. $x^2 = 29$ **12.** $y^2 = 47$ **13.** $w^2 = 8$ **14.** $z^2 = 24$ **15.** $x^2 = 40$

16. $y^2 = 48$ **17.** $w^2 = 50$ **18.** $z^2 = 54$ **19.** $x^2 = 200$ **20.** $y^2 = 250$

For Exercises 21–30, see Example 2.

21. $(x + 1)^2 = 25$ **22.** $(y - 1)^2 = 36$ **23.** $(w + 3)^2 = 16$ **24.** $(z - 4)^2 = 49$

25. $(x-5)^2 = 100$ **26.** $(w+3)^2 = 81$ **27.** $(2x-4)^2 = 64$ **28.** $(2z-6)^2 = 100$

29. $(3y-3)^2 = 81$ **30.** $(4w-2)^2 = 36$

For Exercises 31–36, see Example 3.

31. $x^2 - 2x + 1 = 6$ **32.** $y^2 + 2y + 1 = 5$ **33.** $w^2 + 6w + 9 = 15$

34. $z^2 - 4z + 4 = 11$ **35.** $x^2 + 10x + 25 = 17$ **36.** $y^2 - 8y + 16 = 7$

For Exercises 37–50, see Example 4.

37. $(2x-4)^2 = 20$ **38.** $(2y+6)^2 = 24$ **39.** $(4x+12)^2 = 32$

40. $(4w-8)^2 = 48$ **41.** $(3z-6)^2 = 18$ **42.** $(3y+9)^2 = 27$

43. $(5z+10)^2 = 75$ **44.** $(5w-10)^2 = 50$ **45.** $(2x-1)^2 = 8$

46. $(2x+1)^2 = 12$ **47.** $(3w-2)^2 = 18$ **48.** $(4z+3)^2 = 32$

49. $(5x+1)^2 = 50$ **50.** $(6x-5)^2 = 48$

8.2 Solving Quadratic Equations by Completing the Square

In the examples of Section 8.1 we were able to extract square roots because the expression on the left-hand side of the equation was always a perfect square. For instance, in Example 3 of Section 8.1,

$$y^2 - 6y + 9 = (y-3)^2$$

is a *perfect square trinomial.*

Recall that perfect square trinomials can be expressed as the square of a binomial. For example, $x^2 + 10x + 25 = (x+5)^2$ is a perfect square trinomial. We note that if a perfect square trinomial has a leading coefficient of 1, then its constant term is equal to the square of one-half of the coefficient of the first-degree term. For example, in the perfect square trinomial $y^2 - 20y + 100 = (y-10)^2$, the constant term is 100 and -20 is the coefficient of the first-degree term. Observe that $100 = (-10)^2$ and $-10 = \frac{1}{2}(-20)$.

Example 1. Which of the following are perfect square trinomials?
(a) $x^2 + 8x + 16$. (b) $y^2 - 10y + 100$.

Solution

(a) Half of 8 is 4. Since $4^2 = 16$, this is a perfect square trinomial.

(b) Half of -10 is -5. Since $(-5)^2 = 25$ and $25 \neq 100$, this is not a perfect square trinomial.

In the remainder of this section we show how some quadratic equations can be written in the form

$$\begin{array}{ccc} (x + a)^2 & = & b. \\ \begin{pmatrix} \text{square of} \\ \text{a binomial} \end{pmatrix} & = & \begin{pmatrix} \text{nonnegative} \\ \text{constant} \end{pmatrix} \end{array}$$

In fact, our goal is to write a given trinomial in the form shown above. Once we have written a quadratic equation in this form we can solve it by extracting the roots as explained in Section 8.1. Rewriting a quadratic equation so that the left-hand side is a perfect square trinomial is called *completing the square*.

Example 2. Solve $y^2 + 10y = 20$ by completing the square.

Solution: In order to make the left-hand side a perfect square, we need to add the square of half the coefficient of the first-degree term, $10y$, to each side. Thus we add $\left(\frac{10}{2}\right)^2 = (5)^2 = 25$ to each side. The result is

$$y^2 + 10y + 25 = 20 + 25.$$

But $y^2 + 10y + 25 = (y + 5)^2$, and thus we have

$$(y + 5)^2 = 45.$$

Next we solve this equation by extracting the square roots as in Section 8.1. Therefore,

$$\begin{aligned} y + 5 &= \sqrt{45} \\ &= 3\sqrt{5} \end{aligned} \qquad \text{or} \qquad \begin{aligned} y + 5 &= -\sqrt{45} \\ &= -3\sqrt{5}. \end{aligned}$$

It follows that

$$-5 + 3\sqrt{5} \qquad \text{and} \qquad -5 - 3\sqrt{5}$$

are the solutions to the original equation.

Example 3. Solve $z^2 + 8z + 14 = 0$ by completing the square.

Solution: Since $z^2 + 8z + 14$ is not a perfect square trinomial, we cannot write it as the square of a binomial. We begin by adding -14 to each side of the equation. Thus

$$z^2 + 8z + 14 - 14 = 0 - 14$$

or

$$z^2 + 8z = -14.$$

Now that the constant term has been isolated on the right-hand side, we complete the square of the left-hand side by adding

$$\left[\frac{1}{2}(8)\right]^2 = 4^2 = 16$$

to both sides of the equation. This results in

$$\begin{aligned} z^2 + 8z + 16 &= -14 + 16 \\ &= 2. \end{aligned}$$

But the left-hand side is a perfect square trinomial and thus the equation can be rewritten as

$$(z + 4)^2 = 2.$$

If we extract the square roots of 2, we get

$$z + 4 = \sqrt{2} \qquad \text{or} \qquad z + 4 = -\sqrt{2}.$$

Thus

$$z = -4 + \sqrt{2} \qquad \text{or} \qquad z = -4 - \sqrt{2}.$$

The solutions are $-4 \pm \sqrt{2}$.

The procedures that we have been applying assume that the leading coefficient is 1. *If the leading coefficient of a quadratic equation is not 1, we divide each term of the equation by this coefficient before completing the square.* The next two examples illustrate this procedure.

Example 4. Solve $3x^2 - 12x = 9$ by completing the square.

Solution: First divide each term by the leading coefficient 3. The result of this division is

$$x^2 - 4x = 3.$$

Now complete the square by adding

$$\left(\frac{-4}{2}\right)^2 = (-2)^2 = 4$$

to each side of the equation:

$$x^2 - 4x + 4 = 3 + 4.$$

Factoring the left-hand side, we get

$$(x - 2)^2 = 7.$$

Finally, we find the square roots of 7. Thus

$$x - 2 = \sqrt{7} \quad \text{or} \quad x - 2 = -\sqrt{7}.$$

Therefore,

$$x = 2 + \sqrt{7} \quad \text{or} \quad x = 2 - \sqrt{7}.$$

The reader should check that these are solutions to the original equation.

Example 5. Solve $2y^2 + 3y = 1$ using the method of completing the square.

Solution: Dividing each term by 2, we get

$$y^2 + \frac{3}{2}y = \frac{1}{2}.$$

Next we complete the square by adding

$$\left[\frac{1}{2}\left(\frac{3}{2}\right)\right]^2 = \left(\frac{3}{4}\right)^2 = \frac{9}{16}$$

to each side of the equation.

$$y^2 + \frac{3}{2}y + \frac{9}{16} = \frac{1}{2} + \frac{9}{16}$$

$$\left(y + \frac{3}{4}\right)^2 = \frac{17}{16}. \qquad \left(\frac{1}{2} = \frac{8}{16}\right)$$

Solving the last equation, we get

$$y + \frac{3}{4} = \sqrt{\frac{17}{16}} = \frac{\sqrt{17}}{4} \quad \text{or} \quad y + \frac{3}{4} = -\sqrt{\frac{17}{16}} = \frac{-\sqrt{17}}{4}.$$

Thus the solutions are

$$y = -\frac{3}{4} + \frac{\sqrt{17}}{4} = \frac{-3 + \sqrt{17}}{4} \quad \text{or} \quad y = -\frac{3}{4} - \frac{\sqrt{17}}{4} = \frac{-3 - \sqrt{17}}{4}.$$

Notice that these solutions may be written compactly as

$$\frac{-3 \pm \sqrt{17}}{4}.$$

The following steps summarize the techniques used in the preceding examples for solving a quadratic equation by completing the square.

Step 1. Write the equation so that all variable terms are on one side and all constant terms are on the other side.

Step 2. If the leading coefficient is not 1, divide both sides of the equation by this coefficient.

Step 3. Take half of the coefficient of the first-degree term and square this product. Next, add the result of the squaring to both sides of the equation.

Step 4. One side of the equation is now a perfect square trinomial. Write it as the square of a binomial. Solve the resulting equation by extracting roots.

Step 5. Check your answer in the original equation.

EXERCISES 8.2

For Exercises 1–8, state whether or not the expression is a perfect square trinomial. See Example 1.

1. $x^2 + 8x + 16$ **2.** $y^2 - 6y + 9$ **3.** $w^2 - 14w + 49$

4. $z^2 + 16z + 64$ **5.** $x^2 + 4x - 4$ **6.** $w^2 - 10w - 25$

7. $y^2 + 10y + 100$ **8.** $z^2 - 2z + 2$

For Exercises 9–16, what number should be added to the expression to make it a perfect square trinomial? See Example 2.

9. $x^2 - 10x$ **10.** $y^2 + 2y$ **11.** $w^2 + 16w$ **12.** $z^2 - 12z$

13. $x^2 + 20x$ **14.** $y^2 - 30y$ **15.** $w^2 + 14w$ **16.** $z^2 - 18z$

For Exercises 17–46, solve the equation by completing the square. See Examples 2–5.

17. $x^2 + 6x = 1$ **18.** $y^2 - 8y = 4$ **19.** $w^2 - 4w = 1$

20. $z^2 + 10z = 3$ **21.** $x^2 - 16x = 4$ **22.** $y^2 + 20y = 25$

23. $w^2 + 18w = 9$ **24.** $z^2 - 14z = 1$ **25.** $x^2 - 2x - 3 = 0$

26. $y^2 + 4y - 3 = 0$ **27.** $w^2 + 6w - 11 = 0$ **28.** $z^2 - 8z + 1 = 0$

29. $x^2 - 14x + 1 = 0$ **30.** $y^2 + 18y + 6 = 0$ **31.** $w^2 + 22w + 21 = 0$

32. $z^2 - 24z - 6 = 0$ **33.** $2x^2 + 4x - 6 = 0$ **34.** $4y^2 + 12y - 4 = 0$

35. $3w^2 - 18w + 3 = 0$ **36.** $5z^2 - 30z + 10 = 0$ **37.** $2y^2 + y - 4 = 0$

38. $2w^2 + 3w - 6 = 0$ **39.** $3x^2 + 6x + 3 = 0$ **40.** $4w^2 + 8w - 4 = 0$

41. $4y^2 - 2y - 1 = 0$ **42.** $8y^2 + 4y - 4 = 0$ **43.** $5x^2 + x - 2 = 0$

44. $5z^2 - 3z - 4 = 0$ **45.** $3y^2 - 2y - 4 = 0$ **46.** $2y^2 + 7y + 3 = 0$

8.3 The Quadratic Formula

In this section we derive one of the best-known and most widely used formulas of mathematics, the *quadratic formula*. This formula will enable us to solve any quadratic equation. Because of its importance, the quadratic formula should be committed to memory.

The Quadratic Formula

> If $ax^2 + bx + c = 0$ and $a \neq 0$, then the solutions to the equation are
>
> $$x = \frac{-b + \sqrt{b^2 - 4ac}}{2a} \quad \text{or} \quad x = \frac{-b - \sqrt{b^2 - 4ac}}{2a}.$$

We develop this formula because some quadratics are very difficult to factor and completing the square is often a very tedious procedure. We will establish the quadratic formula by completing the square on the general quadratic equation

$$ax^2 + bx + c = 0, \qquad a \neq 0.$$

Proceeding as we did in Section 8.2, we first add $-c$ to both sides of the equation. This results in

$$ax^2 + bx + c - c = 0 - c$$

or

$$ax^2 + bx = -c.$$

Next, we divide each term by a. The result of this division is

$$x^2 + \frac{b}{a}x = \frac{-c}{a}.$$

In order to complete the square of the left-hand side of the last equation, we add the square of half the coefficient of the first-degree term to both sides. That is, we add

$$\left[\left(\frac{1}{2}\right)\left(\frac{b}{a}\right)\right]^2 = \frac{b^2}{4a^2}$$

to both sides of the equation and get

$$x^2 + \frac{b}{a}x + \frac{b^2}{4a^2} = \frac{-c}{a} + \frac{b^2}{4a^2}.$$

But the left-hand side of this equation is a perfect square trinomial. Thus it can be written as the square of a binomial. That is,

$$\left(x + \frac{b}{2a}\right)^2 = \frac{-c}{a} + \frac{b^2}{4a^2}.$$

Notice that we can add the fractions on the right-hand side of this equation if we first rewrite them with a common denominator of $4a^2$. That is,

$$\begin{aligned}\left(x + \frac{b}{2a}\right)^2 &= \frac{-c}{a} + \frac{b^2}{4a^2}\\ &= \frac{-c}{a} \cdot \frac{4a}{4a} + \frac{b^2}{4a^2}\\ &= \frac{-4ac}{4a^2} + \frac{b^2}{4a^2}\\ &= \frac{b^2 - 4ac}{4a^2}.\end{aligned}$$

Thus we may rewrite $ax^2 + bx + c = 0$ as

$$\left(x + \frac{b}{2a}\right)^2 = \frac{b^2 - 4ac}{4a^2}.$$

Finally, we extract the square roots and obtain

$$\begin{aligned}x + \frac{b}{2a} &= \pm\sqrt{\frac{b^2 - 4ac}{4a^2}}\\ &= \pm\frac{\sqrt{b^2 - 4ac}}{\sqrt{4a^2}}\\ &= \pm\frac{\sqrt{b^2 - 4ac}}{2a}.\end{aligned}$$

Adding $-\dfrac{b}{2a}$ to each side, we get

$$x = \frac{-b}{2a} \pm \frac{\sqrt{b^2 - 4ac}}{2a}.$$

The formula for finding the two solutions may be expressed compactly as

$$x = \frac{-b \pm \sqrt{b^2 - 4ac}}{2a}.$$

In the following examples we will use the quadratic formula to solve quadratic equations.

Example 1. Use the quadratic formula to solve $x^2 - 2x - 5 = 0$.

Solution: For this equation, $a = 1$, $b = -2$, and $c = -5$. Using these values in the quadratic formula, we get

$$x = \frac{-b \pm \sqrt{b^2 - 4ac}}{2a}$$

$$= \frac{-(-2) \pm \sqrt{(-2)^2 - 4(1)(-5)}}{2(1)}$$

$$= \frac{2 \pm \sqrt{4 + 20}}{2}$$

$$= \frac{2 \pm \sqrt{24}}{2}$$

$$= \frac{2 \pm 2\sqrt{6}}{2}$$

$$= 1 \pm \sqrt{6}.$$

We conclude that the solutions are $1 + \sqrt{6}$ and $1 - \sqrt{6}$.

Example 2. Solve $2x^2 = 3x + 1$ by using the quadratic formula.

Solution: In order to use the quadratic formula, we first write the equation in standard form as

$$2x^2 - 3x - 1 = 0.$$

Now that the equation is in standard form, it is easy to see that $a = 2$, $b = -3$, and $c = -1$. Thus

$$x = \frac{-b \pm \sqrt{b^2 - 4ac}}{2a}$$

$$= \frac{-(-3) \pm \sqrt{(-3)^2 - 4(2)(-1)}}{2(2)}$$

$$= \frac{3 \pm \sqrt{17}}{4}.$$

Thus the solutions are

$$\frac{3 + \sqrt{17}}{4} \quad \text{and} \quad \frac{3 - \sqrt{17}}{4}.$$

Example 3. Solve $(x + 4)(x + 5) = 10$ for x.

Solution: Notice that the right-hand side of the given equation does not equal 0. Before we can use the quadratic formula, we must put our equation in standard form. Note that if we multiply the binomials and collect like terms, we get

$$\begin{aligned} (x + 4)(x + 5) &= 10 \\ x^2 + 9x + 20 &= 10 \\ x^2 + 9x + 10 &= 0. \end{aligned}$$

In this form we see that

$$a = 1, \qquad b = 9, \qquad c = 10.$$

Thus

$$\begin{aligned} x &= \frac{-b \pm \sqrt{b^2 - 4ac}}{2a} \\ &= \frac{-9 \pm \sqrt{9^2 - 4(1)(10)}}{2(1)} \\ &= \frac{-9 \pm \sqrt{41}}{2}. \end{aligned}$$

The solutions are

$$\frac{-9 + \sqrt{41}}{2} \quad \text{and} \quad \frac{-9 - \sqrt{41}}{2}.$$

EXERCISES 8.3

For Exercises 1–20, use the quadratic formula to solve the equation. See Example 1.

1. $x^2 + 3x + 2 = 0$ **2.** $y^2 + 4y + 1 = 0$ **3.** $w^2 + 2w - 5 = 0$

4. $z^2 - 3z - 3 = 0$ **5.** $x^2 + 3x + 1 = 0$ **6.** $y^2 - 2y - 5 = 0$

7. $w^2 - 4w - 2 = 0$ **8.** $z^2 - 7z - 5 = 0$ **9.** $x^2 - 10x + 12 = 0$

10. $y^2 + 8y + 6 = 0$ **11.** $3x^2 - x - 1 = 0$ **12.** $2y^2 - 3y - 7 = 0$

13. $2w^2 - w - 2 = 0$ **14.** $5z^2 - 3z - 1 = 0$ **15.** $2x^2 - 3x - 5 = 0$

16. $4y^2 - 3y - 1 = 0$ **17.** $2w^2 + 5w + 2 = 0$ **18.** $5z^2 - 8z + 1 = 0$

19. $7x^2 - 2x - 1 = 0$ **20.** $10w^2 - 5w - 1 = 0$

For Exercises 21–30, rewrite the equation in standard form and then solve using the quadratic formula. See Example 2.

21. $2x^2 - 3x = 2$ **22.** $2y^2 - 3y = 5$ **23.** $w^2 = 2w - 1$ **24.** $2z^2 = 6z - 3$

25. $x^2 + 5x = -2$ **26.** $y^2 = 9y - 5$ **27.** $w^2 - 2w = 3$ **28.** $3z^2 - 4z = 2$

29. $6x^2 = x + 2$ **30.** $7y^2 = 2y + 1$

For Exercises 31–40, first write the equation in standard form and then solve using the quadratic formula. See Example 3.

31. $(x - 2)(x + 3) = 1$ **32.** $(y + 1)(y + 6) = 10$ **33.** $(w + 2)(w - 1) = 1$

34. $(z - 4)(z - 1) = -1$ **35.** $(x + 5)(x + 1) = 10$ **36.** $(y - 4)(y - 2) = 12$

37. $(2w + 1)(w - 1) = 1$ **38.** $(x + 2)(3x - 1) = -1$ **39.** $(2x - 1)(3x + 1) = 2$

40. $(3y + 2)(5y - 4) = -7$

8.4 Complex Numbers

In Section 8.3, each time we used the quadratic formula, the quantity under the radical sign was a nonnegative number. If we apply the quadratic formula to

$$x^2 - x + 1 = 0,$$

the result is

$$x = \frac{-(-1) \pm \sqrt{(1)^2 - 4(1)(1)}}{2(1)}$$

$$= \frac{1 \pm \sqrt{-3}}{2}.$$

Note that the number under the radical sign is -3, a negative number. But we know there is no real number a such that

$$a^2 = -3.$$

Thus, if the equation

$$x^2 - x + 1 = 0$$

has a solution, it is not a real number because $\sqrt{-3}$ is not a real number.

In order to be capable of solving any quadratic equation, we must discuss radicals when the number under a radical sign is negative. To this end we introduce a new set of numbers called *complex numbers*. Complex numbers will allow us to take the square root of a negative real number. We begin with the following two definitions.

> The complex number i has the property $i = \sqrt{-1}$ and therefore $i^2 = -1$.

> Any number that can be written as
>
> $$a + bi,$$
>
> where a and b are real numbers, is called a complex number.

It follows that

$$5 + 2i, \qquad 1 - 3i, \qquad \frac{1}{2} + \frac{4}{3}i$$

are all complex numbers. Notice also that the real number $6 = 6 + 0 \cdot i$ is a complex number. Since any real number a can be written as

$$a = a + 0 \cdot i,$$

every real number is a complex number.

Whenever we encounter a square root of a negative number, we will express it in terms of i. For example,

$$\sqrt{-3} = i\sqrt{3}$$

and

$$\sqrt{-9} = i\sqrt{9} = i \cdot 3 = 3i.$$

In general, we have the following definition.

> If $a > 0$, then $\sqrt{-a} = i\sqrt{a}$.

It is customary to write the factor i on the left of the radical sign so that it is clear that i is not part of the radicand. If a is a perfect square, we can eliminate the radical. In such instances, we usually write i as the second, or right-hand factor. For example,

$$\begin{aligned} \sqrt{-64} &= i\sqrt{64} \\ &= 8i. \end{aligned}$$

Similarly, the number

$$1 + \sqrt{-25}$$

may be written as

$$1 + i\sqrt{25} = 1 + 5i.$$

Example 1. Write each of the following in the form $a + bi$, where a and b are real numbers.
(a) $\sqrt{-100}$. (b) $5 + \sqrt{-36}$. (c) $-\sqrt{-4}$.

Solution

(a) $\sqrt{-100} = i\sqrt{100}$
$= i \cdot 10$
$= 10i.$

Thus $\sqrt{-100} = 0 + 10i$.

(b) $5 + \sqrt{-36} = 5 + i\sqrt{36}$
$= 5 + 6i.$
(c) $-\sqrt{-4} = -i\sqrt{4} = -i2 = -2i.$

Complex numbers are added and subtracted using the ordinary rules of algebra. For example.

$$\begin{aligned} (3 + 5i) + (7 - 3i) &= (3 + 7) + (5i - 3i) && \text{associative property} \\ &= (3 + 7) + (5 - 3)i && \text{distributive property} \\ &= 10 + 2i. && \text{addition} \end{aligned}$$

Therefore,

$$(3 + 5i) + (7 - 3i) = 10 + 2i.$$

Example 2. Subtract $4 + 2i$ from $6 + 5i$.

Solution

$$\begin{aligned} (6 + 5i) - (4 + 2i) &= 6 + 5i - 4 - 2i && \text{subtraction} \\ &= (6 - 4) + (5i - 2i) && \text{associative property} \\ &= (6 - 4) + (5 - 2)i && \text{distributive property} \\ &= 2 + 3i. && \text{addition} \end{aligned}$$

Thus

$$(6 + 5i) - (4 + 2i) = 2 + 3i.$$

When we multiply two complex numbers, we proceed as we did when we multiplied two binomials. However, we *always replace* i^2 *with* -1. Thus

$$\begin{aligned} (3 - 5i)(4 + 2i) &= 3 \cdot 4 + 3 \cdot 2i + (-5i)4 + (-5i)(2i) && \text{FOIL} \\ &= 12 + 6i - 20i - 10i^2 && \text{multiplication} \\ &= 12 + 6i - 20i - 10(-1) && i^2 = -1 \\ &= 22 - 14i && \text{combine like terms} \end{aligned}$$

Therefore,

$$(3 - 5i)(4 + 2i) = 22 - 14i.$$

Example 3. Find the product $3i(2 - 6i)$.

Solution

$$\begin{aligned} 3i(2 - 6i) &= 6i - 18i^2 && \text{distributive property} \\ &= 6i - 18(-1) && i^2 = -1 \\ &= 18 + 6i \end{aligned}$$

Thus

$$3i(2 - 6i) = 18 + 6i.$$

In order to divide complex numbers, we use the complex conjugate of a number.

> The *complex conjugate* of $a + bi$ is $a - bi$.

When dividing a number by a complex number, we represent the quotient as a fraction. In order to express this quotient in the form $a + bi$, we multiply the numerator and denominator by the complex conjugate of the denominator and then simplify if possible. For example, to write the quotient $\frac{5i}{2 + i}$ in the form $a + bi$, we proceed as follows:

$$\frac{5i}{2 + i} \cdot \frac{(2 - i)}{(2 - i)} \qquad \text{multiply numerator and denominator by } 2 - i$$

$$= \frac{10i - 5i^2}{4 - i^2} \qquad \text{multiplication}$$

$$= \frac{10i - 5(-1)}{4 - (-1)} \qquad i^2 = -1$$

$$= \frac{5 + 10i}{5} \qquad \text{simplify}$$

$$= \frac{5}{5} + \frac{10}{5}i$$

$$= 1 + 2i.$$

Thus

$$\frac{3i}{2 + i} = 1 + 2i,$$

where $1 + 2i$ has the form $a + bi$ with $a = 1$ and $b = 2$. Whenever the numerator and denominator of the quotient of two complex numbers is multiplied by the complex conjugate of the denominator, the quotient will have the form $a + bi$.

Example 4. Divide $5 - i$ by $2 + 3i$. Express your answer in the form $a + bi$.

Solution

$$\frac{5 - i}{2 + 3i} = \frac{(5 - i)}{(2 + 3i)} \cdot \frac{(2 - 3i)}{(2 - 3i)} \qquad \text{multiply numerator and denominator by } 2 - 3i$$

$$= \frac{10 - 15i - 2i + 3i^2}{4 - 9i^2} \qquad \text{FOIL}$$

$$= \frac{7 - 17i}{13} \qquad i^2 = -1\text{; combine like terms}$$

$$= \frac{7}{13} - \frac{17}{13}i.$$

EXERCISES 8.4

For Exercises 1–10, write the number in the form a + bi, where a and b are real numbers. See Example 1.

1. $3 + \sqrt{-1}$ **2.** $5 - \sqrt{-1}$ **3.** $6 + \sqrt{-16}$ **4.** $7 - \sqrt{-9}$

5. $-4 + \sqrt{-4}$ **6.** $11 + \sqrt{-25}$ **7.** $-\sqrt{64}$ **8.** $\sqrt{-64}$

9. $\sqrt{-81}$ **10.** $-\sqrt{81}$

For Exercises 11–20, perform the additions and subtractions indicated. See Example 2.

11. $(7 + i) + (2 + 5i)$ **12.** $(4 + 2i) + (5 + i)$

13. $(6 - 5i) + (3 + 2i)$ **14.** $(7 + i) + (4 - 3i)$

15. $(8 + 5i) - (3 + 2i)$ **16.** $(11 + 4i) - (7 + i)$

17. $(3 + 2i) - (5 + i)$ **18.** $(7 + 2i) - (4 + 6i)$

19. $(4 - 3i) + (2 - 4i) + 6i$ **20.** $(12 - 7i) - (7 + 2i) + 3i$

For Exercises 21–30, multiply the two complex numbers. See Example 3 and the discussion preceding it.

21. $(4i)(6 + 3i)$ **22.** $(7 - 2i)(5i)$

23. $(4 + i)(2 + i)$ **24.** $(3 + 2i)(1 + i)$

25. $(3 + 2i)(2 + 3i)$ **26.** $(4 + 3i)(2 + 5i)$

27. $(4 - i)(9 + i)$ **28.** $(3 - i)(2 + i)$

29. $(4 + 3i)(7 - 2i)$ **30.** $(6 - 5i)(2 - 3i)$

For Exercises 31–40, find the given quotient. Express your answers in the form a + bi. See Example 4.

31. $\dfrac{5i}{2 + i}$ **32.** $\dfrac{3i}{4 - i}$ **33.** $\dfrac{-i}{4 + 3i}$ **34.** $\dfrac{-2i}{5 - 3i}$ **35.** $\dfrac{10}{1 + 3i}$

36. $\dfrac{-7}{5 - 2i}$ **37.** $\dfrac{3 + 2i}{2 + 3i}$ **38.** $\dfrac{4 - 3i}{3 + 4i}$ **39.** $\dfrac{7 + 3i}{2 - 3i}$ **40.** $\dfrac{5 + 6i}{3 + 5i}$

Solve Exercises 41 and 42.

41. Find the first eight powers of i. HINT: We know that $i^1 = i$ and $i^2 = -1$; to find i^3, note that $i^3 = i^2 \cdot i$.

42. Based on your answers to number 41, evaluate i^{23}, i^{32}, i^{41}, and i^{46}. Can you generalize to i^n, where n is a positive integer?

8.5 Solving Quadratic Equations: The General Case

We are now ready to solve any quadratic equation $ax^2 + bx + c = 0$, where a, b, and c are real numbers. Recall that if we use the quadratic formula to solve

$$ax^2 + bx + c = 0,$$

the result is

$$x = \frac{-b \pm \sqrt{b^2 - 4ac}}{2a}.$$

Our knowledge of complex numbers will permit us to use the quadratic formula even when the radicand $b^2 - 4ac$ is a negative number. We note that the radicand $b^2 - 4ac$ is called the *discriminant* of the equation $ax^2 + bx + c = 0$.

In this section we consider two examples of quadratic equations with complex solutions. The exercises at the end of the section do not all have complex solutions.

Example 1. Solve $x^2 - 4x + 5 = 0$.

Solution: If we apply the quadratic formula to this equation, we see that $a = 1$, $b = -4$, and $c = 5$. Thus

$$\begin{aligned} x &= \frac{-b \pm \sqrt{b^2 - 4ac}}{2a} \\ &= \frac{(-4) \pm \sqrt{(-4)^2 - 4(1)(5)}}{2(1)} \\ &= \frac{4 \pm \sqrt{16 - 20}}{2} \\ &= \frac{4 \pm \sqrt{-4}}{2} \\ &= \frac{4}{2} \pm \frac{i\sqrt{4}}{2} \\ &= 2 \pm i. \end{aligned}$$

Thus the solutions are $2 + i$ and $2 - i$.

Notice that the solutions to the equation in Example 1 were complex conjugates. Whenever a quadratic equation has a complex solution $a + bi$, the complex conjugate $a - bi$ will also be a solution.

Example 2. Find the solution of $3x^2 - 2x = -1$.

Solution: First we write the equation in standard form as

$$3x^2 - 2x + 1 = 0.$$

Next we apply the quadratic formula with $a = 3$, $b = -2$, and $c = 1$. The result is

$$x = \frac{-(-2) \pm \sqrt{(-2)^2 - 4(3)(1)}}{2(3)}$$

$$= \frac{2 \pm \sqrt{-8}}{6}$$

$$= \frac{2 \pm 2i\sqrt{2}}{6} = \frac{1}{3} \pm \frac{i\sqrt{2}}{3}.$$

Therefore, the solutions are

$$\frac{1}{3} + \frac{i\sqrt{2}}{3} \quad \text{and} \quad \frac{1}{3} - \frac{i\sqrt{2}}{3}.$$

We have learned several methods for solving quadratic equations. The methods we have studied are

1. Factoring.
2. Extraction of roots.
3. Completing the square.
4. Quadratic formula.

The method that we use depends upon the particular equation that is to be solved.

EXERCISES 8.5

For Exercises 1–30, solve the quadratic equation with an appropriate method. Some solutions do not involve i. Express any answers that contain i in the form a + bi or a + ib.

1. $x^2 - 9x + 14 = 0$
2. $x^2 + x - 12 = 0$
3. $x^2 + 6x + 8 = 0$
4. $x^2 - 3x - 18 = 0$
5. $2x^2 + 5x - 3 = 0$
6. $3x^2 + 5x + 2 = 0$
7. $2x^2 - x = 15$
8. $4x^2 - 11x = 3$
9. $x^2 = 16$
10. $x^2 = 81$
11. $x^2 = -16$
12. $x^2 = -81$
13. $(x + 2)^2 = -9$
14. $(x - 1)^2 = -4$
15. $(x - 3)^2 = -49$
16. $(x + 5)^2 = -100$
17. $(x - 1)(x + 3) = 6$
18. $(x + 1)(x - 4) = 5$
19. $(2x - 1)(x + 2) = 20$
20. $(x + 1)(3x - 1) = 12$
21. $x^2 - x + 1 = 0$
22. $x^2 - 3x + 2 = 0$
23. $x^2 + 2x + 5 = 0$
24. $x^2 + 2x + 6 = 0$
25. $x^2 + 3x = -2$
26. $x^2 + x = -4$
27. $2x^2 - 4x = -1$
28. $3x^2 + x = -2$
29. $(2x - 1)(x + 1) = -2$
30. $(3x + 1)(x + 2) = -4$

Solve Exercises 31 and 32.

31. If $3 - i$ is a solution to a quadratic equation, what is the other solution?

32. If $3 + 2i$ is a solution to a quadratic equation, what is the other solution?

Graphing Parabolas 8.6

Earlier in the book, we graphed first-degree inequalities in one variable (Chapter 2) and first-degree equations and inequalities in two variables (Chapter 6). In this section we consider the graphs of second-degree or quadratic equations of the form

$$y = ax^2 + bx + c.$$

We will find that these graphs all have the same general shape.

We begin by sketching the graph of the quadratic equation $y = x^2 + 1$. Note that a point (a, b) is on the graph of $y = x^2 + 1$ if a true statement results when x is replaced by a and y is replaced by b. Thus $(2, 5)$ belongs to the graph of $y = x^2 + 1$ because when we replace x with 2 and y with 5, the result is

$$5 = 2^2 + 1,$$

which is a true statement. Note that $(3, 1)$ does not belong to the graph because

$$1 = 3^2 + 1$$

is not a true statement. We now construct a list of some ordered pairs that satisfy $y = x^2 + 1$. This is done by choosing values for x and then solving for y. For example,

$$\begin{aligned} \text{when} \quad x &= 0, & y &= 0^2 + 1 = 1 \\ x &= 3, & y &= 3^2 + 1 = 10 \\ x &= -3, & y &= (-3)^2 + 1 = 10. \end{aligned}$$

Thus we see that the points $(0, 1)$, $(3, 10)$, and $(-3, 10)$ are all on the graph of $y = x^2 + 1$. If, in the equation $y = x^2 + 1$, we replace x with $-3, -2, -1, 0, 1, 2, 3$, we obtain the following list of seven ordered pairs that correspond to points on the graph of $y = x^2 + 1$.

Ordered pairs for the equation
$y = x^2 + 1$

x	y	(x, y)
-3	10	$(-3, 10)$
-2	5	$(-2, 5)$
-1	2	$(-1, 2)$
0	1	$(0, 1)$
1	2	$(1, 2)$
2	5	$(2, 5)$
3	10	$(3, 10)$

Next we plot these ordered pairs on a rectangular coordinate system as shown in Figure 8.1.

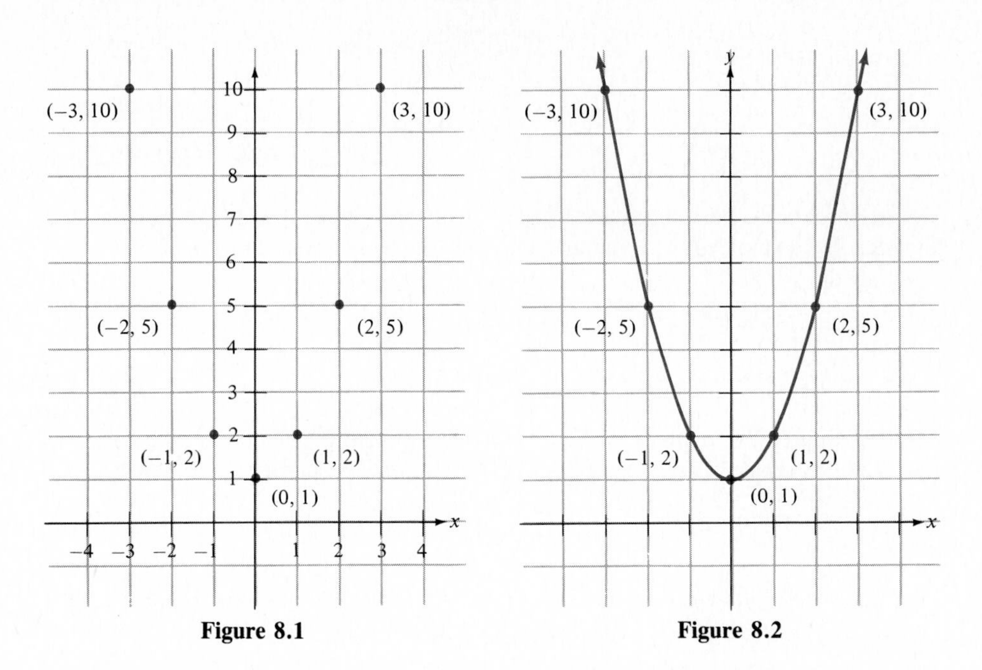

Figure 8.1

Figure 8.2

Before we sketch a smooth curve that connects these points, we make some important observations. First, note that the lowest point on the graph is (0, 1). This is because for any number $a \neq 0$, $a^2 > 0$ and thus $y = a^2 + 1 > 1$. This lowest point is called the *vertex* of the graph. Next, note that when $x = 2$, or $x = -2$, we get the same y-value, namely $y = (\pm 2)^2 + 1 = 4 + 1 = 5$. Similarly, $x = 3$ and $x = -3$ both correspond to $y = 10$. In general, note that if $x = a$ or $x = -a$, we get the same value for y. Finally, note that as we select x-values farther away from the origin, the corresponding y-values are getting larger. Combining these observations with the seven points already plotted, we sketch the smooth curve shown in Figure 8.2, which is the graph of $y = x^2 + 1$. Notice that the curve opens upward. *This curve is called a parabola. In fact, the graph of every quadratic equation* $y = ax^2 + bx + c$ *is a parabola.*

Example 1. Sketch the graph of

$$y = -x^2 + 3.$$

Solution: First we construct a list of some ordered pairs that satisfy the equation. We construct this list by substituting the values −3, −2, −1, 0, 1, 2, 3 for x and solving for y. The result of these computations are as follows:

$y = -x^2 + 3$

x	y	(x, y)
-3	-6	$(-3, -6)$
-2	-1	$(-2, -1)$
-1	2	$(-1, 2)$
0	3	$(0, 3)$
1	2	$(1, 2)$
2	-1	$(2, -1)$
3	-6	$(3, -6)$

Next we plot these points on a rectangular coordinate system and connect the points with a smooth curve. The result is the parabola shown in Figure 8.3. Notice that this parabola opens downward. Also, notice that it has a highest point, namely (0, 3). This highest point is called the vertex of the graph of $y = -x^2 + 3$.

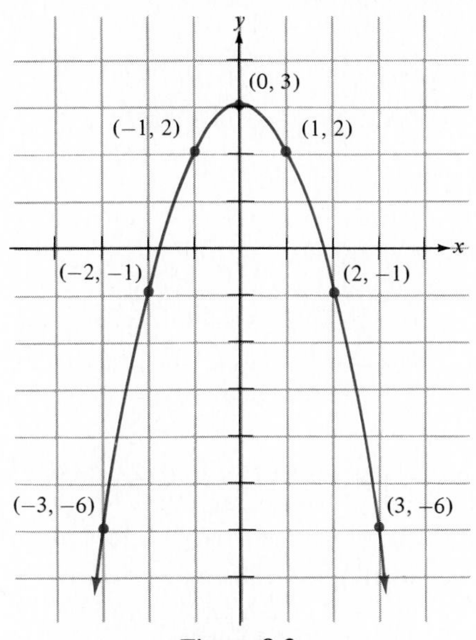

Figure 8.3

Before considering more graphs of quadratic equations, we will use the last two graphs to point out some general characteristics of parabolas. It can be shown that the graph of $y = ax^2 + bx + c$ *opens upward* when *a is positive* and *downward* when *a is negative*. Also, any parabola that opens upward (downward) will have a lowest (highest) point called the *vertex* of the parabola.

Example 2

(a) Does the graph of $y = 2x^2 - 5$ open upward or downward?
(b) Sketch the graph and identify the vertex.

Solution

(a) Since the coefficient of the x^2 term is 2, and 2 is positive, we know that the curve opens upward.

(b) As an aid to sketching the curve, consider the following partial list of ordered pairs that satisfy the equation.

Ordered pairs for the equation

$$y = 2x^2 - 5$$

x	-3	-2	-1	0	1	2	3
y	13	3	-3	-5	-3	3	13

Now we plot the points that correspond to these ordered pairs and connect them as shown in Figure 8.4. We see that the lowest point on the graph is $(0, -5)$. Thus the vertex is the point $(0, -5)$.

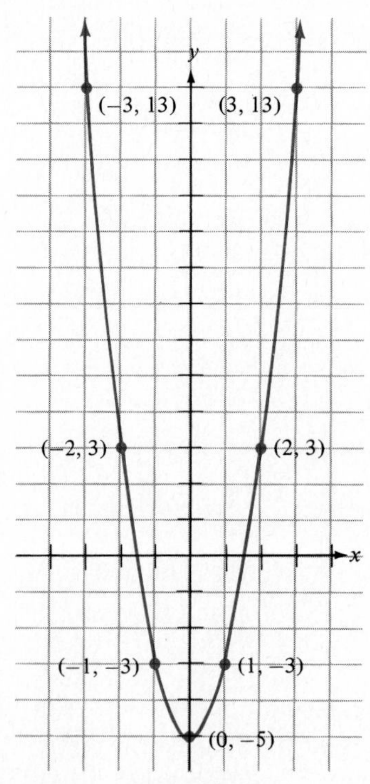

Figure 8.4

Notice that all three of the parabolas that we have sketched are *symmetric with respect to the y-axis*. That is, if the graphs were folded about the y-axis, the two halves would match perfectly. Stated algebraically, a graph is symmetric with respect to the y-axis if both (a, b) and $(-a, b)$ are points on the curve.

All parabolas are symmetric with respect to some line, called the *axis of symmetry*, but this line is not always the y-axis. The sketches of two other parabolas, each with its axis of symmetry and vertex indicated, are shown in Figure 8.5.

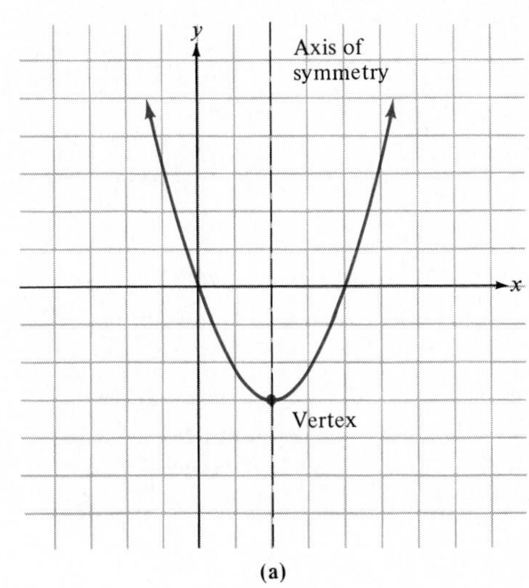

(a)

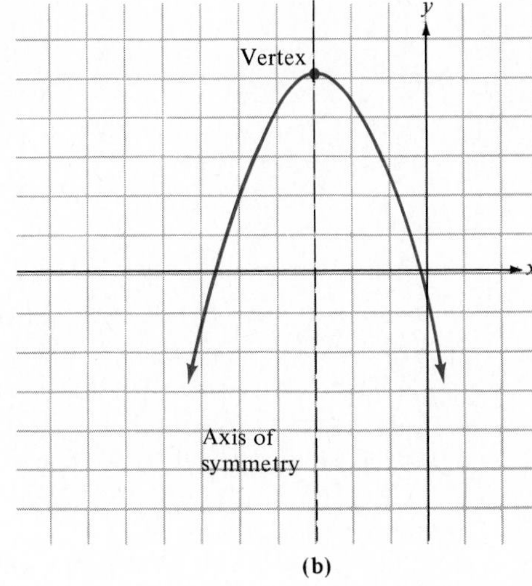

(b)

Figure 8.5

Note that in each case the axis of symmetry divides the curve into two parts that are mirror images of each other. Also note that *the axis of symmetry is the vertical line that passes through the vertex of the parabola*. When sketching a parabola it is helpful to select pairs of x-values that are on opposite sides of, but the same distance from, the axis of symmetry.

We now turn our attention to sketching parabolas that are the graphs of quadratic equations of the form

$$y = (x - a)^2 + b.$$

For example, let us sketch the graph of

$$y = (x - 3)^2 - 7.$$

We begin by locating the vertex of the graph. First, note that $x - 3 = 0$ when $x = 3$. If $x \neq 3$, then $x - 3 \neq 0$ and $(x - 3)^2 > 0$. Thus, when $x \neq 3$, $(x - 3)^2 - 7 > -7$. It follows that the least value of $y = (x - 3)^2 - 7$ occurs when $x = 3$ and that this value is -7. We conclude that the vertex of the parabola is $(3, -7)$.

Since the axis of symmetry of a parabola is the vertical line that passes through the vertex, we know that the line with the equation $x = 3$ is the axis of symmetry. Next, we select pairs of x values that are equidistant from the axis of symmetry, such as 2 and 4, 1 and 5, and 0 and 6 (Figure 8.6).

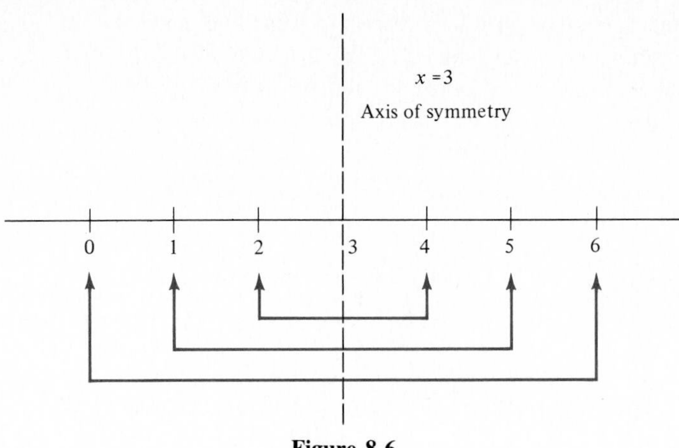

Figure 8.6

Table 8.1 contains these x-values and the computation of the corresponding y-values. The arrows indicate the x-values for which the corresponding y-values are equal. Notice that these x-values are equidistant from the axis of symmetry $x = 3$.

Table 8.1 Ordered Pairs for the Equation

x	$y = (x - 3)^2 - 7$	(x, y)
0	$(0 - 3)^2 - 7 = 9 - 7 = 2$	(0, 2)
1	$(1 - 3)^2 - 7 = 4 - 7 = -3$	(1, −3)
2	$(2 - 3)^2 - 7 = 1 - 7 = -6$	(2, −6)
3	$(3 - 3)^2 - 7 = 0 - 7 = -7$	(3, −7)
4	$(4 - 3)^2 - 7 = 1 - 7 = -6$	(4, −6)
5	$(5 - 3)^2 - 7 = 4 - 7 = -3$	(5, −3)
6	$(6 - 3)^2 - 7 = 9 - 7 = 2$	(6, 2)

Plotting these points on a rectangular coordinate system and connecting them with a smooth curve results in Figure 8.7.

Generalizing the previous discussion, we conclude that the graph of

$$y = (x - a)^2 + b$$

is a parabola that opens upward with **vertex** at (a, b) and having the line $x = a$ as its **axis of symmetry.** Similarly, the graph of

$$y = -(x - a)^2 + b$$

is a parabola that is open downward with **vertex** (a, b) and having the line given by $x = a$ as its **axis of symmetry.**

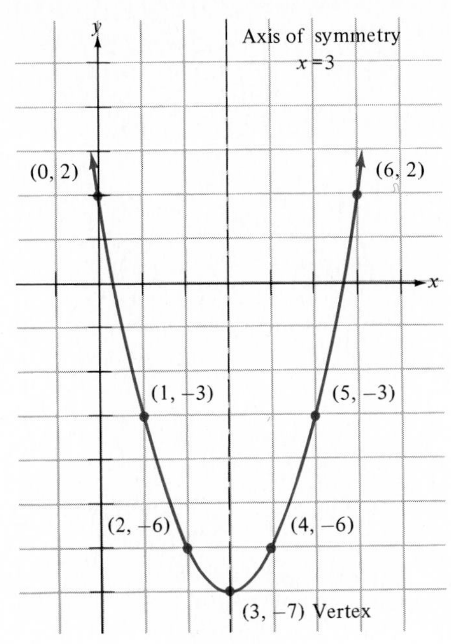

Figure 8.7

Example 3. Sketch the graph of $y = -(x + 1)^2 + 3$. Identify the vertex and the axis of symmetry of the graph.

Solution: Comparing $y = -(x + 1)^2 + 3$ to $y = (x - a)^2 + b$, we see that $a = -1$ and $b = 3$. Thus the vertex is $(-1, 3)$ and the axis of symmetry is the line with the equation $x = -1$. To help sketch the graph, we select some pairs of x-values that are equidistant from the axis of symmetry such as -4 and 2, -3 and 1, and -2 and 0. These x-values and the corresponding y-values are given in the following table.

Ordered pairs for the equation

$y = -(x + 1)^2 + 3$

x	−4	−3	−2	−1	0	1	2
y	−6	−1	2	3	2	−1	−6

We use these points to sketch the graph of $y = -(x + 1)^2 + 3$ as shown in Figure 8.8 on page 316.

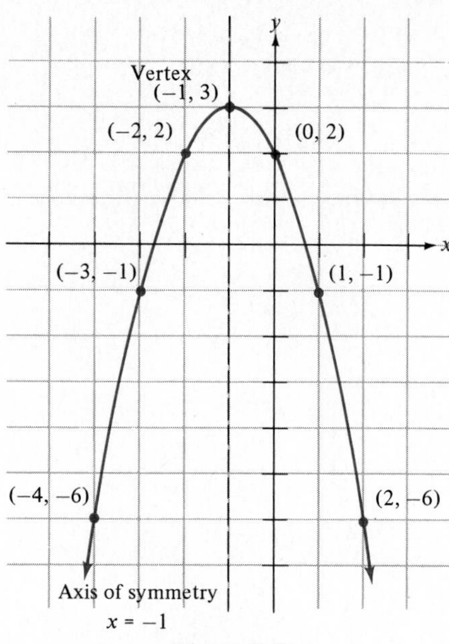

Figure 8.8

EXERCISES 8.6

For Exercises 1–16, determine whether the graph of the equation opens upward or downward. Identify the vertex of the graph and sketch the graph. See Examples 1 and 2.

1. $y = x^2$ **2.** $y = -x^2$ **3.** $y = x^2 + 3$ **4.** $y = x^2 + 2$

5. $y = x^2 - 2$ **6.** $y = x^2 - 4$ **7.** $y = -x^2 + 3$ **8.** $y = -x^2 - 5$

9. $y = 3x^2$ **10.** $y = -3x^2$ **11.** $y = -4x^2 + 1$ **12.** $y = 4x^2 - 3$

13. $y = \frac{1}{3}x^2$ **14.** $y = -\frac{1}{4}x^2$ **15.** $y = \frac{1}{3}x^2 - 2$ **16.** $y = -\frac{1}{4}x^2 + 10$

For Exercises 17–24, graph the parabola. Identify the vertex and axis of symmetry of each parabola. See Example 3.

17. $y = (x - 2)^2$ **18.** $y = (x + 2)^2$ **19.** $y = -(x - 4)^2$

20. $y = -(x + 5)^2$ **21.** $y = -(x - 4)^2 + 2$ **22.** $y = -(x + 5)^2 - 4$

23. $y = (x + 1)^2 - 3$ **24.** $y = (x - 2)^2 + 5$

For Exercises 25–30, first factor the right-hand side of the equation and then sketch the graph. After factoring, see Example 3.

25. $y = x^2 - 2x + 1$ **26.** $y = x^2 + 2x + 1$ **27.** $y = x^2 - 4x + 4$

28. $y = x^2 + 4x + 4$ **29.** $y = x^2 - 10x + 25$ **30.** $y = x^2 - 8x + 16$

Applications 8.7

Quadratic equations are often used to solve practical problems. In this section we consider word problems that require us to find the solutions of a quadratic equation. We begin with problems that involve the use of the Pythagorean theorem.

The *Pythagorean theorem* relates the squares of the lengths of the three sides of a right triangle shown in Figure 8.9. In the figure, the side opposite the right angle is

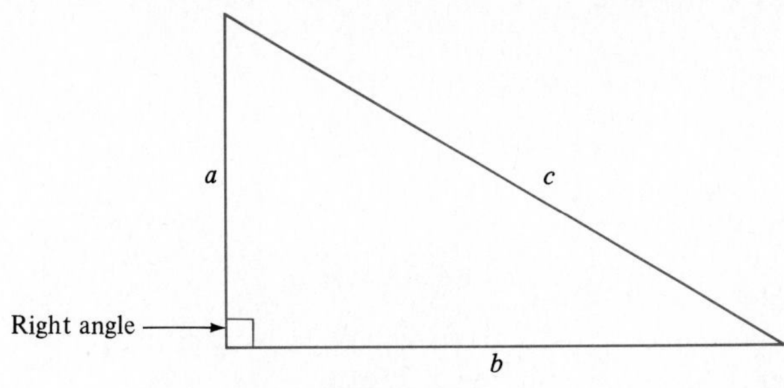

Figure 8.9

called the *hypotenuse*. We represent its length by c. The other two sides are called *legs* of the right triangle. They have length a and b, respectively. The Pythagorean theorem states that

$$a^2 + b^2 = c^2.$$

Example 1. Find the length of the hypotenuse of a right triangle if the lengths of the legs are 5 feet and 12 feet.

Solution: The appropriate diagram is shown in Figure 8.10. The Pythagorean theorem tells us that

$$5^2 + 12^2 = c^2.$$

Thus we need to solve

$$25 + 144 = c^2 \qquad \text{or} \qquad c^2 = 169.$$

Extracting the square roots, we get

$$c = \pm 13.$$

Since length is always positive and c represents a length, we conclude that $c = 13$ feet.

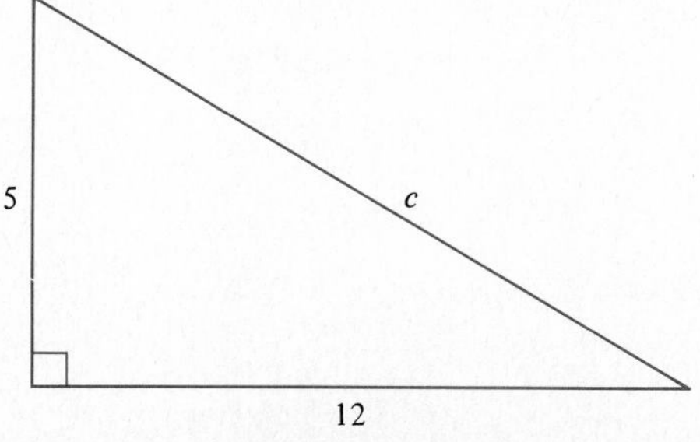

Figure 8.10

Example 2. Mr. Pryce wants to build a corner shelf with two equal sides in the shape of a right triangle as shown in Figure 8.11. If the hypotenuse is to be 36 inches long, how long should each leg be?

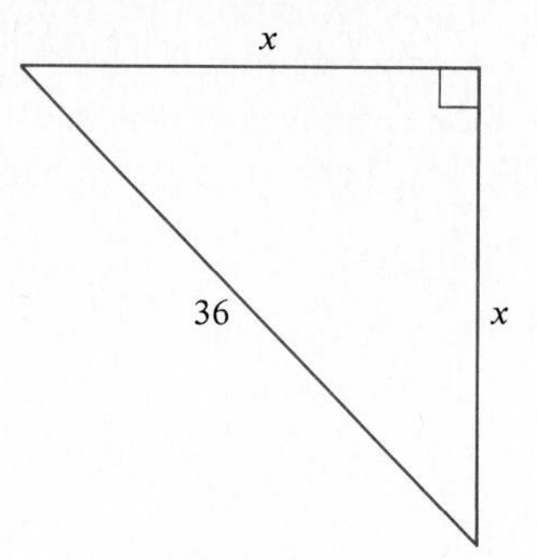

Figure 8.11

Solution: If we let x denote the length of each of the two equal sides, it follows from the Pythagorean theorem that

$$x^2 + x^2 = 36^2$$

or

$$2x^2 = 36^2.$$

Thus

$$x^2 = \frac{36^2}{2}.$$

Extracting the square roots, we get

$$x = \pm\sqrt{\frac{36^2}{2}}$$

$$= \pm\frac{\sqrt{36^2}}{\sqrt{2}}$$

$$= \frac{\pm 36}{\sqrt{2}}. \qquad \sqrt{36^2} = 36$$

Since x is a length, we conclude that $x = \dfrac{36}{\sqrt{2}}$. If we rationalize the denominator of the last expression, we get

$$x = \frac{36\sqrt{2}}{2}$$

$$= 18\sqrt{2}.$$

Finally, if we use 1.414 to approximate $\sqrt{2}$, we see that $x = 18\sqrt{2} \simeq 18(1.414) = 25.45$. Thus

$$x \simeq 25.45 \text{ inches.}$$

Example 3. What is the length of the diagonal of a rectangle with length 10 feet and width 8 feet?

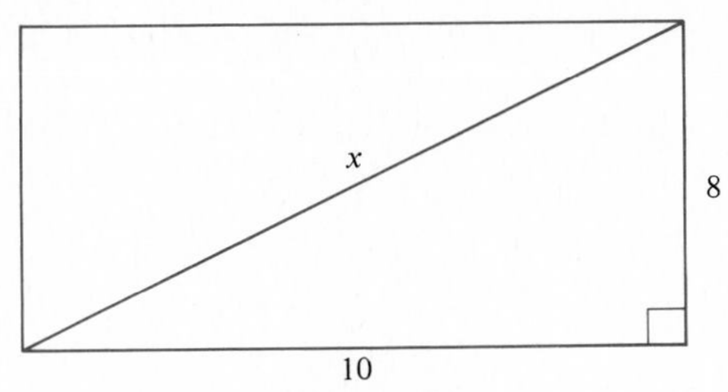

Figure 8.12

Solution: Referring to Figure 8.12, we see that $10^2 + 8^2 = x^2$. Thus

$$x^2 = 164$$

or

$$x = \pm\sqrt{164}$$
$$= \pm 2\sqrt{41}.$$

Again, we only accept a positive answer for length and thus $x = 2\sqrt{41}$ feet. Using a calculator or the table in the back of the book, we can approximate x to two decimal places by

$$x \simeq 2(6.40) \qquad \sqrt{41} \simeq 6.40$$

or

$$x \simeq 12.80 \text{ feet.}$$

In the next two examples we use the quadratic formula to solve a quadratic equation.

Example 4. Mr. Ancrum has just purchased three goats in order to have a fresh supply of what he calls "chemical-free" milk. He has been advised that the goats will require 7500 square feet of grazing space. He has available 400 linear feet of fencing. If he uses all 400 feet of the fencing, what are the dimensions of a rectangular plot that will have an area of 7500 square feet?

Solution: Recall that the perimeter of a rectangle is given by $P = 2L + 2W$, where L and W denote the *length* and *width,* respectively. Since the perimeter of the grazing area will be 400 feet, and W is the width of the plot, the length L may be determined by solving

$$400 = 2L + 2W$$

for L. That is,

$$2L = 400 - 2W$$

or

$$L = 200 - W.$$

This rectangular plot is shown in Figure 8.13.

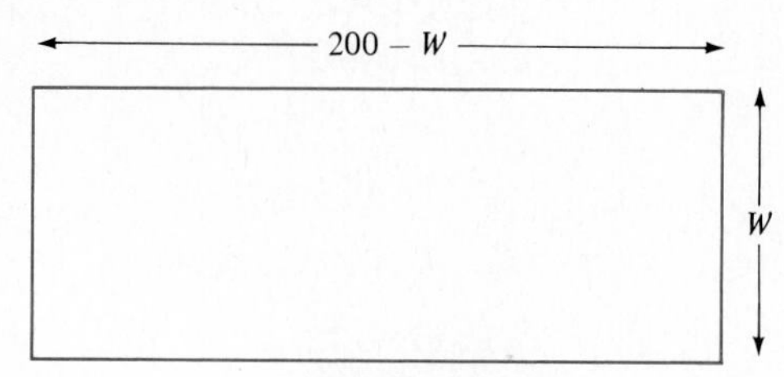

Figure 8.13

Recall that the area of a rectangle is given by A = (length) times (width). Thus the area of this rectangle is $W(200 - W) = 200W - W^2$ square feet. Since we want the area to equal 7500 square feet, we equate $200W - W^2$ and 7500. Thus we must solve the quadratic equation

$$200W - W^2 = 7500$$

or, equivalently,

$$W^2 - 200W + 7500 = 0.$$

To solve this equation, we apply the quadratic formula with $a = 1$, $b = -200$, and $c = 7500$. The result is

$$W = 50 \qquad \text{or} \qquad W = 150.$$

Either way, we see that the dimensions of the rectangle are 150×50.

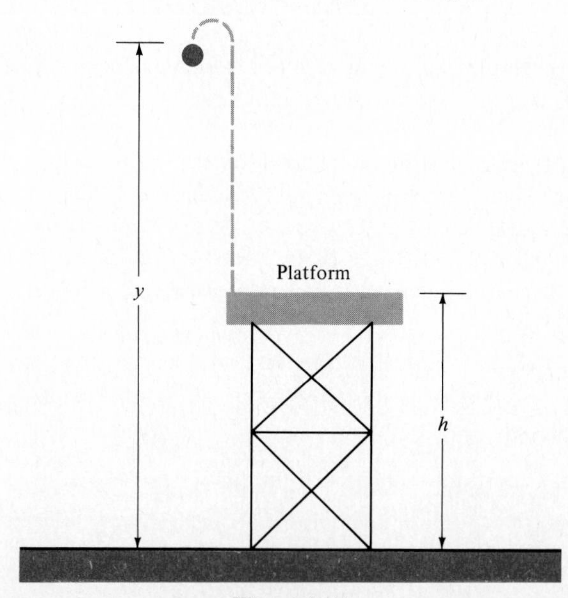

Figure 8.14

Suppose that a ball is thrown directly upward from a platform that is h feet above the ground as shown in Figure 8.14. If the initial velocity of the ball is v feet per second, then its height above the ground after t seconds is given by

$$y = -16t^2 + vt + h.$$

Example 5. A ball is thrown upward from a platform that is 44 feet above the ground. If the initial velocity of the ball is 80 feet per second, when will the ball hit the ground?

Solution: For this problem $h = 44$ and $v = 80$. Substituting these values in the equation

$$y = -16t^2 + vt + h,$$

we get

$$y = -16t^2 + 80t + 44.$$

Next we observe that when the ball hits the ground, its distance above the ground is $y = 0$. Thus we need to solve the equation

$$-16t^2 + 80t + 44 = 0.$$

We can simplify this equation by reducing the size of the coefficients if we divide both sides of the equation by -16. The result of this division is the equivalent equation

$$t^2 - 5t - \frac{11}{4} = 0.$$

We solve this equation using the quadratic formula with $a = 1$, $b = -5$, and $c = -\frac{11}{4}$. Thus

$$\begin{aligned} t &= \frac{-b \pm \sqrt{b^2 - 4ac}}{2a} \\ &= \frac{5 \pm \sqrt{25 - 4(1)\left(-\frac{11}{4}\right)}}{2} \\ &= \frac{5 \pm \sqrt{36}}{2} = \frac{5 \pm 6}{2}. \end{aligned}$$

The possible solutions are

$$t = \frac{11}{2} \quad \text{and} \quad t = \frac{-1}{2}.$$

Since time is positive, we will ignore the negative answer. We conclude that the ball will hit the ground in $\frac{11}{2} = 5\frac{1}{2}$ seconds.

EXERCISES 8.7

Solve the following exercises.

For Exercises 1–8, see Examples 1–3.

1. The lengths of the legs of a right triangle are 7 inches and 24 inches. What is the length of the hypotenuse?
2. The lengths of the legs of a right triangle are 10 feet and 24 feet. What is the length of the hypotenuse?
3. One leg of a right triangle is 10 inches long. The hypotenuse is 15 inches long. How long is the other leg?
4. One leg of a right triangle is 5 feet long. The hypotenuse is 13 feet long. How long is the other leg?
5. Mrs. Jennings is building a triangular corner shelf that is to have two equal sides that meet in the corner. If the other side is to be 50 inches long, how long will each side be?
6. Pat is building a triangular corner closet to hold knick-knacks. The closet will have two equal sides. If the longest side is to be 60 inches in length, determine the length of each of the two equal sides.
7. What is the length of the diagonal of a rectangle with length 12 feet and width 9 feet?
8. The diagonal of a rectangle is 20 inches. The width of the rectangle is 12 inches. What is the length of the rectangle?

For Exercises 9–12, see Example 4.

9. The perimeter of a rectangle is 300 feet. What are the dimensions of the rectangle if its area is 5400 square feet?
10. The area of a rectangle is 162 square inches, the length is twice the width, find the dimensions of the rectangle.
11. The length of a rectangle is 7 inches more than the width. If the area of the rectangle is 78 square inches, find the dimensions of the rectangle.
12. The length of a rectangle is 4 feet longer than twice the width. If the area of the rectangle is 160 feet, find the dimensions of the rectangle.

For Exercises 13–16, see Example 5.

13. A ball is thrown directly upward from a platform that is 36 feet above the ground. If the initial velocity of the ball is 64 feet per second, when will the ball hit the ground?
14. A ball is thrown directly upward from a platform that is 96 feet above the ground. If the initial velocity of the ball is 80 feet per second, when will it hit the ground?

15. A bullet is fired directly upward with an initial velocity of 800 feet per second. When will the bullet hit the ground?

16. A bullet is fired directly upward from the roof of a building that is 816 feet above the ground. If the initial velocity of the bullet is 800 feet per second, when will it strike the ground?

For Exercises 17–22, use the Pythagorean theorem to get a quadratic equation and then solve the equation.

17. A baseball diamond is actually a square with each base at a corner. The distance between first base and second base is 90 feet. What is the straight-line distance from first base to third base? (HINT: Draw a figure.)

18. A 20-foot ladder is leaning against a building with its base 5 feet from the building, as shown in the accompanying figure. How high up the building is the top of the ladder?

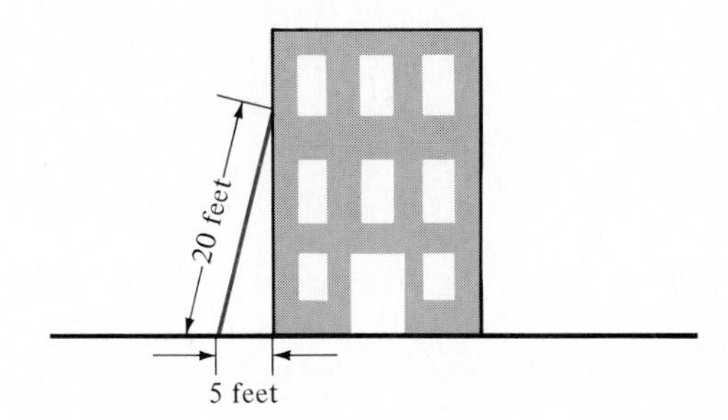

19. The guide wire for a light pole is attached to an iron stake that is 8 feet from the base of the pole. If the wire is attached to the pole at a point 20 feet above the ground, how long is the wire? See the accompanying figure.

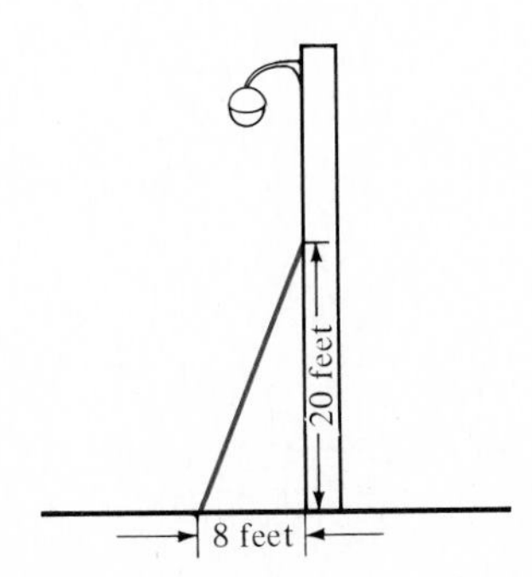

20. The guide wire for a transmission tower is attached to the center of the tower 120 feet above the ground. It is also attached to an iron stake that is 20 feet from the center of the tower base, as shown in the accompanying figure on page 324. How long is the guide wire?

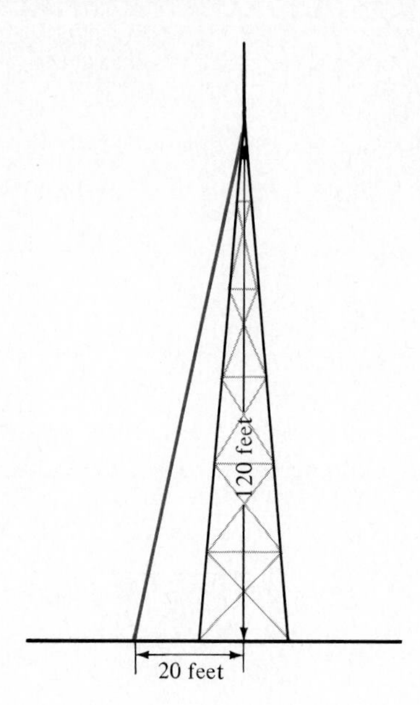

21. Raul knows that he can swim $2\frac{1}{4}$ miles. He has been stranded on an island that is 3 miles from a boat launching site. He also knows that it is 2 miles from the launching site to a point on the shore that is directly opposite the island. Should he try to swim to shore? See the accompanying figure.

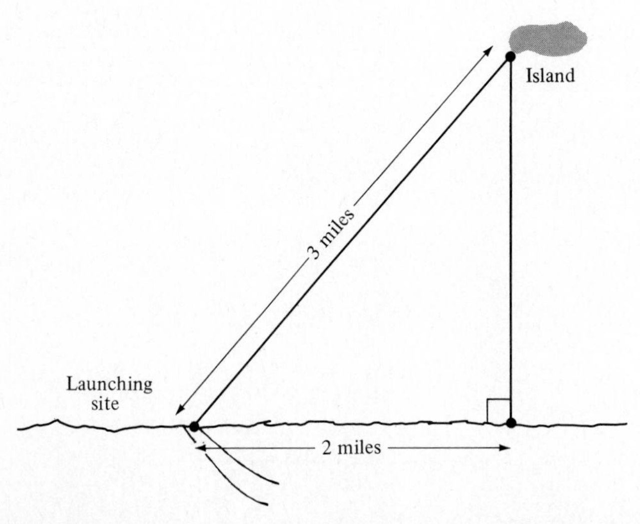

22. Lisa is on an island that is 4 miles from the dock at her summer camp. She knows that it is 2 miles from the island to a point on the shore that is directly opposite the island. If she swims directly to the shore, how far will she have to walk to get back to camp? See the accompanying figure on p. 325.

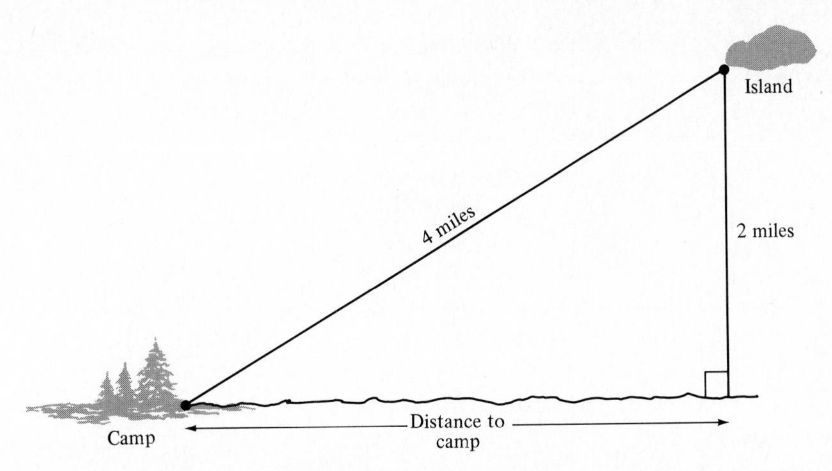

In the rectangle shown, the ratio of the width to the length is equal to the ratio of the length to the width plus the length. That is,

$$\frac{W}{L} = \frac{L}{W + L}.$$

The value $\frac{W}{L}$ *is called the* **Golden Ratio.** *Any rectangle that satisfies the above proportion is called a* **Golden Rectangle.** *Exercises 23–30 involve the Golden Ratio.*

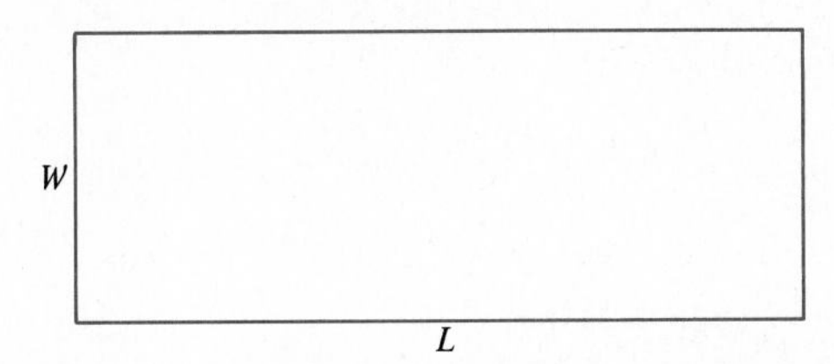

23. Find the value of the Golden Ratio. (HINT: Use the figure with $L = 1$ and solve the equation for positive W.

24. A Golden Rectangle is 10 feet long. How wide is it?

25. A Golden Rectangle is 5 feet wide. How long is it?

26. A display window in a furniture store has the shape of a Golden Rectangle. If the window is 22 feet long, how high (wide) is it?

27. A large painting is to be done on a rectangular canvas that is a Golden Rectangle. If the length of the canvas is 8 feet, what is the width?

28. A rectangular door has the shape of a Golden Rectangle. If the door is 3 feet wide, how high is it?

29. A common rule of thumb is that the Golden Ratio is approximately $\frac{3}{5} = .6$. Use the results of Exercise 23 and either a calculator or the square root tables to justify this approximation.

30. Use the approximation given in Exercise 29 to estimate the width of a Golden Rectangle that is 70 inches long.

Key Words

Axis of symmetry of a parabola
Completing the square
Complex conjugate
Complex number
Extraction of roots
Hypotenuse
Parabola
Perfect square trinomial
Pythagorean theorem
Quadratic equation
Quadratic formula
Vertex of a parabola

CHAPTER 8 TEST

For Problems 1–4, solve the equation by finding the square roots of both sides of the equation. Simplify all radicals and check your answers.

1. $x^2 = 49$ **2.** $y^2 = 80$ **3.** $(x + 2)^2 = 36$ **4.** $(2z + 6)^2 = 100$

For Problems 5 and 6, state whether or not the expression is a perfect square trinomial.

5. $y^2 - 10y - 25$ **6.** $y^2 + 18y + 81$

For Problems 7–10, solve the equation by completing the square.

7. $x^2 + 10x = 11$ **8.** $w^2 - 8w + 12 = 0$

9. $2y^2 + 5y + 2 = 0$ **10.** $5z^2 - 3z = 4$

For Problems 11–18, perform the operations indicated. Express your answer in the form $a + bi$ or $a + ib$.

11. $(3 + \sqrt{-100}) + (5 - \sqrt{-16})$ **12.** $(4 - 3i) - (2 + i)$ **13.** $3i(7 + i)$

14. $2i(3 - 4i)$ **15.** $(4 + i)(3 - i)$ **16.** $(5 + 2i)(2 + 3i)$

17. $\dfrac{4i}{2 - i}$ **18.** $\dfrac{5 - 3i}{4 + 3i}$

For Problems 19 and 20, solve the quadratic equation using the quadratic formula.

19. $x^2 + 2x + 5 = 0$ **20.** $3x^2 + 6x + 14 = 0$

For Problems 21–24, graph the quadratic equation. Identify the vertex and the axis of symmetry of each parabola.

21. $y = x^2 + 2$ **22.** $y = -3x^2 + 1$ **23.** $y = \frac{1}{5}x^2 - 4$ **24.** $y = -(x + 1)^2 + 4$

Solve Problems 25–27 by writing and solving an appropriate equation.

25. The lengths of the legs of a right triangle are 8 feet and 12 feet, as shown in the accompanying figure. What is the length of the hypotenuse?

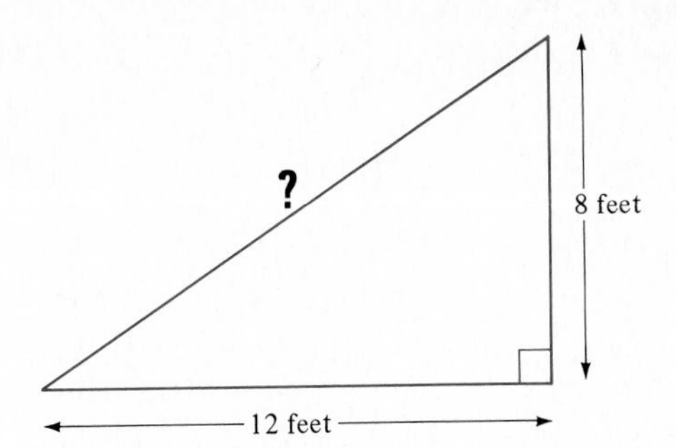

26. The diagonal of a square is 10 inches long, as shown in the accompanying figure.

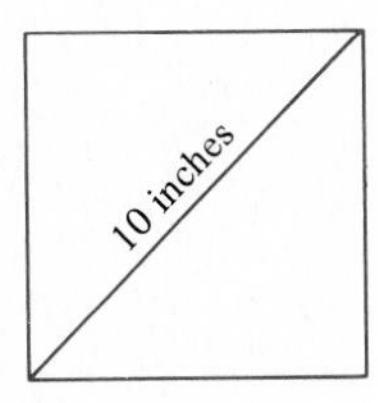

(a) What is the length of any side of the square?
(b) What is the perimeter of the square?
(c) What is the area of the square?

27. One end of a guide wire for a utility pole is to be fastened to the pole 20 feet above the ground. The other end is to be fastened to a ground stake that is 6 feet from the base of the pole (see the accompanying figure). How long will the wire have to be to meet these requirements?

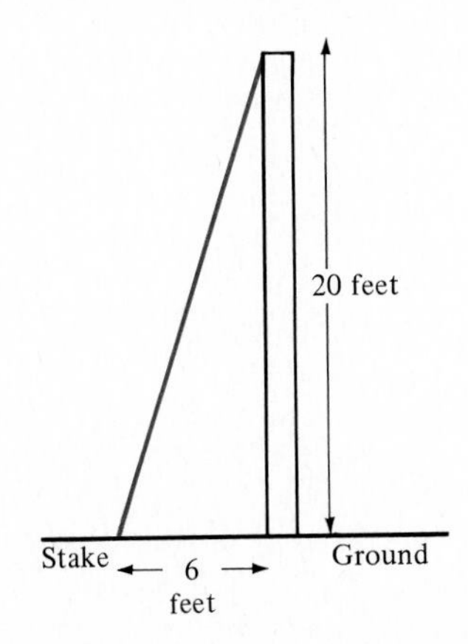

CHAPTER 9

Selected Topics

A farm agency knows that the loss, L, in crop value per acre for wheat is proportional to the average number, W, of weed plants per square yard. This tells the farmer that the ratio $\frac{L}{W}$ is constant. Thus, if he knows the number of weeds per square yard, he can estimate his loss in crop value.

U.S. Department of Agriculture

CHAPTER 9
CHAPTER 9
CHAPTER 9
CHAPTER 9
CHAPTER 9
CHAPTER 9
CHAPTER 9
CHAPTER 9
CHAPTER 9
CHAPTER 9
CHAPTER 9
CHAPTER 9
CHAPTER 9
CHAPTER 9
CHAPTER 9
CHAPTER 9
CHAPTER 9
CHAPTER 9
CHAPTER 9
CHAPTER 9
CHAPTER 9
CHAPTER 9
CHAPTER 9
CHAPTER 9
CHAPTER 9
CHAPTER 9
CHAPTER 9
CHAPTER 9
CHAPTER 9
CHAPTER 9
CHAPTER 9
CHAPTER 9
CHAPTER 9

In this chapter we present a collection of topics that will help to advance your knowledge of the elements of algebra. These topics can also serve as an introduction to the beginnings of intermediate algebra or precalculus. All of the sections can be studied independently of one another, except Section 9.3. Section 9.3, which deals with equations of straight lines, assumes that the reader is familiar with the contents of Section 9.2 on slopes of straight lines.

9.1 Functions

The concept of a function plays an important role in much of mathematics, science, and everyday life. Functions are usually stated in terms of mathematical equations. Let us take an example from everyday life where the function concept is used. Suppose that a person pays \$2 a gallon for gasoline for his car. Then the cost of the gasoline is a function of the number of gallons purchased. That is,

$$\text{cost} = (\$2) \times (\text{number of gallons}).$$

We can express this cost function with the equation

$$C = 2g.$$

The variable C represents the total cost of the gasoline and the variable g represents the number of gallons purchased. For $g = 15$ the cost is $C = 2(15) = \$30$. Also, for each specific value of g the equation produces *exactly* one value of C. Thus, for the equation $C = 2g$, we see that the value of the variable C (cost) *depends* on the variable g (gallons purchased). We call C the *dependent variable* and g the *independent variable*. Furthermore, we say that the rule $C = 2g$ defines C as a *function* of g.

In mathematics, when we define a function, we generally use y for the dependent variable and x for the independent variable. In this book the values of the variables x and y will be real numbers.

In general,

> A *function* is a rule that assigns to each value x in one set one and only one value y in another set.

The set of values that x may assume is called the *domain* of the function. The set of all corresponding y values is called the *range* of the function.

The simplest way to define y as a function of x is to let y equal an algebraic expression in x.

Example 1. Does the rule given by the equation $y = 3x + 2$ define y as a function of x when the set of values x can assume is $\{-1, 0, 1\}$? If the rule defines a function, identify the domain and range.

Solution: Notice that when

$$x = -1, \quad y = -1; \qquad x = 0, \quad y = 2; \qquad x = 1, \quad y = 5.$$

As a result, we see that each possible value of x produces exactly one value for y. Therefore, by definition, $y = 3x + 2$ defines y as a function of x. The domain of the function is the set $\{-1, 0, 1\}$. The set of corresponding y values $\{-1, 2, 5\}$ is the range of the function.

Example 2. Does the equation $y = x^2$ define y as a function of x when x assumes values from the set $\{-2, -1, 0, 1, 2\}$? If the equation defines a function, identify the domain and range.

Solution: Using the given rule $y = x^2$, we make up a table of corresponding x and y values. Thus we have

x	-2	-1	0	1	2
y	4	1	0	1	4

It might appear that the given equation does not define y as a function of x, because $x = 2$, and $x = -2$ both correspond to the same y value 4. But this is permitted by the definition of a function. What a rule for a function *must not* do is assign a number such as 2 to two or more y values. Here we see that 2 corresponds to *only one* y value, 4. Thus we see that $y = x^2$ defines y as a function of x. The domain is the set $\{-2, -1, 0, 1, 2\}$. The range is the set of corresponding y values $\{0, 1, 4\}$.

Example 3. Does the rule $y^2 = x$ define y as a function of x when x may be any positive real number? If the rule defines a function, what is the domain and range?

Solution: The rule $y^2 = x$ *does not* define y as a function of x. Notice, for example, that when $x = 9$, we have

$$y^2 = 9$$

or

$$y = \pm 3.$$

Thus, when $x = 9$, there are two corresponding values of y, $+3$ and -3. This violates the definition of a function which states that a given x value, in this case 9, must be assigned to exactly one y value.

If a rule defining y as a function of x is given without specifically stating what values x can assume, then the domain of the function is assumed to be *all real* values x for which there is *exactly one* corresponding *real* value y. Note that when a function is specified by an algebraic expression, we exclude from the domain any numbers that would result in either an imaginary number or division by zero. The following examples illustrate how to determine the domain of y as a function of x when the domain is not specified.

Example 4. Specify the domain of the function $y = \dfrac{1}{x - 3}$.

Solution: Notice that if we replace x with 3, we obtain 0 for a denominator. But division by 0 is not permitted. Thus the domain is

$$\text{domain} = \{x \mid x \text{ is real and } x \neq 3\}.$$

Example 5. Specify the domain of the function $y = \sqrt{x - 1}$.

Solution: The square root of a negative number is not a real number, it is an imaginary number. Thus we wish to assign values to x that make the expression under the radical sign positive or zero. As a result, we must have $x - 1 \geq 0$ or $x \geq 1$. The domain of the function is

$$\text{domain} = \{x \mid x \text{ is real and } x \geq 1\}.$$

If a function is defined by a rule such as $y = 3x + 2$, then we can conveniently use a letter such as f to refer to the function. Usually, the letters f, g, h (or F, G, H) are used to name functions. Then the notation $f(x)$, which is read "the value of the function f at x," or simply "f of x," means the *value of y in the range* that corresponds to the *domain value x*. Thus $f(x)$ is the same as y. Thus if $y = 3x + 2$, we may write

$$f(x) = 3x + 2.$$

The symbol $f(2)$ represents the value of the function f at 2. To find $f(2)$, we replace x with 2 wherever we see x in the algebraic expression defining the function. With this understanding we see that for $f(x) = 3x + 2$ we have

$$f(2) = 8 \qquad \text{because} \qquad f(2) = 3(2) + 2 = 8$$

Also,

$$f(3) = 11 \qquad \text{because} \qquad f(3) = 3(3) + 2 = 11$$
$$f(-1) = -1 \qquad \text{because} \qquad f(-1) = 3(-1) + 2 = -1$$
$$f(0) = 2 \qquad \text{because} \qquad f(0) = 3(0) + 2 = 2.$$

Notice that since $y = 3x + 2$ and $f(x) = 3x + 2$ are the same, we may write

$$y = f(x),$$

which states compactly that "y is a function of x." The notation $f(x)$ is referred to as *functional notation*. Note that $f(x)$ does *not* mean a number f times a number x.

Example 6. Given $g(x) = x^2 + 1$, find
(a) $g(0)$. (b) $g(2)$. (c) $g(-4)$.

Solution: Remember that the number in parentheses replaces x in the formula for the function g. Consequently,
(a) $g(0) = 1$ because $g(0) = 0^2 + 1 = 1$.
(b) $g(2) = 5$ because $g(2) = 2^2 + 1 = 5$.
(c) $g(-4) = 17$ because $g(-4) = (-4)^2 + 1 = 17$.

Example 7. If $h(x) = 5x + 1$, find
(a) $h\left(\frac{1}{5}\right)$, (b) $h(-2)$, (c) $h(b)$.

Solution
(a) $h\left(\frac{1}{5}\right) = 5\left(\frac{1}{5}\right) + 1 = 1 + 1 = 2.$
(b) $h(-2) = 5(-2) + 1 = -10 + 1 = -9.$
(c) $h(b) = 5(b) + 1 = 5b + 1.$

The *graph* of a function $y = f(x)$ on a rectangular coordinate system is the set of all points (x, y), where x is a value in the domain of the function and y is the corresponding range value.

Let us consider the graph of the function $f(x) = 2x + 6$. Since both y and $f(x)$ represent the value of the function at x, we will write $f(x) = 2x + 6$ as $y = 2x + 6$. Thus the graph of $f(x) = 2x + 6$ is the graph of the equation $y = 2x + 6$. Recall from Chapter 6 that $y = 2x + 6$ is a linear equation in two variables whose graph is the straight line shown in Figure 9.1.

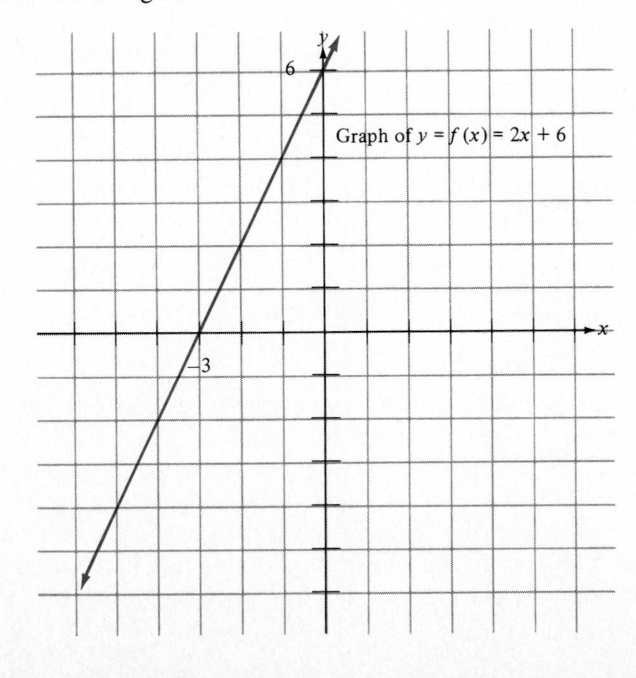

Figure 9.1

Now suppose that we are given the graph of an equation in x and y without knowing the equation. We can use the *vertical-line test* to determine whether the graph represents y as a function of x. By the definition of a function, if we are given a value of x there is exactly one corresponding value of y. If, for some equation, there is more than one y value for a given x value, then the equation does *not* define y as a function of x. Graphically, this means that a vertical line will intersect the graph in two or more points. Thus we have the following rule:

Vertical Line Test

> If a vertical line intersects a graph in two or more points, then the graph *is not* the graph of y as a function of x.

The following example illustrates the application of the vertical-line test.

Example 8. Decide if the graphs given in Figure 9.2 represent y as a function of x.

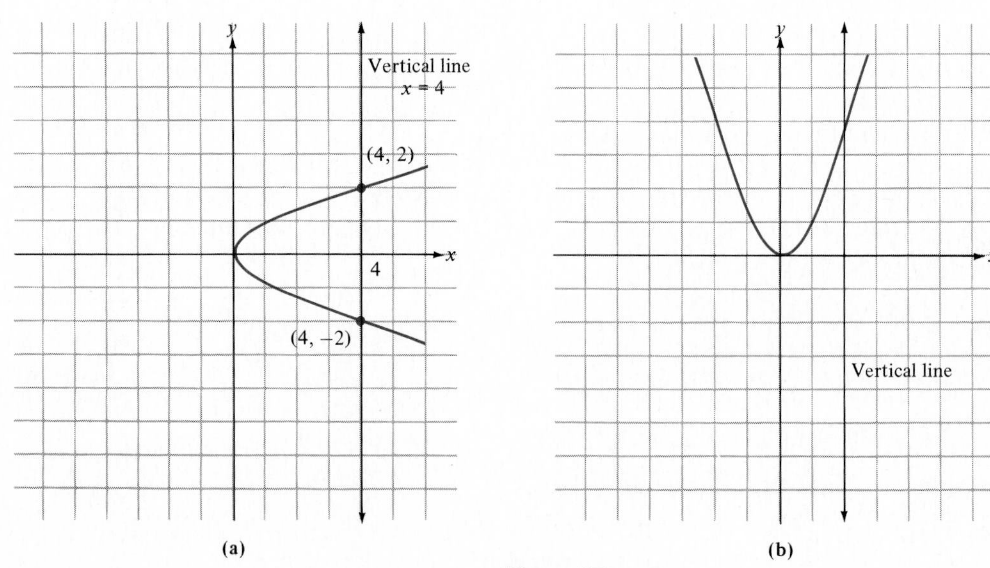

Figure 9.2

Solution

(a) Figure 9.2(a) does not represent y as a function of x, since the vertical line $x = 4$ intersects the graph at two points (4, 2) and (4, −2). This test shows that on this graph $x = 4$ corresponds to two y-values 2 and −2, which contradicts the definition of y as a function of x. In fact, the graph is the graph of the equation $y^2 = x$, which we saw in Example 3 does not define y as a function of x.

(b) Figure 9.2(b) does represent the graph of y as a function of x, since *any* vertical line intersects the graph in exactly one point.

EXERCISES 9.1

For Exercises 1–16, determine whether or not the equation defines y as a function of x with the given set of x values. State the domain and range of all functions. See Examples 1–3.

1. $y = 3x$; $\{0, 1, 2, 3\}$
2. $y = 5x$; $\{-2, -1, 0, 1, 2\}$
3. $y = 2x + 5$; $\{-1, 0, 1\}$
4. $y = -4x + 1$; $\{-2, -1, 1, 2\}$
5. $y - x = 0$; $\{x \mid x \text{ is real}\}$
6. $2y - 4x = 0$; $\{x \mid x \text{ is real}\}$
7. $y = 3x^2$; $\{-3, -2, -1, 0, 1, 2, 3\}$
8. $y = x^2 - 1$; $\{-2, -1, 0, 1, 2\}$
9. $y^2 = 4x$; $\{0, 1, 4\}$
10. $y^2 = 9x$; $\{0, 4, 9\}$
11. $|y| = x$; $\{0, 1, 2, 3\}$
12. $y = |x|$; $\{-2, -1, 1, 2, 3\}$
13. $y = x^3$; $\{-3, -2, -1, 0, 1, 2, 3\}$
14. $y^2 = x^2$; $\{0, 1, 4, 9, 25\}$
15. $y = 4$; $\{x \mid x \text{ is real}\}$
16. $y = \sqrt{x}$; $\{0, 1, 4, 9, 25\}$

For Exercises 17–34, determine the domain of the given function of x. See Examples 4 and 5.

17. $y = \dfrac{2}{x}$
18. $y = \dfrac{3}{x - 1}$
19. $y = \sqrt{x}$
20. $y = \sqrt{x - 2}$
21. $y = \dfrac{1}{(x - 1)(x - 2)}$
22. $y = \dfrac{1}{(x - 3)(x + 5)}$
23. $y = x^2 - 1$
24. $y = x^2 - 4$
25. $y = \sqrt{2x - 1}$
26. $y = \dfrac{1}{x^3}$
27. $y = \dfrac{1}{x^2 - 1}$
28. $y = \dfrac{1}{|x - 1|}$
29. $y = \dfrac{1}{\sqrt{x - 2}}$
30. $y = \dfrac{1}{(x - 3)^2}$
31. $y = \dfrac{1}{|x| - 1}$
32. $y = \dfrac{1}{|x| + 1}$
33. $y = \dfrac{2x - 1}{x - 5}$
34. $y = \dfrac{3x - 1}{x - 7}$

For Exercises 35–46, find the indicated function values. See Examples 6 and 7.

35. Given $f(x) = 2x + 1$, find **(a)** $f(0)$; **(b)** $f(3)$; **(c)** $f(-2)$.

36. Given $f(x) = 3x + 2$, find **(a)** $f(-1)$; **(b)** $f(4)$; **(c)** $f(-2)$.

37. Given $g(x) = x^2 + 3$, find **(a)** $g(0)$; **(b)** $g(\sqrt{2})$; **(c)** $g(-2)$.

38. Given $g(x) = 2x^2 + 1$, find **(a)** $g(0)$; **(b)** $g(3)$; **(c)** $g(\sqrt{3})$.

39. Given $h(x) = \sqrt{x + 4}$, find **(a)** $h(0)$; **(b)** $h(-3)$; **(c)** $h(5)$.

40. Given $h(x) = \dfrac{1}{x}$, find **(a)** $h\left(\dfrac{1}{2}\right)$; **(b)** $h(-1)$; **(c)** $h(a)$.

41. Given $F(x) = \dfrac{2x - 1}{x - 4}$, find **(a)** $F(0)$; **(b)** $F(2)$; **(c)** $F(3)$.

42. Given $F(x) = \dfrac{3x + 1}{x + 5}$, find **(a)** $F(0)$; **(b)** $F(-1)$; **(c)** $F(2)$.

43. Given $G(x) = \sqrt{25 - x^2}$, find **(a)** $G(0)$; **(b)** $G(3)$; **(c)** $G(-3)$.

44. Given $G(x) = \sqrt{169 - x^2}$, find **(a)** $G(5)$; **(b)** $G(-5)$; **(c)** $G(0)$.

45. Given $H(x) = 5$, find **(a)** $H(0)$; **(b)** $H(-1)$; **(c)** $H(20)$.

46. Given $H(x) = -6$, find **(a)** $H(0)$; **(b)** $H(-2)$; **(c)** $H(30)$.

For Exercises 47–54, use the vertical-line test to determine whether or not the graphs represent functions of x. See Example 8.

47.

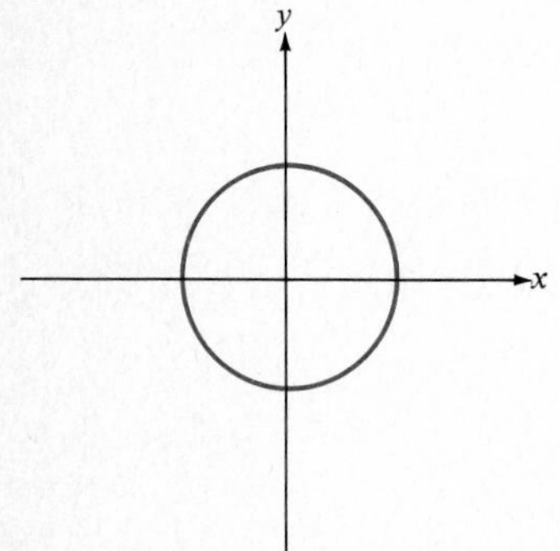

48.

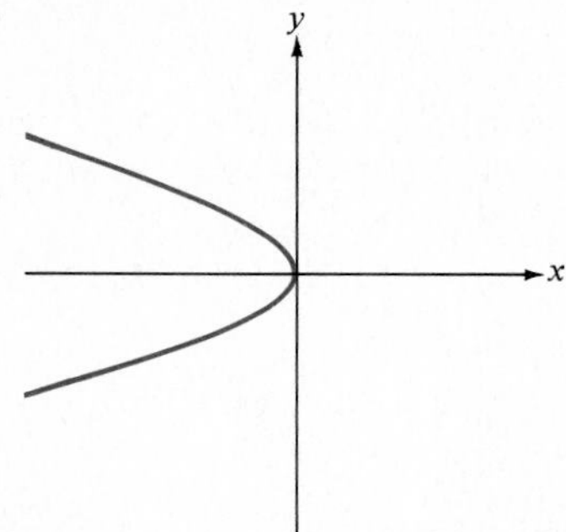

49.

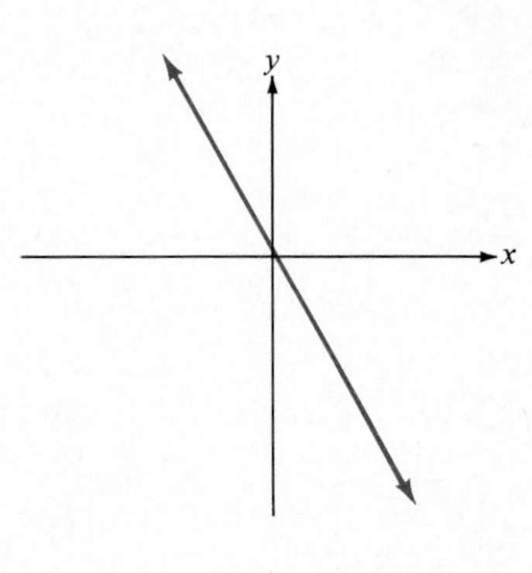

50.

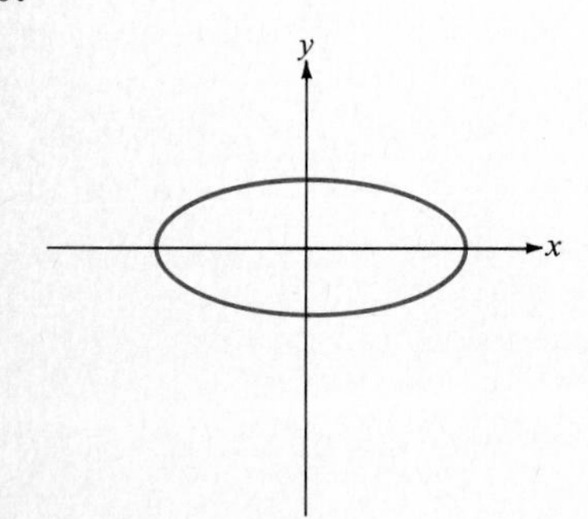

51.

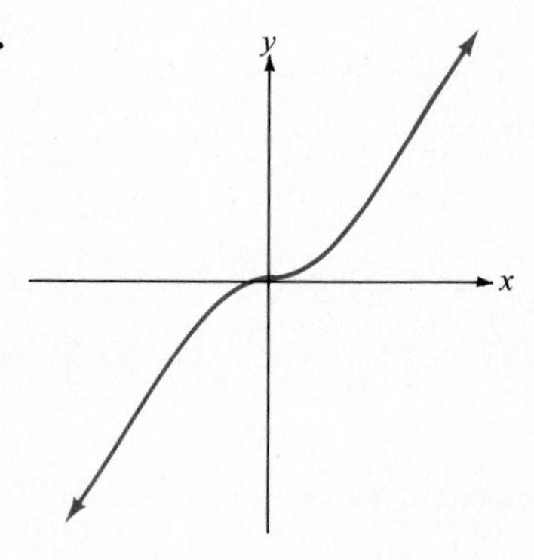

52.

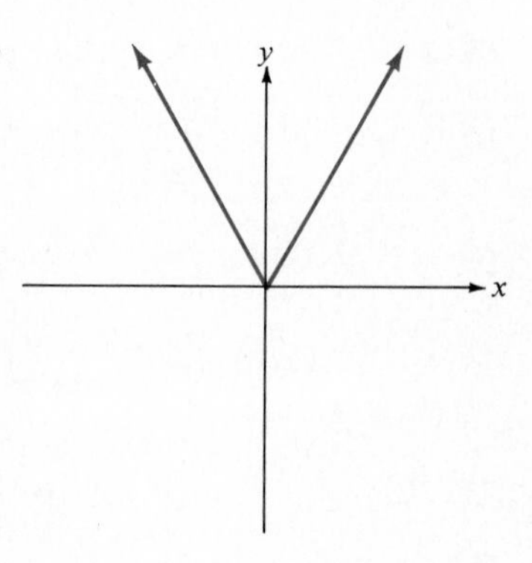

53.

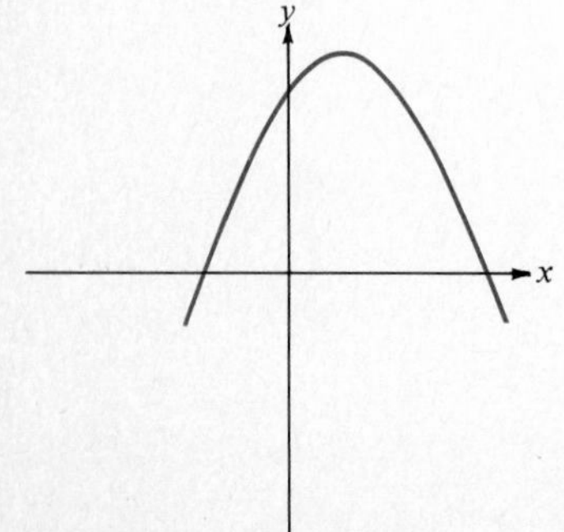

54.

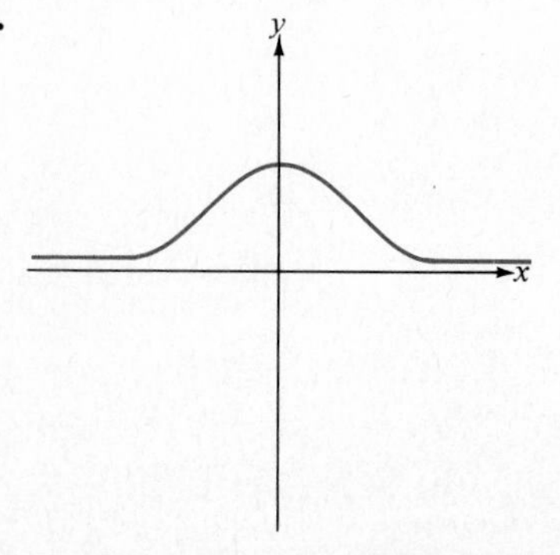

Slope of a Straight Line 9.2

We saw in Section 9.1 that the graph of the function

$$f(x) = 2x + 6$$

is a straight line. This function can also be written as $y = 2x + 6$, and in this form it is easily seen to represent a linear equation in two variables. Recall from Section 6.3 that the graph of any linear equation in two variables is a straight line. Conversely, for any given straight line on a rectangular coordinate system, there is a corresponding linear equation in two variables. In the next section we discuss how to find the equation of a straight line. But to do this we need to introduce an important concept related to the equation of a line called the *slope* of a line. The concept of the slope of a line is connected with measuring the "steepness" of a line.

Consider Figure 9.3, with two straight lines labeled L_1 (read, L sub-one) and L_2 (read, L sub-two).

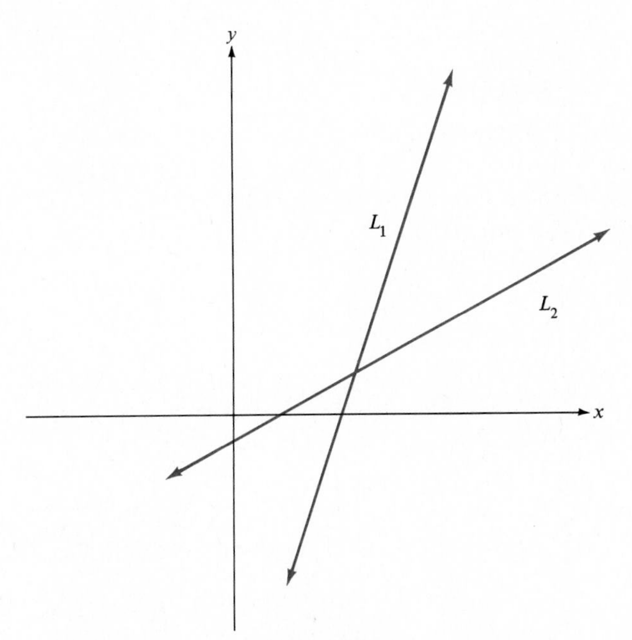

Figure 9.3

We see that relative to the x-axis, the line labeled L_1 is "steeper" than the line labeled L_2. In order to measure the "steepness" of a line we associate with the line a number called its *slope*.

To define the slope of a line, we consider Figure 9.4 on page 338, where points $P(x_1, y_1)$ and $Q(x_2, y_2)$ represent any two points on the line. $P(x_1, y_1)$ represents a point P whose coordinates are x_1 and y_1. Similarly, $Q(x_2, y_2)$ represents a second point Q whose coordinates are x_2 and y_2.

Note that as we move on the line from P to Q, the y variable changes from y_1 to y_2.

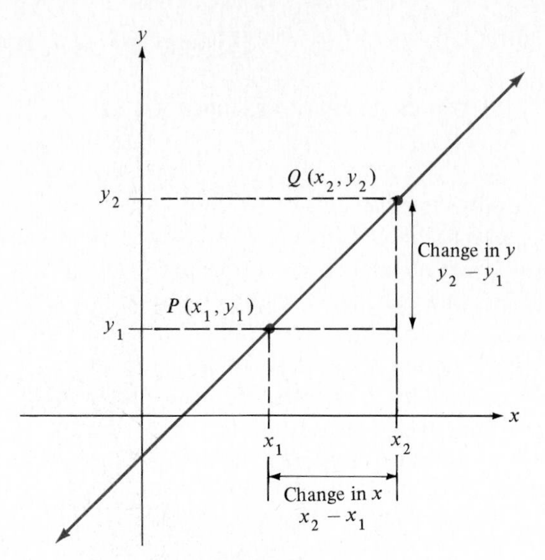

Figure 9.4

The amount of change is given by the quantity $(y_2 - y_1)$. The corresponding change in the x-variable is given by the quantity $(x_2 - x_1)$. If the line rises steeply, the change in y is greater than the change in x. If the line rises slowly, the change in y is less than the change in x. We illustrate these observations graphically in Figure 9.5. Notice that both lines pass through the same point, $P(1, 2)$.

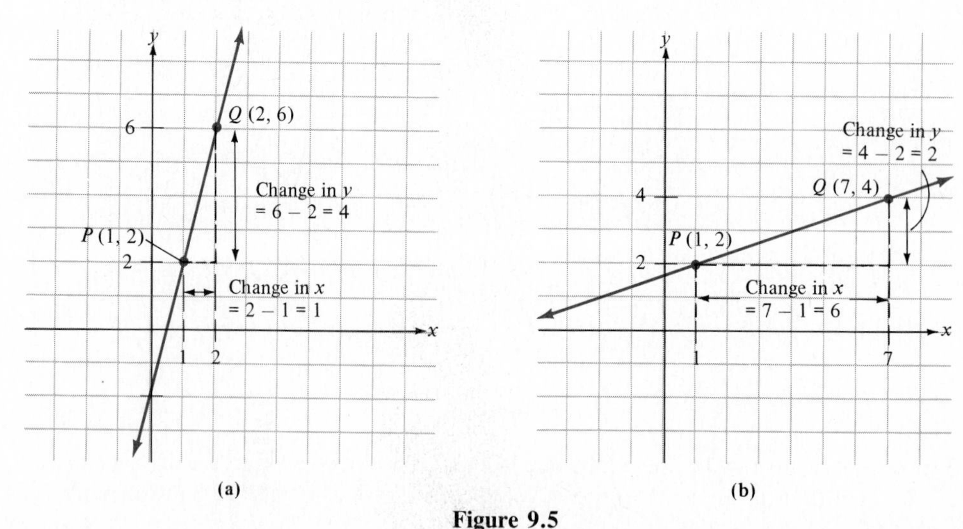

Figure 9.5

Now, if we form the *ratio* of the *change in y* to the *change in x* for the steeper line given in Figure 9.5(a), we obtain the following number:

$$\frac{\text{change in } y \text{ from } P \text{ to } Q}{\text{change in } x \text{ from } P \text{ to } Q} = \frac{6 - 2}{2 - 1} = \frac{4}{1} = 4.$$

Forming the same ratio for the line that rises slowly in Figure 9.5(b), we obtain

$$\frac{\text{change in } y \text{ from } P \text{ to } Q}{\text{change in } x \text{ from } P \text{ to } Q} = \frac{4 - 2}{7 - 1} = \frac{2}{6} = \frac{1}{3}.$$

Thus we see that the ratio for the *steeper* line,

$$\frac{\text{change in } y}{\text{change in } x} = \frac{4}{1} = 4$$

is a *larger* number than the corresponding ratio for the line that is *rising slowly,* which is

$$\frac{\text{change in } y}{\text{change in } x} = \frac{1}{3}.$$

In general, if P and Q are two points on a line, the ratio

$$\frac{\text{change in } y \text{ from } P \text{ to } Q}{\text{change in } x \text{ from } P \text{ to } Q}$$

is a number that enables us to measure the steepness of a line. It can be shown that the ratio for a given line would be the same for any two points that we choose on the line.

The previous discussion motivates the following definition for measuring the steepness of a nonvertical line.

Slope of a Line

If $P(x_1, y_1)$ and $Q(x_2, y_2)$ are any two points on a line, then the slope, m, of the line is

$$m = \frac{y_2 - y_1}{x_2 - x_1} = \frac{\text{change in } y \text{ from } P \text{ to } Q}{\text{change in } x \text{ from } P \text{ to } Q}.$$

Note that the formula remains valid if we write

$$m = \frac{y_1 - y_2}{x_1 - x_2}.$$

The last formula is valid because it can be obtained from the original formula,

$$m = \frac{y_2 - y_1}{x_2 - x_1},$$

by multiplying numerator and denominator by -1.

In the following examples we illustrate how to compute the slope of a line given the coordinates of two points on the line. We do not have to graph the line; all we need is the formula for the slope, m. However, we show the graphs in each example in order to associate the sign of the slope with the way the line slants.

Example 1. Find the slope of the line passing through the points (1, 4) and (3, 9).

Solution: The given line is shown in Figure 9.6.

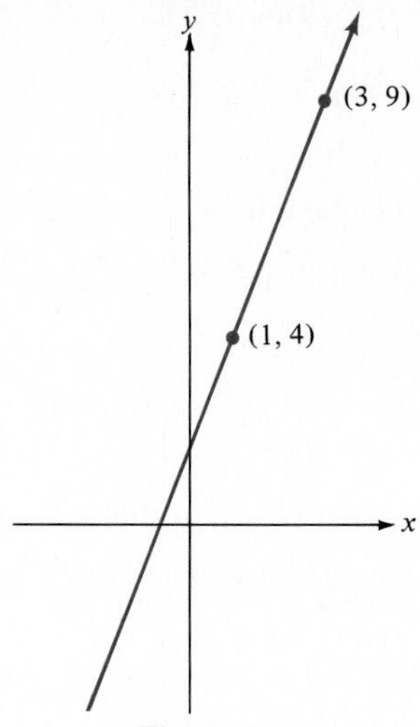

Figure 9.6

It does not matter which coordinates we label (x_2, y_2) and (x_1, y_1) as long as we subtract consistently. Thus we set

$$(x_2, y_2) = (3, 9) \quad \text{or} \quad x_2 = 3, y_2 = 9$$

and

$$(x_1, y_1) = (1, 4) \quad \text{or} \quad x_1 = 1, y_1 = 4.$$

The formula for the slope, m, gives us

$$m = \frac{y_2 - y_1}{x_2 - x_1} = \frac{9 - 4}{3 - 1} = \frac{5}{2}.$$

The slope, m, is $\frac{5}{2}$. This means that if x changes by 2 units, the corresponding change in y is 5 units.

Notice that if we let

$$(x_2, y_2) = (1, 4) \quad \text{or} \quad x_2 = 1, y_2 = 4$$
$$(x_1, y_1) = (3, 9) \quad \text{or} \quad x_1 = 3, y_1 = 9,$$

we obtain the same result. That is,

$$m = \frac{y_2 - y_1}{x_2 - x_1} = \frac{4 - 9}{1 - 3} = \frac{-5}{-2} = \frac{5}{2}.$$

Example 1 illustrates that lines that *rise* from *left to right* have *positive* slopes. Also, lines with *positive* slopes are lines that will *rise* from *left to right*.

Example 2. Find the slope of the line passing through the points (3, −2) and (7, −5).

Solution: The given line is shown in Figure 9.7. Let

$$(x_2, y_2) = (7, -5)$$

and

$$(x_1, y_1) = (3, -2).$$

Substituting in the slope formula, we obtain

$$m = \frac{y_2 - y_1}{x_2 - x_1} = \frac{-5 - (-2)}{7 - 3} = \frac{-5 + 2}{+4} = \frac{-3}{+4}.$$

Thus a change of +4 units in x (increase of 4 units) as we move from one point on the line to another point is accompanied by a decrease of 3 units (−3) in y.

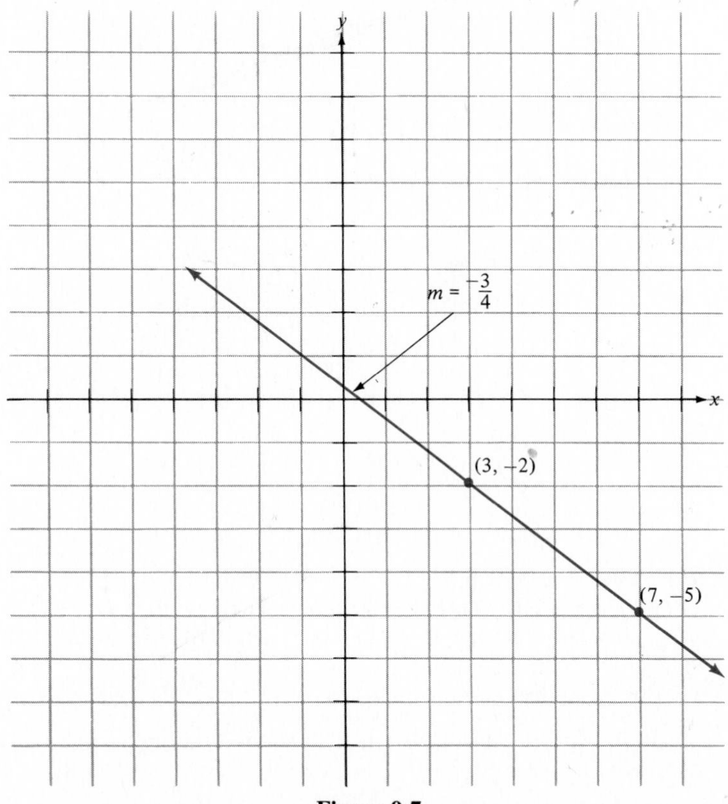

Figure 9.7

In Example 2 the line *fell* from *left to right* and had a *negative* slope. In general, lines that *fall* from *left* to *right* have *negative* slopes, and vice versa.

In the next example we show that horizontal lines have a slope of zero.

Example 3. Find the slope of the horizontal line passing through the points $(-2, 3)$ and $(4, 3)$.

Solution: Let

$$(x_2, y_2) = (-2, 3) \qquad \text{and} \qquad (x_1, y_1) = (4, 3).$$

Then

$$m = \frac{y_2 - y_1}{x_2 - x_1} = \frac{3 - 3}{-2 - 4} = \frac{0}{-6} = 0.$$

The slope of the line is 0 (see Figure 9.8).

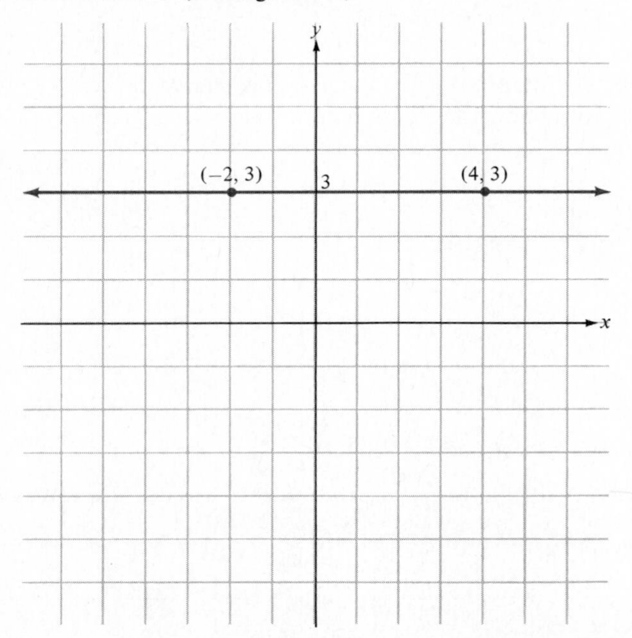

Figure 9.8

Notice that as illustrated in Example 3, any two points on a horizontal line must have the same y-coordinates. Therefore, the numerator (change in the y-coordinates) in the slope formula will be zero. The resulting slope m equals zero. As a result we state that

> All horizontal lines have a slope $m = 0$.

The next example illustrates that vertical lines do not have slopes defined for them.

Example 4. Find the slope of the vertical line passing through the points $(2, 3)$ and $(2, -1)$.

Solution: The graph of the given line is shown in Figure 9.9. Let

$$(x_2, y_2) = (2, 3)$$

and

$$(x_1, y_1) = (2, -1)$$

Therefore,

$$m = \frac{y_2 - y_1}{x_2 - x_1} = \frac{3 - (-1)}{2 - 2} = \frac{3 + 1}{0} = \frac{4}{0} \text{ is not defined.}$$

Remember that division by zero is not defined.

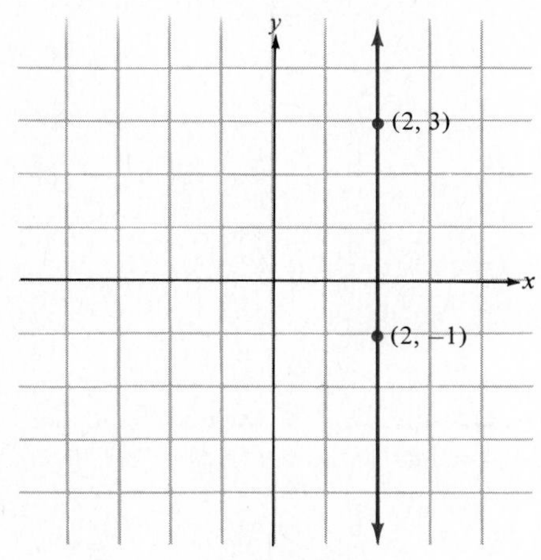

Figure 9.9

As Example 4 illustrated, the change in x for a vertical line will always be zero. Therefore, the denominator in the slope formula (change in x) will be zero. But division by zero is not defined. Consequently,

> The slope of a vertical line is not defined.

A summary of the previous discussion about slopes is shown in Figure 9.10.

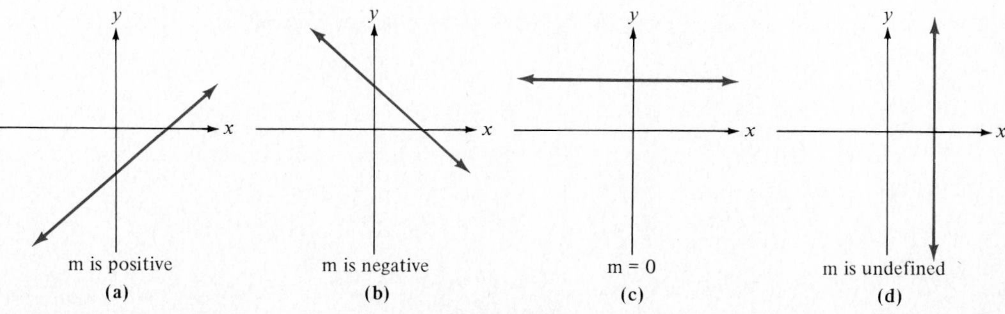

Figure 9.10

Example 5. Find the slope of the line in Figure 9.11.

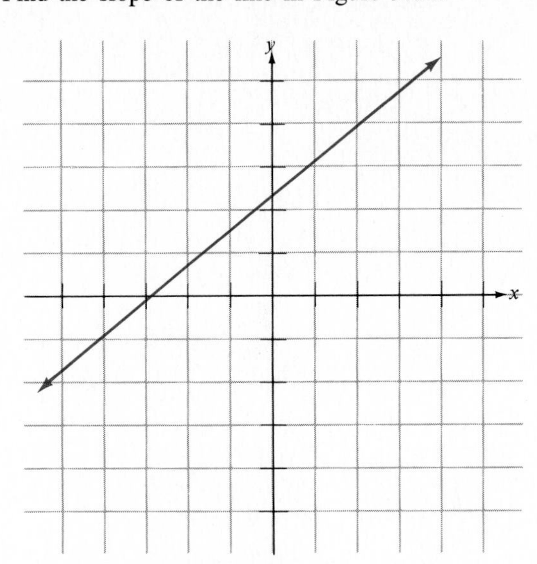

Figure 9.11

Solution: To compute the slope of the line, we choose any two points on the line. Let us choose the points with coordinates $(-3, 0)$ and $(2, 4)$. First let

$$(x_2, y_2) = (2, 4) \qquad \text{and} \qquad (x_1, y_1) = (-3, 0).$$

Therefore, the slope is

$$m = \frac{y_2 - y_1}{x_2 - x_1} = \frac{4 - 0}{2 - (-3)} = \frac{4}{5}.$$

EXERCISES 9.2

For Exercises 1–20, find the slope of the line passing through the given pair of points. See Example 1–4.

1. $(3, 4), (5, 7)$
2. $(2, 6), (8, 12)$
3. $(7, 10), (9, 12)$
4. $(5, 11), (11, 15)$
5. $(-3, 9), (5, 5)$
6. $(10, 3), (4, 4)$
7. $(5, 12), (7, 12)$
8. $(5, -9), (5, -13)$
9. $(4, -8), (4, -12)$
10. $(-3, 12), (4, 12)$
11. $(-6, -1), (-9, -5)$
12. $(-2, -3), (-4, -7)$
13. $(-9, -6), (-13, -2)$
14. $(-8, -10), (-6, -12)$
15. $(-3, 0), (0, 6)$
16. $(-7, 0), (0, 7)$
17. $(4, 0), (0, 12)$
18. $(5, 0), (0, 10)$
19. $(a, 0), (0, b)$
20. $(m, 0), (0, n)$

For Exercises 21–26, find the slope of the line. See Example 5.

21.

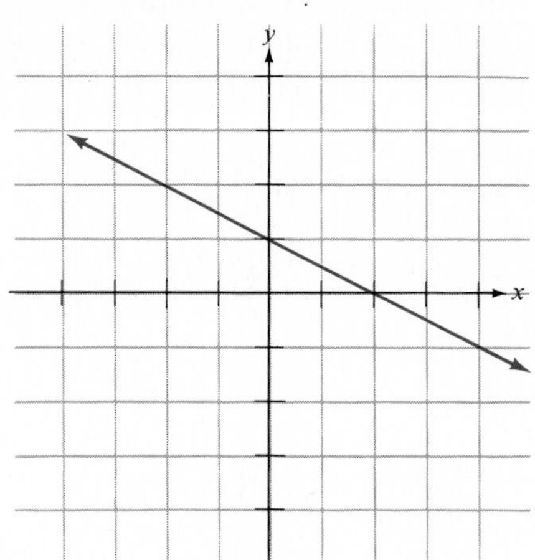

22.

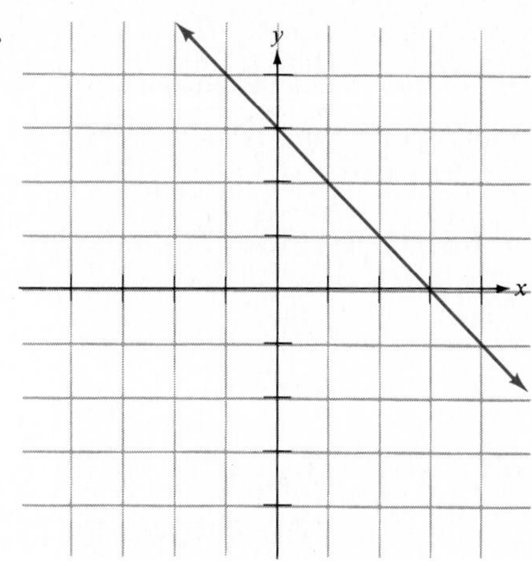

23.

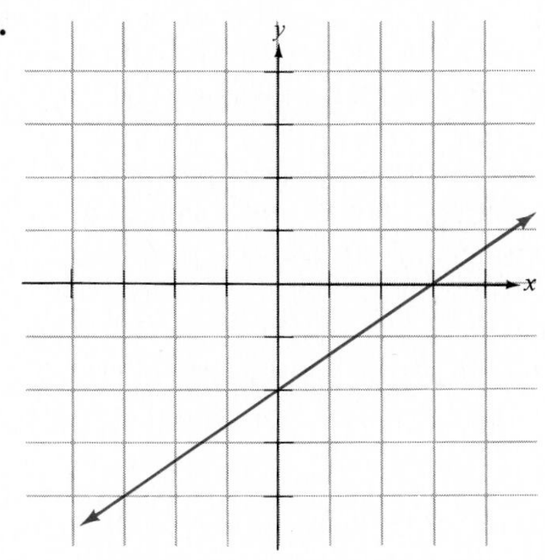

24.

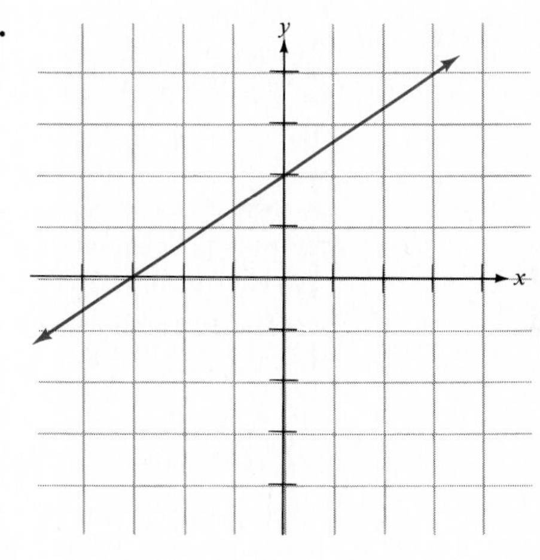

25.

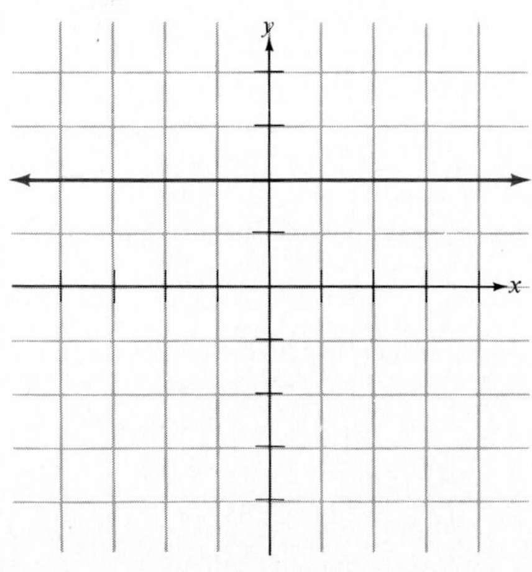

26.

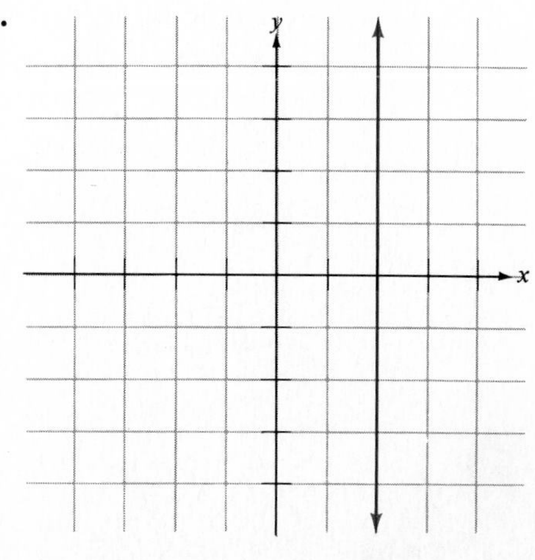

For Exercises 27–31, determine whether the statement is true or false.

27. Horizontal lines do not have slopes.

28. A line with a negative slope passing through the point (0, 3) must intersect the positive side of the x-axis.

29. All lines passing through the same point must have the same slope.

30. A line that has slope $\frac{1}{2}$ and passes through the origin, (0, 0), must pass through the point (2, 1).

31. Some lines do not have slopes defined for them.

9.3 Point-Slope and Slope-Intercept Form of the Equation of a Straight Line

In this section we discuss how to find the equation of a line from information that is given about the line. An equation of a line may be written in the form $ax + by = c$ and has the property that the coordinates (x, y) of *any* point on the line satisfy the equation. Also, any point whose coordinates satisfy the equation is on the line. The equation $ax + by = c$ is called the *standard form* of the equation of a line. As we will see, there are other forms of the equation of a line, provided that the line is nonvertical. We will discuss the equation of a vertical line at the end of this section.

What sort of information do we need about a line to write its equation? To answer this question, let us consider line L in Figure 9.12.

This figure indicates that we know the following information about line L.

1. It passes through point P with coordinates (2, 3).
2. The slope of the line, m, is 4.

Point $Q(x, y)$ in the figure represents any other point on the line with coordinates (x, y). To write the equation of line L means writing an equation relating coordinates x and y of any point on the line. To determine this equation, we use point (2, 3) and slope $m = 4$, together with the slope formula

$$\frac{y_2 - y_1}{x_2 - x_1} = m.$$

Remember that (x_2, y_2) and (x_1, y_1) can be any two points on the line. Thus we let

$$m = 4, \qquad (x_1, y_1) = (2, 3), \qquad (x_2, y_2) = (x, y),$$

where (x, y) is any other point on the line. Substituting in the formula for the slope, we have

$$\frac{y - 3}{x - 2} = 4.$$

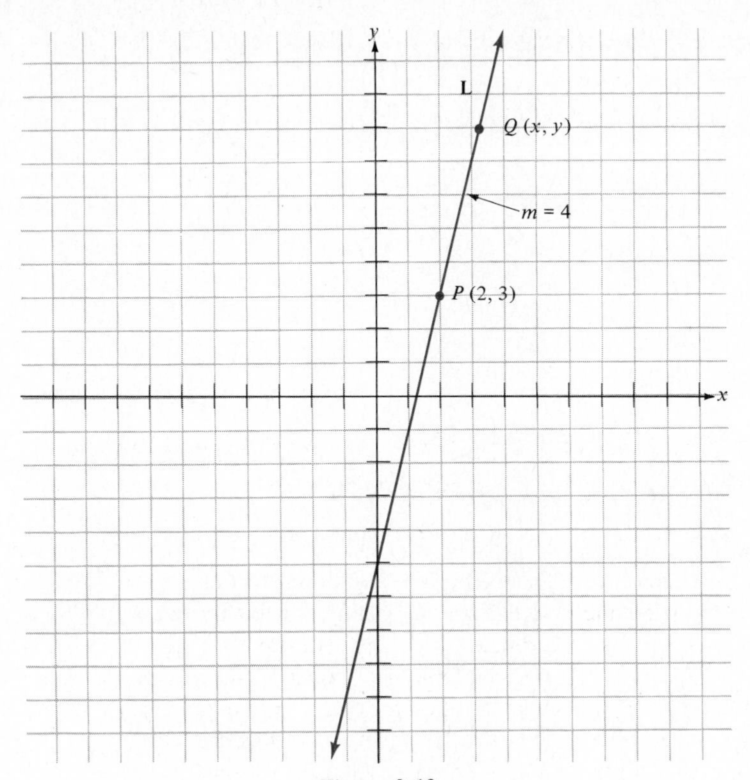

Figure 9.12

Multiplying both sides of the last equation by $(x - 2)$ gives

$$(x - 2)\frac{y - 3}{x - 2} = 4(x - 2)$$

$$y - 3 = 4(x - 2).$$

The last equation is called the *point-slope form* of the equation of the line. The coordinates of every point on the line will satisfy the equation. Also, every point whose coordinates satisfy the equation is on the line.

In general, if we know

1. The coordinates (x_1, y_1) of a point on the line, and
2. The slope m of the line,

then the line has an equation of the form

Point-Slope Form

$$y - y_1 = m(x - x_1).$$

Example 1. Find the equation of the line that contains the point (2, 1) and has a slope of 5.

Solution: We use the point-slope form of the equation of a straight line. First we let

$$(x_1, y_1) = (2, 1) \qquad \text{or} \qquad x_1 = 2, \quad y_1 = 1$$

and

$$m = 5.$$

Substituting in the formula

$$y - y_1 = m(x - x_1),$$

we obtain

$$y - 1 = 5(x - 2). \tag{1}$$

We can write the last equation in a simpler form as

$$y - 1 = 5x - 10 \qquad \text{distributive law}$$

$$y = 5x - 9. \qquad \text{add 1 to both sides} \tag{2}$$

Either equation (1) or equation (2) is the desired equation. However, equation (2) is the preferred form.

Example 2. Find the equation of the line passing through the point (−2, 5) and having slope $-\frac{3}{2}$.

Solution: Let

$$(x_1, y_1) = (-2, 5) \qquad \text{and} \qquad m = -\frac{3}{2}.$$

Using the point-slope form of the equation of a straight line,

$$y - y_1 = m(x - x_1),$$

we see that

$$y - 5 = -\frac{3}{2}\left[x - (-2)\right], \qquad x_1 = -2, \quad y_1 = 5$$

or

$$y - 5 = -\frac{3}{2}(x + 2).$$

To simplify the equation we multiply both sides by 2. Thus we see that

$$
\begin{aligned}
2(y - 5) &= -3(x + 2) && \\
2y - 10 &= -3x - 6 && \text{distributive property} \\
2y &= -3x + 4 && \text{add 10 to both sides} \\
y &= -\tfrac{3}{2}x + 2. && \text{divide both sides by 2}
\end{aligned}
$$

Suppose that we are given the point $(0, b)$, where a line intersects the y-axis and the slope of the line, m, as shown in Figure 9.13. Recall that b is called the *y-intercept* of

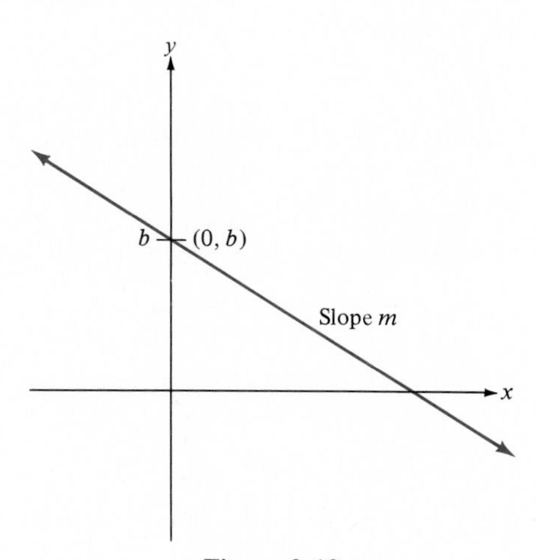

Figure 9.13

the line. Given the point $(0, b)$ and slope m, we can use the point-slope form to write a very simple and useful form of the equation of a straight line. To this end we let

$$(x_1, y_1) = (0, b) \quad \text{or} \quad x_1 = 0 \text{ and } y_1 = b.$$

Also, remember that we know the slope m. Applying the point-slope form,

$$y - y_1 = m(x - x_1),$$

we get

$$y - b = m(x - 0), \qquad x_1 = 0, \quad y_1 = b$$

If we solve for y, we obtain

$$
\begin{aligned}
y - b &= mx \\
y &= mx + b.
\end{aligned}
$$

The last equation is called the *slope-intercept form* of a linear equation.

Summarizing, if a line has slope m, and y-intercept b, then its equation is given by

Slope-Intercept Form

$$y = mx + b.$$

Notice that the coefficient of x in this form is the slope of the line and the constant term is the y-intercept.

Example 3. Find the equation of the line with slope 3 and y-intercept -4.

Solution: We are given that

$$m = 3 \quad \text{and} \quad b = -4.$$

Replacing m with 3 and b with -4 in the slope-intercept form,

$$y = mx + b,$$

we obtain

$$y = 3x + (-4)$$

or

$$y = 3x - 4.$$

This is the equation of the line with slope 3, y-intercept -4, in slope-intercept form.

Example 4. Find the equation of the line with slope $-\frac{7}{2}$ and y-intercept 5.

Solution: We use the slope-intercept form

$$y = mx + b$$

with $m = -\frac{7}{2}$ and $b = 5$. Substituting for m and b in the formula above, we get the desired equation,

$$y = -\frac{7}{2}x + 5.$$

Suppose that we are given the standard equation, $ax + by = c$, of a nonvertical line and wish to determine the slope of this line. We accomplish this by rewriting the equation in slope-intercept form. We illustrate how to deal with this type of problem in the next example.

Example 5. Find the slope of the line represented by the equation

$$-8x + 2y = 10.$$

Solution: We rewrite the equation $-8x + 2y = 10$ so that it is in slope-intercept form. To do this, we solve for y as follows:

$$\begin{aligned} -8x + 2y &= 10 \\ 2y &= 8x + 10 && \text{add } 8x \text{ to both sides} \\ y &= 4x + 5. && \text{divide both sides by 2} \end{aligned}$$

The last equation is the slope-intercept form,

$$y = mx + b,$$

of the original equation. In this form we see that the slope, m, is 4. Notice also that the y-intercept is 5.

The next example shows how to find the equation of a line if we are given two distinct points on the line.

Example 6. Find the equation of the line passing through the points (1, 2) and (3, 7).

Solution: First we use the two points to compute the slope m of the line. Then, using the slope, together with either one of the two given points, we apply the point-slope form to find the equation of the line. To compute the slope, we let

$$(x_1, y_1) = (1, 2) \qquad \text{and} \qquad (x_2, y_2) = (3, 7).$$

Thus

$$m = \frac{y_2 - y_1}{x_2 - x_1} = \frac{7 - 2}{3 - 1} = \frac{5}{2}.$$

Next, using the point-slope form,

$$y - y_1 = m(x - x_1)$$

with $(x_1, y_1) = (1, 2)$ and $m = \frac{5}{2}$, we have

$$y - 2 = \frac{5}{2}(x - 1).$$

We put the last equation in a simpler form by rewriting it in slope-intercept form. As a result, we have

$$y = \frac{5}{2}x - \frac{1}{2}.$$

If we had used the point (3, 7) in the point-slope form, we would obtain

$$y - 7 = \frac{5}{2}(x - 3).$$

This also simplifies to $y = \frac{5}{2}x - \frac{1}{2}$. Thus we see that no matter which point we select, we get the same slope-intercept equation.

We now consider the equations of horizontal and vertical lines, which appear to have special forms.

Recall from Section 9.2 that all horizontal lines have a slope of $m = 0$. Suppose that a horizontal line passed through the point (a, b). If we were to use the point-slope form of the equation of a line, we would obtain the following general rule.

> An equation of the horizontal line through (a, b) is
>
> $$y = b.$$

We also recall from Section 9.2 that vertical lines do *not* have slopes defined for them. However, vertical lines do have equations. In general,

> An equation of the vertical line through (a, b) is
>
> $$x = a.$$

Example 7. Find the equation of the lines passing through the given points. (a) $(-3, 5)$, $(7, 5)$. (b) $(2, 1)$, $(2, 6)$.

Solution

(a) If we were to graph the line passing through points $(-3, 5)$ and $(7, 5)$, we would see that it is a horizontal line. Note, therefore, that two distinct points with the same y-coordinates determine a horizontal line. Thus the equation of the line is

$$y = 5.$$

(b) The line passing through points $(2, 1)$ and $(2, 6)$ is a vertical line. Note that two distinct points with the same x-coordinates determine a vertical line. Therefore, the equation of the line is

$$x = 2.$$

EXERCISES 9.3

For Exercises 1–18, find the equation of the line passing through the given point with slope m. See Examples 1 and 2.

1. (3, 5); $m = 4$ **2.** (1, 1); $m = 2$ **3.** (1, 6); $m = 3$

4. (3, 4); $m = 1$ **5.** (1, −2); $m = 5$ **6.** (2, −7); $m = 8$

7. (4, −2); $m = -1$ **8.** (5, −1); $m = -3$ **9.** (−3, 7); $m = 2$

10. (−2, 4); $m = 3$ **11.** (2, 4); $m = \frac{5}{2}$ **12.** (6, 1); $m = \frac{2}{3}$

13. (9, 3); $m = -\frac{1}{3}$ **14.** (8, 2); $m = -\frac{7}{4}$ **15.** (−1, −5); $m = 9$

16. (−2, −2); $m = 2$ **17.** (10, 3); $m = 0$ **18.** (12, 7); $m = 0$

For Exercises 19–30, find the equation of the line with the given slope m and y-intercept b. See Examples 3 and 4.

19. $m = 7$; $b = 2$ **20.** $m = 8$; $b = 4$ **21.** $m = 11$; $b = 3$

22. $m = 9$; $b = 5$ **23.** $m = -2$; $b = 5$ **24.** $m = -3$; $b = 1$

25. $m = -5$; $b = 7$ **26.** $m = 13$; $b = 6$ **27.** $m = -\frac{5}{3}$; $b = 8$

28. $m = 1$; $b = -1$ **29.** $m = 0$; $b = -6$ **30.** $m = -4$; $b = 8$

For Exercises 31–40, find the slope of the line represented by the equation. See Example 5.

31. $-2x + y = 3$ **32.** $-3x + y = 4$ **33.** $3x - y = 2$ **34.** $5x - y = 1$

35. $12x - 3y = 6$ **36.** $15x + 5y = 30$ **37.** $2x - 5y = 20$ **38.** $4x - 3y = 9$

39. $y = 7$ **40.** $3x - 2y = 10$

For Exercises 41–52, find the equation of the line passing through the given points. See Examples 6 and 7.

41. (3, 5), (4, 7) **42.** (3, 3), (5, 9) **43.** (1, −2), (3, 6)

44. (2, −5), (3, 0) **45.** (0, 2), (3, 0) **46.** (0, 5), (1, 0)

47. (3, 7), (−2, 7) **48.** (8, −3), (9, −3) **49.** (−4, −1), (−4, 3)

50. (5, 7), (5, −6) **51.** (2, −7), (5, −7) **52.** (−1, 10), (−2, 10)

Variation: Direct and Inverse 9.4

Many real-world situations are modeled using the concept of variation. For example, the pressure that water exerts on an object varies directly with the depth of the object. Also, the intensity with which light strikes an object varies inversely as the square of the distance between the object and the light source. In this section concepts of direct and inverse variation will be developed. Applications that involve variation will also be considered. We begin by discussing direct variation.

If one quantity is a constant multiple of another quantity, we say that the quantities *vary directly*. Thus we have the following definition.

Direct Variation

If k is a constant and

$$y = kx,$$

then y varies directly as x.

If we divide both sides of the equation $y = kx$ by x, the result is

$$\frac{y}{x} = k,$$

where k is a constant. Thus, if two quantities, x and y, vary directly, we see that their ratio remains constant. The constant k is called the *constant of variation*. If we know x and y vary directly, we may write either $y = kx$ or $\frac{y}{x} = k$. Choose whichever form you prefer. If y varies directly with x, we also say that y *is proportional to* x and call k *the constant of proportionality*.

For example, suppose we know that y varies directly as x and that $y = 21$ when $x = 3$. What is the constant of variation and what is the value of y when $x = 4$? Since y varies directly as x, we may write $y = kx$. But $y = 21$ when $x = 3$; thus $21 = k \cdot 3$ or $k = 7$. Therefore, the constant of variation is 7. Now that we know the constant of variation, we may write

$$y = 7x.$$

Thus, when $x = 4$, we get $y = 7(4) = 28$.

Example 1. If y varies directly as x and $y = 15$ when $x = 6$, what is the value of y when $x = 10$?

Solution: Since y varies directly as x, we know that for some constant k,

$$y = kx.$$

But $y = 15$ when $x = 6$; thus

$$15 = k \cdot 6.$$

Solving this last equation for k results in

$$k = \frac{5}{2}.$$

Thus $y = \frac{5}{2}x$. Therefore, when $x = 10$ we get

$$y = \frac{5}{2}(10) \qquad \text{or} \qquad y = 25$$

Thus far in our examples, if two variables varied directly, we wrote one variable as a constant multiple of the other variable. The next example illustrates how to solve a direct variation problem by expressing the ratio of the two variables as a constant.

Example 2. Suppose that D varies directly as the square of t. If $D = 36$ when $t = 2$, what is the value of D when $t = 5$?

Solution: Since D varies directly as the square of t, we know that the ratio of D and t^2 is a constant. That is,

$$\frac{D}{t^2} = k\,.$$

Therefore,

$$\frac{36}{2^2} = k \qquad \text{or} \qquad k = 9$$

and for all D and t, we have $\frac{D}{t^2} = 9$. Thus, when $t = 5$, we get

$$\frac{D}{5^2} = 9 \qquad \text{or} \qquad D = 225.$$

In Example 1, y varied directly as x; that is, $y = kx$. The concept of direct variation can be extended to include powers of a variable. Thus, in Example 2, *D varied directly as the square of t*; that is,

$$D = kt^2. \qquad k \text{ is a constant}$$

Additional applications that involve direct variation follow.

The volume of a sphere *varies directly as the cube of the radius*. Thus

$$V = kr^3. \qquad (k = \tfrac{4}{3}\pi)$$

When a guitar string is plucked, its frequency of vibration f *varies directly as the square root of the tension t* on the string. Thus

$$f = k\sqrt{t}. \qquad (k \text{ constant})$$

Applied problems that involve direct variation are given in Example 3 and Exercises 36–40 at the end of this section.

Example 3. The pressure on an object submerged in a liquid varies with the depth of the object (Figure 9.14). An object 6 feet below the surface of the liquid experiences a pressure of 300 pounds per square foot. What is the pressure on the object when it is at a depth of 15 feet?

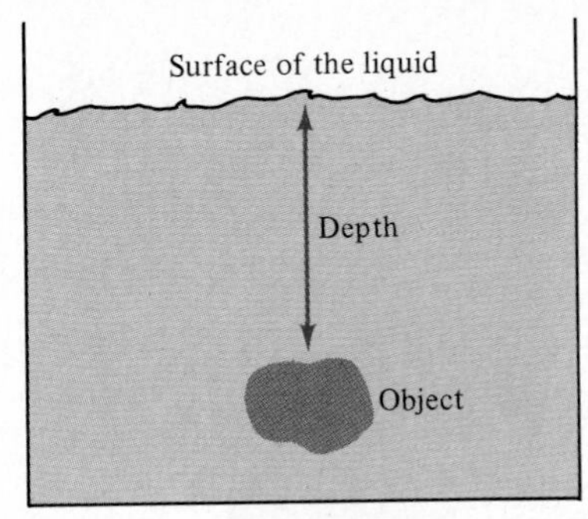

Figure 9.14

Solution: Let P denote pressure and d denote depth. Since P varies directly with d, we know that

$$P = kd.$$

Since $P = 300$ when $d = 6$, we have

$$300 = k \cdot 6$$

or

$$k = 50.$$

It follows that $P = 50d$. Consequently, when $d = 15$, we have

$$P = 50(15) \qquad \text{or} \qquad P = 750.$$

We conclude that the pressure is 750 pounds per square foot when the object is at a depth of 15 feet.

When two quantities *vary inversely,* their product is a constant. Thus "y varies inversely as x" is written algebraically as

$$xy = k.$$

If we divide both sides of this equation by x, we get

$$y = \frac{k}{x}$$

or

$$y = k\left(\frac{1}{x}\right).$$

Thus y varies inversely as x if y is a constant multiple of the reciprocal of x. The previous discussion is summarized by the following definition.

Indirect Variation

> If k is a constant and
>
> $$y = \frac{k}{x},$$
>
> then y varies inversely as x.

We also say that *y is inversely proportional to x*. As with direct variation, k is called the *constant of variation*. If x and y vary inversely, we may write either $y = \frac{k}{x}$ or $xy = k$. Both equations are correct.

Example 4. If y varies inversely as x and $y = 20$ when $x = 4$, find the value of y when $x = 10$.

Solution: Since y varies inversely as x, we know that

$$y = k\left(\frac{1}{x}\right).$$

As we did with direct variation problems, we will first determine the value of the constant of variation. Since $y = 20$ when $x = 4$, we have $20 = k\left(\frac{1}{4}\right)$ and in particular, $k = 80$. Thus

$$y = 80\left(\frac{1}{x}\right).$$

Now, when $x = 10$, we have

$$y = 80\left(\frac{1}{10}\right) \quad \text{or} \quad y = 8.$$

We see that $y = 8$ when $x = 10$.

As with direct variation, there are many applications that involve inverse variation of a variable raised to a power. For example, the weight W, of an object varies inversely as the *square of its distance d from the center of the earth*. Thus

$$W = \frac{k}{d^2}.$$

Another interesting application of inverse variation is the fact that the diameter d of the lens opening of a camera *varies inversely as the square root* of the exposure time t required to photograph an object:

$$d = \frac{k}{\sqrt{t}}. \qquad (k \text{ constant})$$

Example 5. The intensity of a source of light, such as a candle, may be measured with a light meter. As we move away from the light source, the light is distributed over a larger area and the light becomes less intense (Figure 9.15). In fact, the light intensity, denoted by I, varies inversely as the square of the distance d from the source of light. If $I = 200$ when $d = 2$ units, what is I when $d = 4$ units?

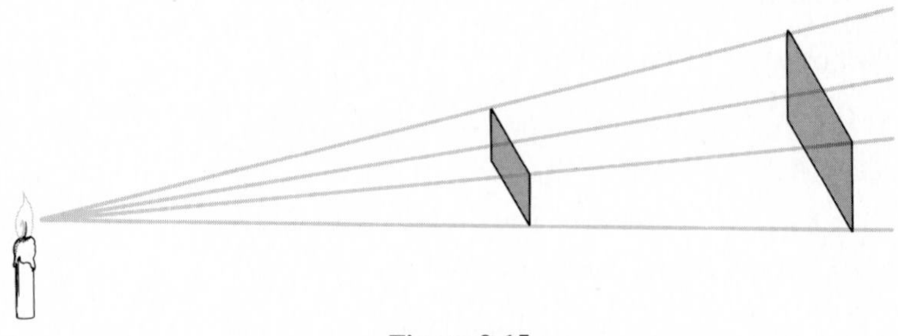

Figure 9.15

Solution: From the information given we know that

$$I = \frac{k}{d^2}.$$

To solve this problem, we will first determine the constant of variation k. Since $I = 200$ when $d = 2$, we have

$$200 = \frac{k}{2^2}.$$

It follows that $k = 800$ and

$$I = \frac{800}{d^2}.$$

We can now solve for I when $d = 4$. Substituting $d = 4$, we have

$$I = \frac{800}{4^2}$$

$$= \frac{800}{16} = 50.$$

EXERCISES 9.4

For Exercises 1–10, assume that y varies directly as x and solve for y. See Example 1.

1. If $y = 18$ when $x = 6$, find y when $x = 9$.

2. If $y = 20$ when $x = 5$, find y when $x = 7$.

3. If $y = 42$ when $x = 21$, find y when $x = 33$.

4. If $y = 42$ when $x = 7$, find y when $x = 11$.

5. If $y = 44$ when $x = 4$, find y when $x = 7$.

6. If $y = 54$ when $x = 6$, find y when $x = 8$.

7. If $y = 6$ when $x = 9$, find y when $x = 15$.

8. If $y = 18$ when $x = 12$, find y when $x = 18$.

9. If $y = 4$ when $x = 10$, find y when $x = 25$.

10. If $y = 15$ when $x = 25$, find y when $x = 15$.

For Exercises 11–15, assume that y varies directly as the square of x. See Example 2.

11. If $y = 45$ when $x = 3$, what is y when $x = 2$?

12. If $y = 63$ when $x = 3$, what is y when $x = 5$?

13. If $y = 75$ when $x = 5$, what is y when $x = 3$?

14. If $y = 96$ when $x = 12$, what is y when $x = 15$?

15. If $y = 15$ when $x = 5$, what is y when $x = 10$?

For Exercises 16–20, answer the following questions about direct variation. See Example 3 and the discussion that precedes it.

16. If Q varies directly as the square root of t and $Q = 10$ when $t = 4$, what is the value of Q when $t = 16$?

17. If Q varies directly as the square root of t and $Q = 6$ when $t = 16$, what is the value of Q when $t = 36$?

18. If V varies directly as the cube of r and $V = 32$ when $r = 2$, what is the value of V when $r = 3$?

19. If V varies directly as the cube of r and $V = 18$ when $r = 3$, what is the value of V when $r = 6$?

20. If y varies directly as the cube root of x and $y = 3$ when $x = 8$, what is the value of y when $x = 27$?

For Exercises 21–30, assume that y varies inversely as x. See Example 4.

21. If $y = 10$ when $x = 3$, find y when $x = 5$.

22. If $y = 8$ when $x = 5$, find y when $x = 10$.

23. If $y = 5$ when $x = 10$, find y when $x = 2$.

24. If $y = 8$ when $x = 8$, find y when $x = 32$.

25. If $y = 20$ when $x = 5$, find y when $x = 25$.

26. If $y = 10$ when $x = 9$, find y when $x = 15$.

27. If $y = 25$ when $x = \frac{2}{3}$, find y when $x = \frac{10}{3}$.

28. If $y = \frac{5}{2}$ when $x = 5$, find y when $x = 20$.

29. If $y = \frac{20}{3}$ when $x = \frac{10}{3}$ find y when $x = 2$.

30. If $y = 15$ when $x = \frac{5}{4}$, find y when $x = \frac{25}{2}$.

Answer Exercises 31–35 about inverse variation. See Example 5 and the discussion that precedes it.

31. Suppose that I varies inversely as the square of d. If $I = 25$ when $d = 2$, what is the value of I when $d = 5$?

32. Suppose that z varies inversely as the square of t. If $z = 10$ when $t = 4$, what is the value of z when $t = 2$?

33. Suppose that S varies inversely as the square root of q. If $S = 10$ when $q = 81$, what is the value of S when $q = 25$?

34. Suppose that T varies inversely as the square root of s. If $T = \frac{10}{9}$ when $s = 100$, what is the value of T when $s = \frac{25}{9}$?

35. Suppose that Y varies inversely as the cube of X. If $Y = 125$ when $X = 2$, what is the value of Y when $X = 10$?

Solve Exercises 36–45. See Examples 3 and 5.

36. The pressure on an object submerged in a liquid varies directly with the depth of the object. An object at a depth of 4 feet is experiencing a pressure of 260 pounds per square foot. What is the pressure on the object when it is 6 feet below the surface? (See Example 3.)

37. The distance, D (in miles), that we can see across a large body of water varies directly as the square root of our height, h (in feet), above the surface of the water. If $D = 11.25$ miles when $h = 25$ feet, what is D when $h = 49$ feet?

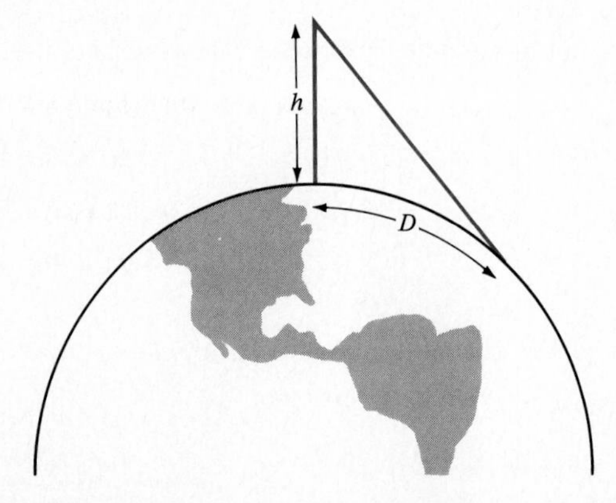

38. Tension in a spring is measured in pounds. The tension, T, varies directly as the distance, x, that the spring has been elongated from its original length. If $T = 60$ pounds when $x = 3$ inches, what will T be when $x = 5$ inches? See the accompanying figure at the top of page 361.

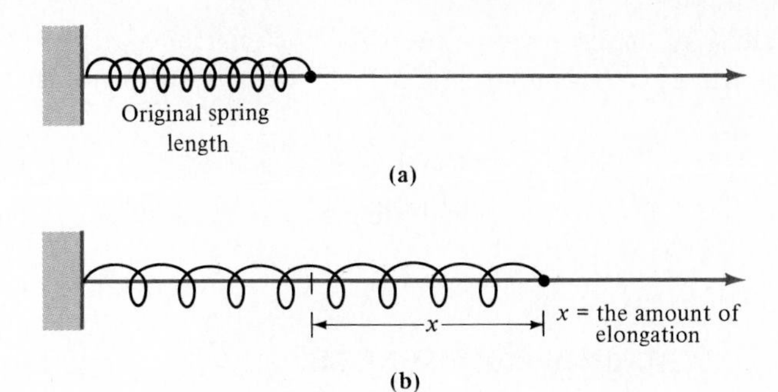

39. The weight of a uniformly shaped steel beam, W, varies directly as the length of the beam, L. If $W = 1200$ pounds when $L = 8$ feet, what is the weight of a 14-foot beam?

40. The loss in crop value per acre for wheat, L, varies directly with the average number of weed plants per square yard, W. If $L = \$80$ per acre when $W = 16$ plants per square yard, what is L when $W = 36$?

41. The intensity of a source of light, I, varies inversely as the square of the distance from the source of light, d. If $I = 40$ when $d = 5$ units, what is I when $d = 10$? (See Example 5.)

42. The weight of an object, W, varies inversely with the square of the distance between the object and the center of the earth. An object that weighs 200 pounds is on the surface of the earth, which is 4000 miles from the center of the earth. How much will the object weigh when it is 2000 miles above the surface of the earth?

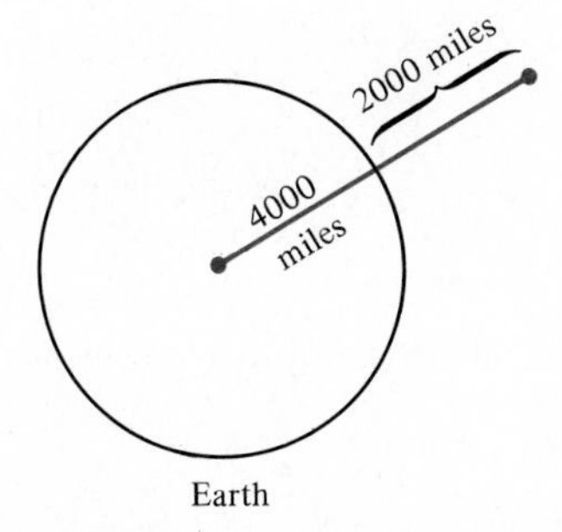

43. An interior designer wants to highlight a large plain wall by embedding reflecting tiles in the wall in a uniform pattern. The number of tiles, N, varies inversely with the square of the distance between the tiles, d. If $N = 100$ when $d = 2$ feet, what is N when $d = 5$ feet?

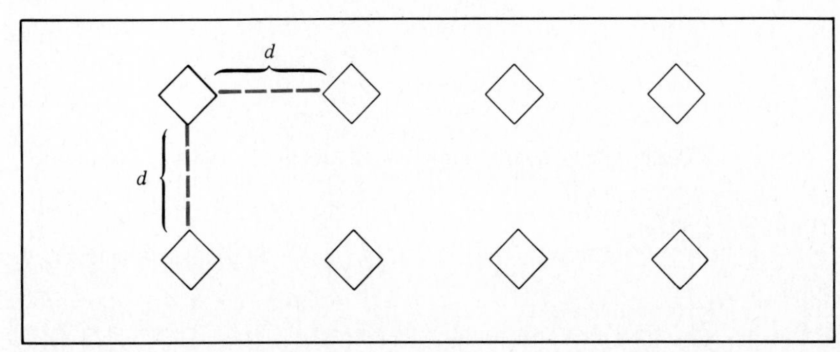

44. A box is filled with N marbles of uniform size and weight. The number of marbles in a pound varies inversely as the cube of the diameter of the marbles. If $N = 32$ when $d = 1$ inch, what is N when $d = 2$ inches?

45. The strength, S, of a wooden plank of fixed width and thickness varies inversely as the length of the plank. If $S = 200$ pounds when $L = 6$ feet, what is the strength of an 8-foot plank?

9.5 Rational Exponents

In this section we extend our knowledge of expressions such as x^m to include the case when m is a rational number p/q (p, q are integers, $q \neq 0$). In Section 3.1, we learned how to interpret x^m when the exponent m is an integer such as 4. Recall that if x is a real number, then

$$x^4 = x \cdot x \cdot x \cdot x.$$

In Section 3.2, we learned that if $x \neq 0$, then

$$x^0 = 1, \qquad x^{-1} = \frac{1}{x}, \qquad x^{-3} = \frac{1}{x^3}.$$

In order to develop an interpretation for expressions with rational exponents such as $x^{1/3}$, let us consider the expression

$$(x^{1/3})^3.$$

Rule 7, the power rule (Section 3.2), states that

$$(x^m)^n = x^{m \cdot n}.$$

If we assume that this rule is valid for expressions that contain rational exponents, it follows that

$$(x^{1/3})^3 = x^{(1/3)3} = x^1 = x.$$

Thus

$$(x^{1/3})^3 = x.$$

However, since $a^3 = b$ means $a = \sqrt[3]{b}$, we conclude that $x^{1/3}$ denotes the cube root of x. That is,

$$x^{1/3} = \sqrt[3]{x}.$$

Generalizing this example, we have the following definition.

If b is a positive real number and n is a positive integer, then

$$b^{1/n} = \sqrt[n]{b}.$$

That is, $b^{1/n}$ is another way of writing the nth root of b.

Example 1. Evaluate each expression.
(a) $16^{1/2}$. (b) $27^{1/3}$. (c) $10{,}000^{1/4}$.

Solution
(a) $16^{1/2} = \sqrt{16} = 4$ because $4^2 = 16$.
(b) $27^{1/3} = \sqrt[3]{27} = 3$ because $3^3 = 27$.
(c) $10{,}000^{1/4} = \sqrt[4]{10{,}000} = 10$ because $10^4 = 10{,}000$.

Using the definition above and Rule 7, we can write

$$(\sqrt[3]{x})^2 = (x^{1/3})^2 = x^{(1/3)2} = x^{2/3}.$$

For example, when $x = 125$, we get

$$\begin{aligned} 125^{2/3} &= (\sqrt[3]{125})^2 \\ &= 5^2 \\ &= 25. \end{aligned}$$

The following definition generalizes the above result. It gives a meaning to rational exponents that agrees with our previous definitions and properties of integer exponents. In the definition, and for the rest of the section, we assume that the base b is a positive real number.

Positive Rational Exponents

If m and n are positive integers and $b > 0$, then

$$b^{m/n} = (\sqrt[n]{b})^m.$$

Notice that the definition tells us that $b^{m/n}$ means to first take the nth root of the base b and then raise that result to the mth power. For example,

$$9^{3/2} = (\sqrt{9})^3 = 3^3 = 27.$$

Example 2. Evaluate each expression.
(a) $8^{2/3}$. (b) $4^{5/2}$. (c) $64^{2/3}$.

Solution: Using the previous definition, we get
(a) $8^{2/3} = (\sqrt[3]{8})^2 = 2^2 = 4$.
(b) $4^{5/2} = (\sqrt{4})^5 = 2^5 = 32$.
(c) $64^{2/3} = (\sqrt[3]{64})^2 = 4^2 = 16$.

From the definition of $b^{m/n}$ we know that

$$b^{m/n} = (\sqrt[n]{b})^m.$$

But since we are assuming all the rules for working with integer exponents may be applied to rational exponents, we may write

$$b^{m/n} = b^{m\cdot(1/n)} = (b^m)^{1/n} = \sqrt[n]{b^m}.$$

Therefore, it is also correct to write

$$b^{m/n} = \sqrt[n]{b^m}.$$

That is, it is also correct to first raise the base b to the mth power and then take the nth root of the result. However, it is usually easier to first extract the nth root and then raise this root to the mth power. For example,

$$49^{3/2} = (\sqrt{49})^3 = 7^3 = 343$$

is easier to do than

$$49^{3/2} = \sqrt{49^3} = \sqrt{117{,}649} = 343.$$

If the exponent is a negative rational number, we proceed as we did with negative integer exponents.

Negative Rational Exponents

If $b > 0$ and $m/n > 0$, then

$$b^{-m/n} = \frac{1}{b^{m/n}}.$$

Example 3. Evaluate each expression.
(a) $25^{-1/2}$. (b) $8^{-2/3}$. (c) $27^{-4/3}$.

Solution: Using the definition of negative rational exponents, we get

$$\text{(a)}\quad 25^{-1/2} = \frac{1}{25^{1/2}} = \frac{1}{\sqrt{25}} = \frac{1}{5}.$$

$$\text{(b)}\quad 8^{-2/3} = \frac{1}{8^{2/3}} = \frac{1}{(\sqrt[3]{8})^2} = \frac{1}{2^2} = \frac{1}{4}.$$

$$\text{(c)}\quad 27^{-4/3} = \frac{1}{27^{4/3}} = \frac{1}{(\sqrt[3]{27})^4} = \frac{1}{3^4} = \frac{1}{81}.$$

As stated previously, all rules that involve integer exponents can be used with rational exponents. The following examples demonstrate the use of some of the rules with expressions that contain rational exponents. We assume that any variable represents a positive real number.

Example 4. Use Rule 1 to simplify each expression.
(a) $7^{1/2} \cdot 7^{3/2}$. (b) $x^{2/3} \cdot x^{1/2}$.

Solution: Rule 1 states that

$$b^m \cdot b^n = b^{m+n}.$$

Applying this rule to the two expressions above, we get
(a) $7^{1/2} \cdot 7^{3/2} = 7^{(1/2)+(3/2)} = 7^{4/2} = 7^2 = 49.$
(b) $x^{2/3} \cdot x^{1/2} = x^{(2/3)+(1/2)} = x^{(4/6)+(3/6)} = x^{7/6}.$

Example 5. Use Rule 2 to simplify each expression.
(a) $8^{1/2} \cdot 2^{1/2}$. (b) $y^{3/4} \cdot z^{3/4}$.

Solution: Rule 2 states that

$$a^m \cdot b^m = (ab)^m.$$

Applying this rule to the expressions above, we get
(a) $8^{1/2} \cdot 2^{1/2} = (8 \cdot 2)^{1/2} = 16^{1/2} = 4.$
(b) $y^{3/4} \cdot z^{3/4} = (yz)^{3/4}.$

Example 6. Use Rule 3 to simplify each expression.
(a) $\dfrac{10^{13/5}}{10^{3/5}}$. (b) $\dfrac{W^{9/7}}{W^{2/7}}$.

Solution: Rule 3 states that

$$\frac{b^m}{b^n} = b^{m-n}.$$

Applying this rule to the two expressions above, we get

(a) $\dfrac{10^{13/5}}{10^{3/5}} = 10^{(13/5)-(3/5)} = 10^{10/5} = 10^2 = 100.$

(b) $\dfrac{W^{9/7}}{W^{2/7}} = W^{(9/7)-(2/7)} = W^{7/7} = W^1 = W.$

Example 7. Use Rule 7 to simplify each expression.
(a) $(8^{1/2})^{2/3}$. (b) $(x^{3/5})^{2/3}$.

Solution: Rule 7 states that

$$(b^m)^n = b^{m \cdot n}.$$

Applying this rule to the expressions above, we get
(a) $(8^{1/2})^{2/3} = 8^{(1/2)\cdot(2/3)} = 8^{1/3} = \sqrt[3]{8} = 2.$
(b) $(x^{3/5})^{2/3} = x^{(3/5)\cdot(2/3)} = x^{2/5} = (\sqrt[5]{x})^2.$

As the next two examples show, the rules for working with exponents may be used to simplify expressions with rational exponents.

Example 8. Simplify the expression $25^{1/4}$.

Solution

$$25^{1/4} = (5^2)^{1/4} = 5^{2(1/4)} = 5^{1/2} = \sqrt{5}.$$

Notice that $\sqrt{5}$ is easier to understand or estimate than $25^{1/4} = \sqrt[4]{25}$.

Example 9. Use the rules of exponents to simplify $(x^{2/3}y^{-2/5}z^{1/10})^{15}$.

Solution: There are several methods that could be used to simplify this expression. Two methods are shown below.

Method 1

$$\begin{aligned}(x^{2/3}y^{-2/5}z^{1/10})^{15} &= \left(\frac{x^{2/3}z^{1/10}}{y^{2/5}}\right)^{15} && \text{Rule 5}\\ &= \frac{(x^{2/3})^{15}(z^{1/10})^{15}}{(y^{2/5})^{15}} && \text{Rules 2 and 6}\\ &= \frac{x^{10}z^{3/2}}{y^6}. && \text{Rule 7}\end{aligned}$$

Method 2

$$\begin{aligned}(x^{2/3}y^{-2/5}z^{1/10})^{15} &= (x^{2/3})^{15}(y^{-2/5})^{15}(z^{1/10})^{15} && \text{Rule 2}\\ &= x^{10}y^{-6}z^{3/2} && \text{Rule 7}\\ &= \frac{x^{10}z^{3/2}}{y^6}. && \text{Rule 5}\end{aligned}$$

Rational exponents are extremely useful in advanced courses in mathematics and science. One reason for their usefulness is simplicity of notation. For example, although

$$x^{3/5} \quad \text{and} \quad (\sqrt[5]{x})^3$$

are equal, the first expression is less cluttered than the second. The following exercises will give you practice with expressions that involve rational exponents.

EXERCISES 9.5

For Exercises 1–20, determine the value of the expression. See Examples 1 and 2.

1. $100^{1/2}$ **2.** $144^{1/2}$ **3.** $125^{1/3}$ **4.** $729^{1/3}$ **5.** $625^{1/4}$

6. $400^{1/2}$ **7.** $625^{3/4}$ **8.** $400^{3/2}$ **9.** $16^{3/2}$ **10.** $25^{3/2}$

11. $8^{4/3}$ **12.** $27^{4/3}$ **13.** $1^{7/2}$ **14.** $1^{2/11}$ **15.** $27^{2/3}$

16. $1000^{2/3}$ **17.** $16^{3/4}$ **18.** $81^{3/4}$ **19.** $32^{2/5}$ **20.** $100{,}000^{2/5}$

For Exercises 21–32, use the definition of negative exponents to evaluate the expression. See Example 3.

21. $16^{-1/2}$ **22.** $49^{-1/2}$ **23.** $36^{-1/2}$ **24.** $64^{-1/2}$

25. $27^{-2/3}$ **26.** $8^{-5/3}$ **27.** $16^{-3/4}$ **28.** $81^{-3/4}$

29. $9^{-3/2}$ **30.** $36^{-3/2}$ **31.** $32^{-2/5}$ **32.** $1000^{-2/3}$

For Exercises 33–44, use Rule 1 to simplify the expression. Assume that all variables are positive. See Example 4.

33. $6^{1/2} \cdot 6^{3/2}$ **34.** $5^{1/2} \cdot 5^{3/2}$ **35.** $9^{2/3} \cdot 9^{4/3}$ **36.** $11^{2/3} \cdot 11^{4/3}$

37. $9^{5/3} \cdot 9^{1/3}$ **38.** $12^{3/7} \cdot 12^{4/7}$ **39.** $x^{1/5} \cdot x^{2/3}$ **40.** $y^{2/5} \cdot y^{1/3}$

41. $z^{3/2} \cdot z^{1/4}$ **42.** $a^{3/4} \cdot a^{1/2}$ **43.** $x^{2/3} \cdot x^{3/4}$ **44.** $y^{3/5} \cdot y^{2/3}$

For Exercises 45–52, use Rule 2 to simplify the expression. Assume that all variables are positive. See Example 5.

45. $3^{1/2} \cdot 12^{1/2}$ **46.** $2^{1/2} \cdot 32^{1/2}$ **47.** $4^{1/3} \cdot 2^{1/3}$ **48.** $4^{1/3} \cdot 250^{1/3}$

49. $x^{2/3} \cdot y^{2/3}$ **50.** $a^{4/5} \cdot b^{4/5}$ **51.** $y^{5/3} \cdot z^{5/3}$ **52.** $p^{3/7} \cdot q^{3/7}$

For Exercises 53–64, use Rule 3 to simplify the expression. Assume that all variables are positive. See Example 6.

53. $\dfrac{5^{8/3}}{5^{2/3}}$ **54.** $\dfrac{9^{11/5}}{9^{1/5}}$ **55.** $\dfrac{12^{7/5}}{12^{2/5}}$ **56.** $\dfrac{11^{9/7}}{11^{2/7}}$ **57.** $\dfrac{2^{11/2}}{2^{5/2}}$ **58.** $\dfrac{3^{15/4}}{3^{3/4}}$

59. $\dfrac{W^{8/3}}{W^{5/3}}$ **60.** $\dfrac{x^{7/5}}{x^{2/5}}$ **61.** $\dfrac{y^{11/3}}{y^{2/3}}$ **62.** $\dfrac{z^{13/4}}{z^{5/4}}$ **63.** $\dfrac{x^{5/2}}{x^{1/4}}$ **64.** $\dfrac{y^{2/3}}{y^{1/2}}$

For Exercises 65–72, use Rule 7 to simplify the expression. Assume that all variables are positive. See Example 7.

65. $(9^{1/5})^{5/2}$ **66.** $(25^{1/3})^{3/2}$ **67.** $(7^{1/3})^{6}$ **68.** $(4^{1/5})^{10}$

69. $(a^{2/3})^{1/2}$ **70.** $(b^{3/2})^{1/3}$ **71.** $(x^{3/5})^{10/3}$ **72.** $(y^{3/7})^{14/3}$

For Exercises 73–80, use the rules for exponents to simplify the expression. See Examples 8 and 9.

73. $36^{1/4}$ **74.** $100^{1/4}$ **75.** $81^{1/4}$ **76.** $121^{1/4}$

77. $(x^{2/5}y^{-1/3}z^{2/10})^{15}$ **78.** $(x^{3/4}y^{-1/2}z^{2/3})^{12}$

79. $(a^{-1/4}b^{2/5}c^{3/10})^{20}$ **80.** $(a^{4/3}b^{1/2}c^{-2/3})^{12}$

Systems of Linear Inequalities in Two Variables 9.6

In Chapter 6 we studied methods for solving systems of linear equations. We also learned how to graph linear equations, linear inequalities, and systems of linear equations. In this section we show how to determine the solution set of a *system of linear inequalities* by graphing.

A *system of linear inequalities* in two variables consists of two or more linear inequalities, such as

$$\begin{aligned} x + y &\le 10 \\ 3x + 4y &\le 36. \end{aligned}$$

The *solution set* of a system of linear inequalities is the set of all ordered pairs (x, y) that satisfy all the inequalities of the system. As shown in Section 6.4, the graph of a linear inequality is a half-plane or a half-plane with its boundary line. To find the solution set of a system of linear inequalities, we use a rectangular coordinate system and the following three steps.

1. Graph each linear inequality on the same rectangular coordinate system.
2. Determine the intersection of all the graphs in step 1. This is the graph of the solution set of the system of linear inequalities.
3. Sketch a graph that shows the solution set only.

The next example demonstrates these three steps.

Example 1. Graph the following system of linear inequalities.

$$\begin{aligned} 2x + y &\ge 6 \\ x - y &> -3. \end{aligned}$$

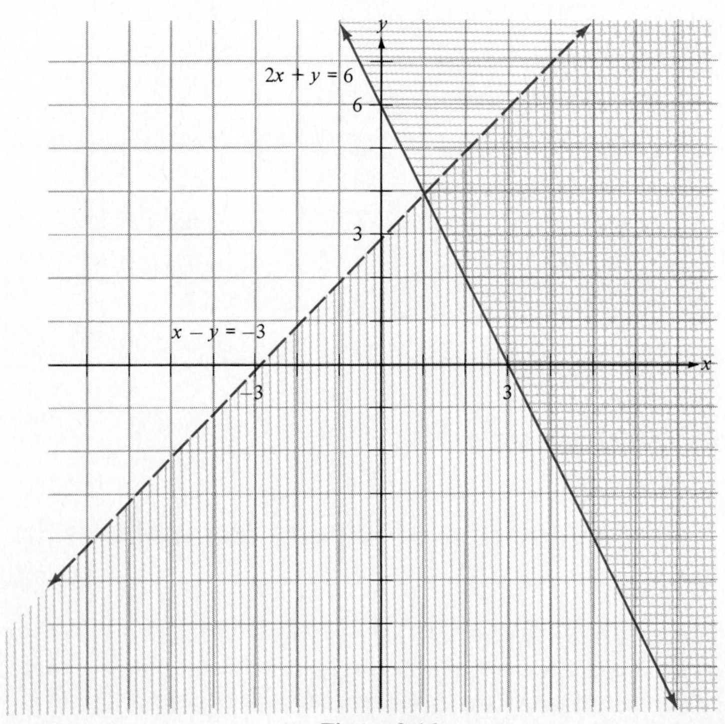

Figure 9.16

Solution: As shown in Chapter 6, we first sketch the boundary line of each inequality. If the inequality is $\leq$ or $\geq$, draw a solid line. If the inequality is $<$ or $>$, draw a dashed line. Next, by selecting convenient test points, we determine the half-plane that belongs to the solution set of each inequality. Finally, we determine the intersection of all of these half-planes. The end result is shown in Figure 9.16. The half-plane that is shaded horizontally is the graph of $2x + y > 6$. Thus the graph of $2x + y \geq 6$ is that half-plane together with the line $2x + y = 6$. The half-plane that is shaded vertically is the graph of $x - y > -3$. The solution set of the system of linear inequalities is the region of overlap of these two graphs. For clarity, a graph of the solution set only is shown in Figure 9.17.

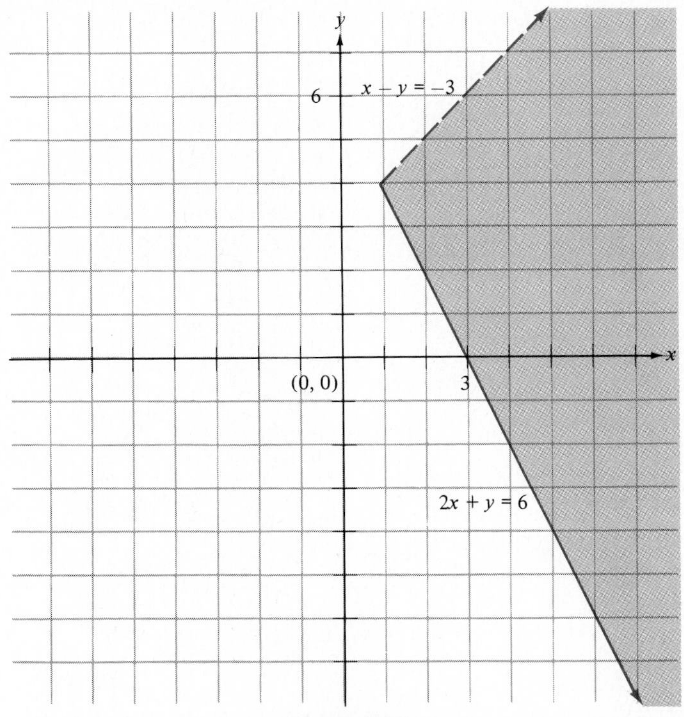

Figure 9.17

Many applications from business and economics involve linear inequalities in which all the variables are nonnegative. Requiring all variables to be greater than or equal to zero tells us that our solution set is restricted to the first quadrant (quadrant I) together with the positive x and y axes and the origin. The next example illustrates such a system of linear inequalities.

Example 2. Graph the solution set of the following system of linear inequalities:

$$x \geq 0$$
$$y \geq 0$$
$$2x + 3y \leq 12.$$

Solution: Using the techniques described in Section 6.4, we sketch the graph of $2x + 3y \leq 12$ as shown in Figure 9.18. The two inequalities $x \geq 0$ and $y \geq 0$ tell us that the set of points (x, y) that satisfy the system of inequalities must be such that both x and y are nonnegative. Thus the desired graph of the solution set is restricted to the first quadrant or the nonnegative part of the x and y axes, as shown in Figure 9.19.

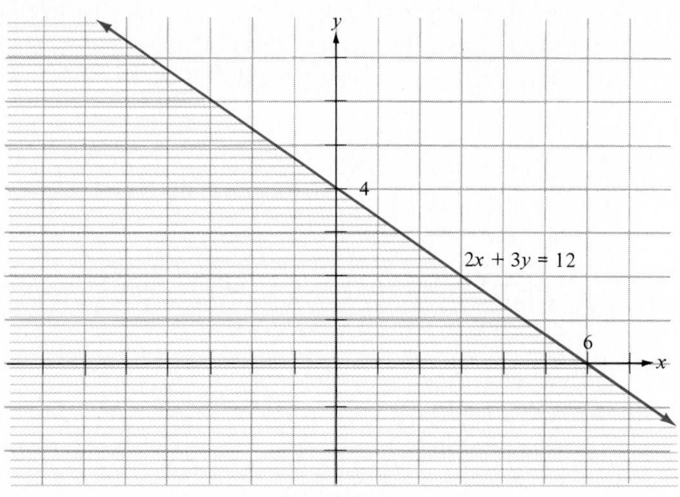

Figure 9.18

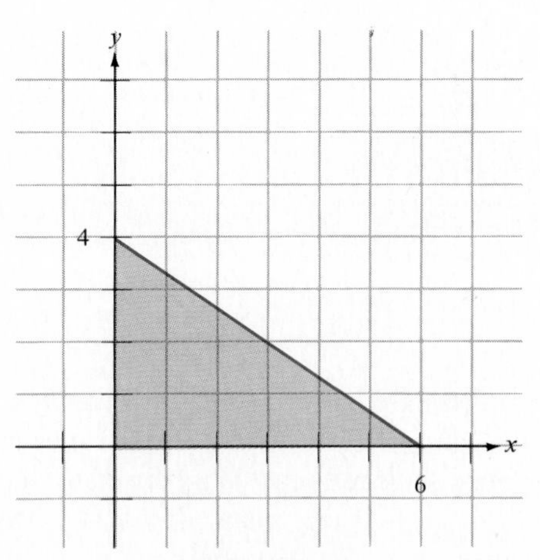

Figure 9.19

The following example shows that the graph of a system of linear inequalities may be bounded by parallel lines.

Example 3

(a) Graph the following system of inequalities.

$$2x + y \leq 8$$
$$2x + y \geq 4.$$

(b) Graph the system given in part (a) when both x and y are positive or zero.

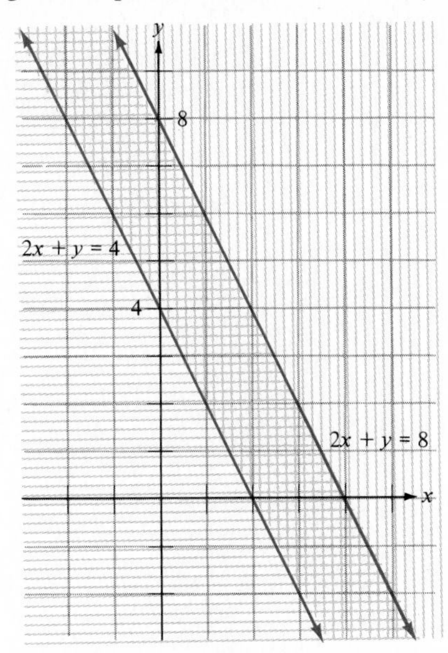

Figure 9.20

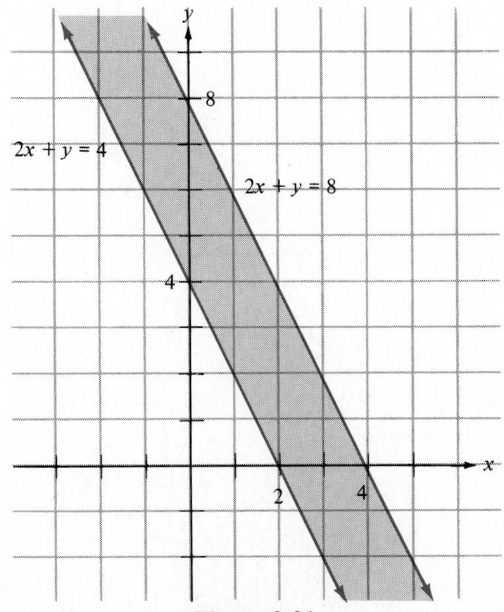

Figure 9.21

Solution

(a) Notice that the left-hand sides of both inequalities are the same. When we graph the boundary of each inequality, we see that they are parallel. In Figure 9.20 on page 371, the graph of $2x + y \leq 8$ is denoted using horizontal lines. The graph of $2x + y \geq 4$ is denoted with vertical lines. The solution set is the intersection of the two half-planes. Notice that both boundary lines belong to the solution set (Figure 9.21, on page 371).

(b) Adding the restrictions that $x \geq 0$ and $y \geq 0$ to the system of inequalities in part (a), we get the solution set shown in Figure 9.22.

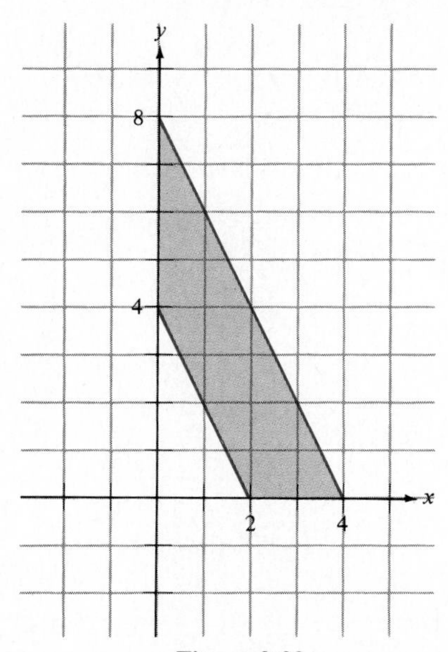

Figure 9.22

EXERCISES 9.6

For Exercises 1–10, graph the system of linear inequalities. See Example 1.

1. $x + y \leq 4$
$2x + y \geq 8$

2. $x + y \geq 6$
$2x + y \leq 8$

3. $3x + 5y \geq 15$
$3x - 2y \leq 6$

4. $2x + y \geq 8$
$2x - y \leq -2$

5. $3x + 4y \leq 12$
$x - y \geq -2$

6. $4x + 5y \leq 20$
$x - y \geq -2$

7. $4x + 3y > 12$
$-x + y \geq 2$

8. $5x + 4y > 20$
$-x + y \geq 2$

9. $x + y \leq 7$
$3x - 2y < -6$

10. $2x + y \leq 12$
$2x + 3y < 18$

For Exercises 11–20, sketch the system of inequalities. See Example 2.

11. $x \geq 0, \quad y \geq 0$
$x + y \leq 4$
$3x + y \leq 6$

12. $x \geq 0, \quad y \geq 0$
$x + y \leq 10$
$2x + y \leq 12$

13. $x \geq 0, \quad y \geq 0$
$3x + 2y \leq 6$
$2x + 3y \leq 6$

14. $x \geq 0, \quad y \geq 0$
$4x + 3y \leq 12$
$3x + 4y \leq 12$

15. $x \geq 0, \quad y \geq 0$
$3x + 2y > 3$
$3x + y \leq 6$

16. $x \geq 0, \quad y \geq 0$
$4x - 3y > 12$
$6x + 4y \geq 12$

17. $x \geq 0, \quad y \geq 0$
$6x + 5y < 30$
$3x - 2y \geq 6$

18. $x \geq 0, \quad y \geq 0$
$5x + 6y \leq 30$
$3x - 2y > 6$

19. $x \geq 0, \quad y \geq 0$
$x + 2y < 16$
$x - y > -1$

20. $x \geq 0, \quad y \geq 0$
$3x + 5y < 15$
$7x + 3y < 21$

For Exercises 21–26, sketch the system of linear inequalities. See Example 3.

21. $3x + 2y \leq 12$
$3x + 2y \geq 6$

22. $2x + 5y \leq 20$
$2x + 5y \geq 10$

23. $2x + y < 6$
$2x + y \geq 2$

24. $x + 3y \leq 9$
$x + 3y > 3$

25. $x \geq 0, \quad y \geq 0$
$4x + 2y \leq 8$
$4x + 2y \geq 4$

26. $x \geq 0, \quad y \geq 0$
$7x + 3y < 42$
$7x + 3y > 21$

Key Words

Constant of proportionality
Constant of variation
Dependent variable
Directly proportional
Domain
Function
Independent variable
Indirectly proportional
Point-slope form
Range
Rational exponent
Slope
Slope-intercept form
System of linear inequalities
Varies directly
Varies indirectly

CHAPTER 9 TEST

For Problems 1 and 2, determine whether or not the equation defines y as a function of x with the given set of x values. State the domain and range of any function.

1. $y = 2x^3; \{-2, 0, 2\}$ **2.** $y^2 = x - 1; \{0, 5, 10\}$

For Problems 3–6, determine the domain of the function of x.

3. $y = \dfrac{5}{x - 4}$ **4.** $y = x^4 + 1$ **5.** $y = \sqrt{3x + 6}$ **6.** $y = \dfrac{x - 2}{(x + 1)(x - 6)}$

For Problems 7 and 8, find the indicated function values.

7. $f(x) = 5x + 1$: **(a)** $f(0)$; **(b)** $f(2)$; **(c)** $f(-2)$.

8. $g(x) = \sqrt{3x + 1}$: **(a)** $g(0)$; **(b)** $g(5)$; **(c)** $g(8)$.

9. Compute the slopes of the lines passing through the given pair of points.
(a) (5, 3), (7, 9) **(b)** (1, −2), (3, −8)

Solve Problems 10–16.

10. Find the equation of the line that contains the point (6, 5) and has slope $\frac{2}{3}$.

11. Find the equation of the line with a slope of −7 and y-intercept 4.

12. Find the equation of the line passing through points (−2, 1) and (0, 5).

13. If y varies directly as x and $y = 15$ when $x = 12$, find the value of y when $x = 8$.

14. If y varies inversely as x and $y = 20$ when $x = \frac{5}{3}$, find the value of y when $x = \frac{5}{2}$.

15. If Q varies directly as the square root of t and $Q = 12$ when $t = 16$, what is the value of Q when $t = 64$?

16. The pressure on an object is 420 pounds per square foot when the object is at a depth of 12 feet. What is the pressure on the object when it is at a depth of 20 feet? Recall that pressure is directly proportional to the depth of the object.

For Problems 17 and 18, determine the value of the expression.

17. $16^{3/4}$ **18.** $27^{-2/3}$

For Problems 19–23, simplify the expression.

19. $7^{3/2} \cdot 7^{1/2}$ **20.** $\dfrac{4^{5/3}}{4^{2/3}}$ **21.** $\dfrac{x^{7/4}}{x^{3/4}}$ **22.** $(5^{1/4})^8$

23. $(x^{2/3}y^{-3/4}z^{1/6})^{12}$

24. Sketch the solution set of the system of linear inequalities.

$$\begin{aligned} x + y &\le 20 \\ 3x + y &\le 36 \end{aligned}$$

ANSWERS

Odd-Numbered Exercises and Chapter Tests

CHAPTER 1

Exercises 1.1

1. 12 **3.** 2 **5.** 14 **7.** 32 **9.** 5
11. 27 **13.** 6 **15.** 3 **17.** 1 **19.** 2
21. True **23.** True **25.** False **27.** True
29. True **31.** False **33.** True
35. The sum of 3 and 6 equals (or is) 9.
37. The difference of 9 and 3 is 6.
39. The product of 4 and 5 is 20.
41. The quotient of 20 by 5 is 4.
43. The sum of x and 2 is 3.
45. The product of 2 and x is less than 10.
47. The difference of x and y is 25.
49. The sum of x and 2 is greater than 3.
51. The product of x and y is less than or equal to 1, or the product xy is less than or equal to 1.
53. The sum of the quotient $\frac{x}{5}$ and 1 is greater than or equal to 2.
55. $<$ or $\leq$ **57.** $>$ or $\geq$ **59.** $<$ or $\leq$
61. $12 + 13 = 25$ **63.** $5(11) = 55$ or $5 \cdot 11 = 55$
65. $2x = 16$ **67.** $\frac{z}{3} = 12$ **69.** $3 + 4y = 15$
71. $w \cdot x \geq 4$ or $wx \geq 4$ **73.** $w + 3 > 12$

ANSWERS ANSWERS

Exercises 1.2

1. 16 **3.** 8 **5.** 1 **7.** 10,000 **9.** 64 **11.** 15 **13.** 26
15. 35 **17.** 8 **19.** 5 **21.** 9 **23.** 11 **25.** 24 **27.** 32
29. 5 **31.** 24 **33.** 2 **35.** 2 **37.** 1 **39.** 21 **41.** 13
43. 12 **45.** 5 **47.** 5 **49.** $\frac{2}{5}$ **51.** $x(x + y)$ **53.** $6 - xy$
55. $x^2 - y^2$ **57.** $\frac{x}{y} - 2$ **59.** $(x + y)(x - y)$ **61.** $x^3 + y^3$

Exercises 1.3

1. **(a)** $-4°$ **(b)** $-10°$ **3.** **(a)** -15 **(b)** $+2$

5.

7.

9. **(a)** $\{2\}$ **(b)** $\{-5, -3, 2\}$ **(c)** $\{-5, -3, .5, \frac{3}{2}, 2\}$ **(d)** $\{\sqrt{2}, \pi\}$ **(e)** $\{-5, -3, .5, \frac{3}{2}, \sqrt{2}, 2, \pi\}$
11. **(a)** Rational **(b)** Irrational **(c)** Rational **(d)** Rational **(e)** Irrational **(f)** Rational
13. **(a)** Integer or rational **(b)** Rational **(c)** Rational **(d)** Rational **(e)** Irrational **(f)** Irrational **(g)** Integer or rational **(h)** Integer or rational **(i)** Irrational
15. **(a)** True **(b)** False **(c)** True **(d)** True **(e)** True

Exercises 1.4

1. **3.**

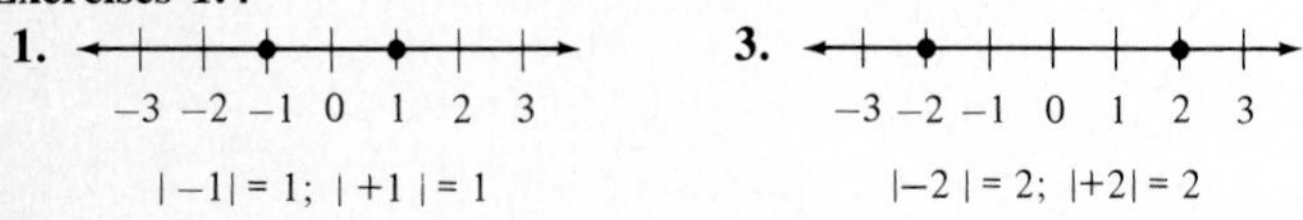

$|-1| = 1; \ |+1| = 1$ $|-2| = 2; \ |+2| = 2$

5.

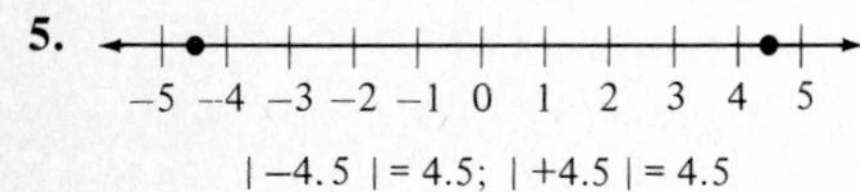

$|-4.5| = 4.5; \ |+4.5| = 4.5$

7.

$|-3\frac{1}{4}| = 3\frac{1}{4}; \ |+3\frac{1}{4}| = 3\frac{1}{4}$

9.

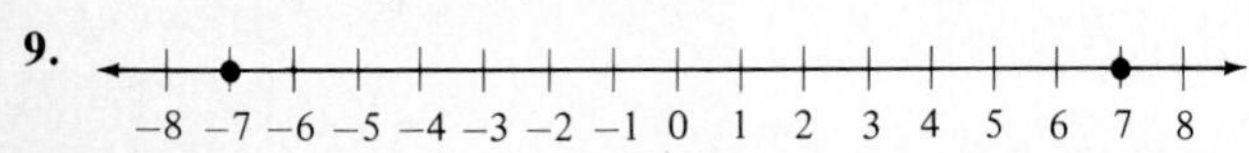

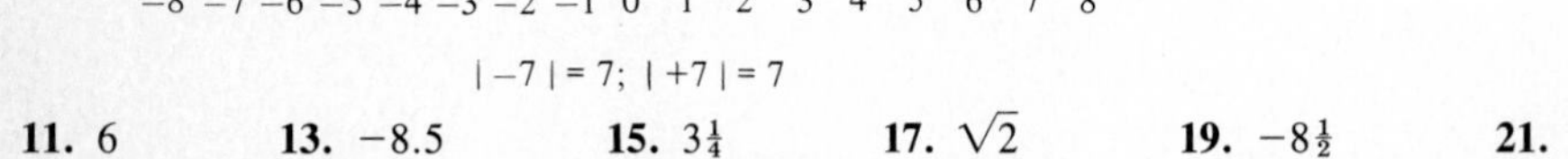

$|-7| = 7; \ |+7| = 7$

11. 6 **13.** -8.5 **15.** $3\frac{1}{4}$ **17.** $\sqrt{2}$ **19.** $-8\frac{1}{2}$ **21.** 8 **23.** $+81$
25. -7 **27.** $+\pi$ **29.** $-2\frac{1}{2}$ **31.** -32 **33.** $+2$ **35.** 2 **37.** -8
39. -15 **41.** -5 **43.** -2 **45.** -5 **47.** x is less than or equal to -3.
49. x is positive. **51.** x is negative. **53.** True **55.** False **57.** True **59.** False
61. True **63.** True **65.** True

Exercises 1.5

1.

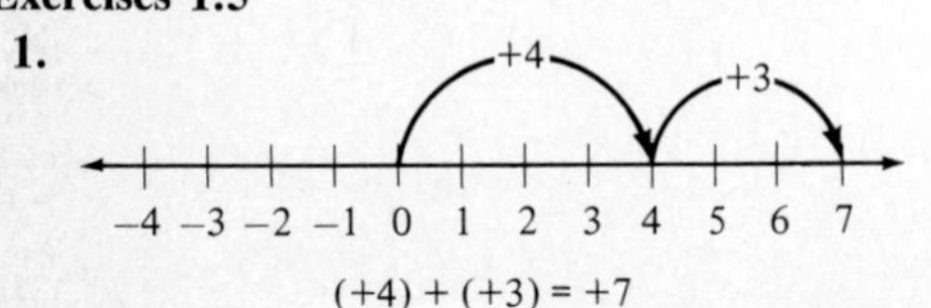

$(+4) + (+3) = +7$

3.

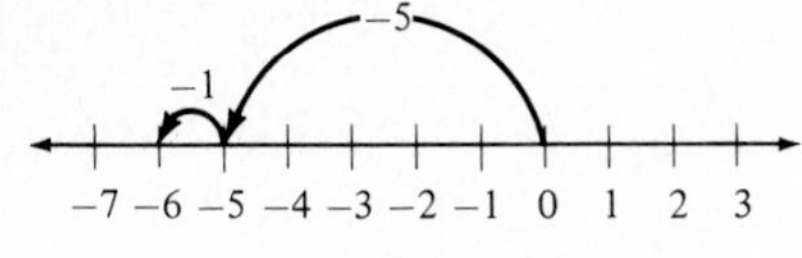

$(-5) + (-1) = -6$

5.

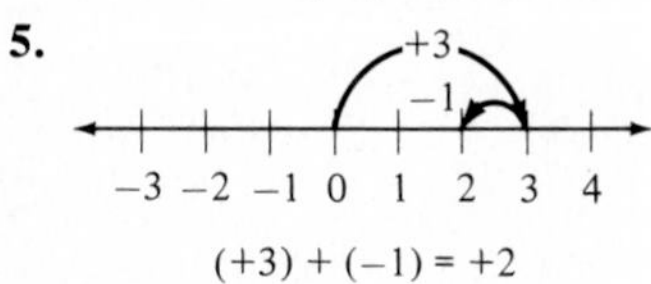

$(+3) + (-1) = +2$

7.

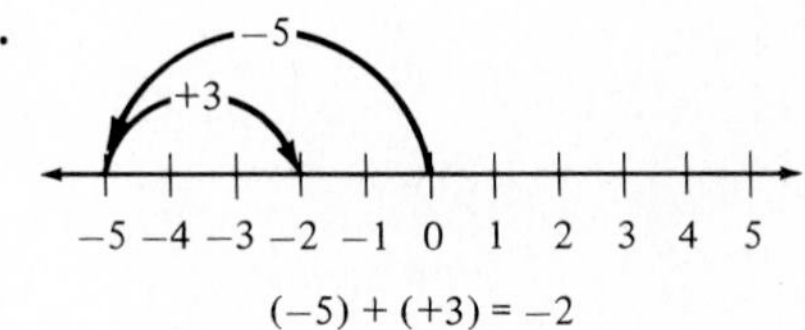

$(-5) + (+3) = -2$

9.

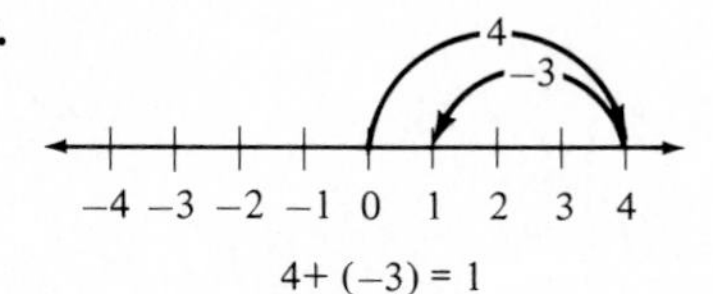

$4 + (-3) = 1$

11. 13 **13.** −10 **15.** −8 **17.** 10 **19.** 1 **21.** 5 **23.** $-\frac{2}{3}$
25. 0 **27.** .50 **29.** $4\frac{1}{2}$ **31.** $7\frac{1}{2}$ **33.** −.5 **35.** 0 **37.** 3.2
39. 3.2 **41.** True **43.** True **45.** False **47.** False **49.** True
51. **(a)** +15 **(b)** 10° **53.** **(a)** −2 **(b)** −6 because $(-2) + (-4) = -6$
55. **(a)** +3, +6, −5 **(b)** $(+3) + (+6) + (-5) = +9 + (-5) = +4$; a total gain of 4 yards **(c)** 6 yards

Exercises 1.6

1. −5 **3.** 8 **5.** −13 **7.** −12 **9.** 15 **11.** −9 **13.** 12
15. −7 **17.** 0 **19.** −24 **21.** 5.5 **23.** −10.30 **25.** −2 **27.** 2
29. −3 **31.** −14 **33.** −9 **35.** −5 **37.** 12 **39.** −13 **41.** 14
43. −4 **45.** −5° **47.** −2 **49.** **(a)** −7520 **(b)** +6325 **(c)** $6325 - (-7520) = 13{,}845$ feet

Exercises 1.7

1. Commutative **3.** Associative **5.** Commutative **7.** Associative **9.** Commutative
11. Additive inverse **13.** Zero property **15.** Associative **17.** Multiplicative inverse
19. Additive identity **21.** Distributive **23.** Distributive **25.** True **27.** False
29. False **31.** True **33.** True **35.** False **37.** True **39.** True
41. True **43.** 2 **45.** −15 **47.** 14 **49.** −18 **51.** −2 **53.** $5x + 10$
55. $2x - 4$ **57.** $r \cdot 3 + 6$ **59.** $(xy)z + (xy)w$ **61.** $4x - 4y$ **63.** $x \cdot 2 - 6$
65. $12 - y \cdot 3$

Exercises 1.8

1. −14 **3.** −20 **5.** 30 **7.** 33 **9.** 20 **11.** −4.8 **13.** -5π
15. $\frac{2}{3}$ **17.** $\frac{1}{12}$ **19.** 9.0 **21.** x **23.** 6 **25.** −6 **27.** $-r$ **29.** 15
31. 5 **33.** 0 **35.** −8 **37.** 16 **39.** −9 **41.** −2 **43.** $-5x + 15$
45. $-x - y + z$ **47.** $-x - y$ **49.** $-x + w - s$ **51.** $-x - y + z$ **53.** $5x - 2z - 1$
55. $8x - y - w$ **57.** **(a)** −49.50 **(b)** −198.00 **(c)** −230.20

Exercises 1.9

1. −10 **3.** −5 **5.** 5 **7.** 4 **9.** −9 **11.** −5 **13.** 0
15. Not possible **17.** 1 **19.** Not possible **21.** $\frac{-1}{4}, \frac{1}{-4}$ **23.** $-\frac{3}{5}, \frac{3}{-5}$

25. $\frac{-4}{5}, -\frac{4}{5}$ **27.** $\frac{-3}{7}, \frac{3}{-7}$ **29.** $\frac{-7}{9}, -\frac{7}{9}$ **31.** $\frac{13}{-11}, -\frac{13}{11}$ **33.** -2
35. $\frac{1}{3}$ **37.** $\frac{14}{3}$ **39.** $-\frac{2}{3}$ **41.** 24 **43.** -3 **45.** 4 **47.** -2
49. 3 **51.** -1 **53.** -2 **55.** -4 **57.** 3 **59.** -1 **61.** $-1°$

Chapter 1 Test

1. $\frac{x}{y} = 2$ **2.** $5y + 6 = 31$ **3.** $x + 15 < 18$ **4.** 12 **5.** 37 **6.** 15
7. 20 **8.** 1 **9.** 12 **10.** 12 **11.** 25 **12.** 2
13. (number line) −5 −4 −3 −2 −1 0 1 2 3 4 5
14. $\{-7, -1, 2\}$ **15.** $\{-7, -1, \frac{22}{7}, 2, .3\}$ **16.** $\{\sqrt{2}, \pi\}$ **17.** 16 **18.** 2
19. True **20.** False **21.** False **22.** 14 **23.** -13 **24.** 0 **25.** -7
26. 2 **27.** -12 **28.** -20 **29.** $-7°$ **30.** C **31.** B **32.** E
33. D **34.** $6x + 12$ **35.** $yx - y$ **36.** True **37.** True **38.** False
39. True **40.** -10 **41.** 0 **42.** 4 **43.** 0 **44.** -5 **45.** -3
46. -120 **47.** 24 **48.** 1 **49.** -1 **50.** 3

CHAPTER 2

Exercises 2.1

1. $11x$ **3.** $5z$ **5.** $7x$ **7.** $-9q$ **9.** $-7x$ **11.** $11x$ **13.** $6z$
15. $2w$ **17.** $7x$ **19.** 0 **21.** $14z - 5$ **23.** $4q + 4$ **25.** $9v - 6$
27. $2x + 8$ **29.** $2z + 6$ **31.** $6x + 4$ **33.** $10z - 11$ **35.** $x - 7$
37. $-8x + 53$ **39.** $4x - 51$ **41.** $x + 4$ **43.** $.17x$ **45.** $.18y$ **47.** $-\frac{1}{3}x + \frac{3}{10}$
49. $\frac{11}{12}z - \frac{1}{2}$

Exercises 2.2

1. 1 **3.** 3 **5.** 2 **7.** 1 **9.** 0 **11.** 11 **13.** -4 **15.** 9
17. 7 **19.** -1 **21.** 4 **23.** 10 **25.** 1 **27.** -2 **29.** -2
31. 34 **33.** 0 **35.** 6 **37.** 3 **39.** 2 **41.** 8 **43.** 17 **45.** 16
47. 8 **49.** 13 **51.** 1 **53.** 4 **55.** 10
57. After simplifying, the result is a false statement such as $3 = 0$.
59. After simplifying, the result is a true statement such as $0 = 0$.

Exercises 2.3

1. 3 **3.** 5 **5.** -7 **7.** -9 **9.** 8 **11.** $\frac{1}{2}$ **13.** $\frac{5}{3}$ **15.** 2
17. $\frac{9}{2}$ **19.** $-\frac{7}{2}$ **21.** $\frac{-4}{3}$ **23.** $\frac{12}{7}$ **25.** $\frac{17}{12}$ **27.** -1 **29.** -5
31. -3 **33.** -3 **35.** $-\frac{3}{4}$ **37.** 3 **39.** -3

Exercises 2.4

1. 15 **3.** 10 **5.** -14 **7.** -24 **9.** 12 **11.** 6 **13.** -18
15. -22 **17.** 10 **19.** -3 **21.** -2 **23.** 9 **25.** 33 **27.** 9
29. -10 **31.** 7 **33.** 9 **35.** 5 **37.** 4 **39.** 13

Exercises 2.5

1. -2 **3.** 6 **5.** -3 **7.** -7 **9.** 4 **11.** 1 **13.** 2 **15.** -9
17. -1 **19.** -4 **21.** -2 **23.** -6 **25.** -11 **27.** 4 **29.** -1

Exercises 2.6

1. 18 **3.** 17 **5.** 12 **7.** 8 **9.** 60 **11.** 16 **13.** 17, 18, and 19
15. 24 **17.** 14 and 33 **19.** 6

Exercises 2.7

1. \$165 **3.** \$1000 **5.** $A = 64\pi$ square feet $\simeq 200.96$ square feet **7.** 40%
9. **(a)** \$1000 **(b)** 60% **11.** 4 feet **13.** $P = 50$ feet, $A = 144$ square feet
15. **(a)** $P = 60$ units, $A = 150$ square units **(b)** $P = 30$ units, $A = 30$ square units **17.** $A = 50$ square units
19. 808 square feet **21.** 800π square inches or $\simeq 2512$ square inches **23.** 6 hours
25. Width = 12 inches; height = 12 inches

Exercises 2.8

1. 20 times **3.** 3 mph **5.** 4 liters **7.** 1.8 gallons **9.** 2500 items
11. $W = 24$ inches; $L = 36$ inches **13.** 40% **15.** 2 hours **17.** 2 hours

Exercises 2.9

1. $x \geq 7$
0 2 4 6 7 8

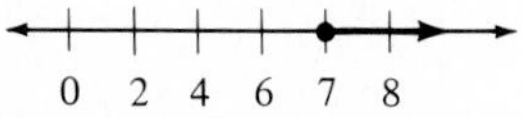

3. $x \leq 5$
0 2 4 5 6 8

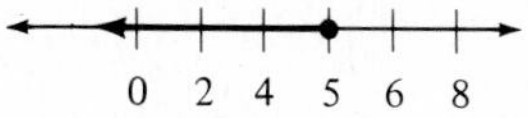

5. $x < -1$
−2 −1 0 1 2 3

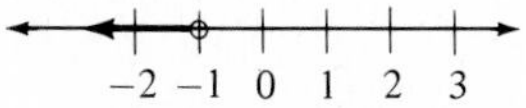

7. $x > 2$
−2 −1 0 1 2 3

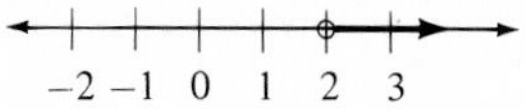

9. $x \leq -4$
−5 −4 −3 −2 −1 0 1

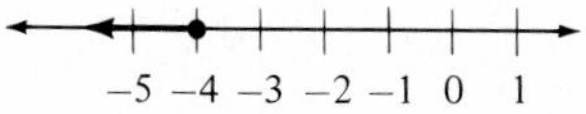

11. $x > -11$ 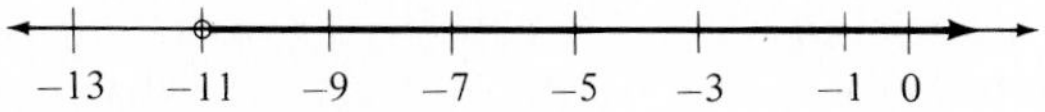
−13 −11 −9 −7 −5 −3 −1 0

13. $x \geq 4$
−4 −2 0 2 4 6

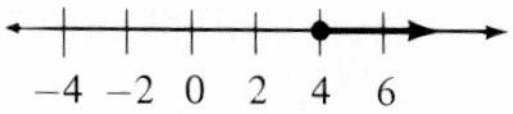

15. $x < -2$
−4 −2 0 2 4 6

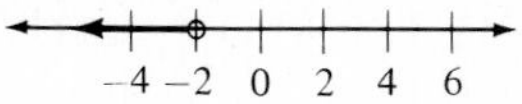

17. $x \geq -\frac{25}{2} = -12.5$
−15 −12.5 −10

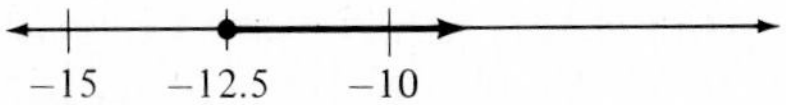

19. $x < -4$
−6 −4 −2 0 1 2

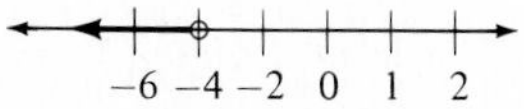

21. $x \geq 1.5$
−1 0 1 2 3
1.5

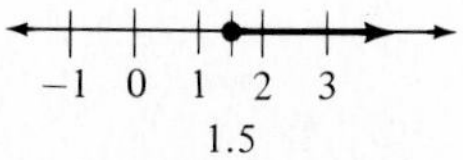

23. $x > -8$
−12 −10 −8 −6 −4

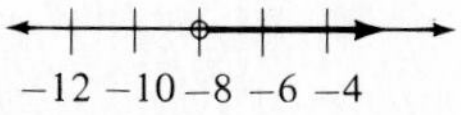

25. $x \geq -4$ (number line: −6 −4 −2 0 2)

27. $x > -2$ (number line: −4 −2 0 2)

29. $x > -1$ (number line: −2 −1 0 1 2)

31. $x < -3$ (number line: −5 −4 −3 −2 −1 0)

33. $x \leq 9$ (number line: 7 8 9 10 11)

35. $x > -3$ (number line: −4 −3 −2 −1 0 1)

37. $x > 1$ (number line: −2 −1 0 1 2 3)

39. If x denotes the number of buses, then we want $45x \geq 390$. Thus we need $x \geq 8\frac{2}{3}$. Therefore, we need 9 buses.

Chapter 2 Test

1. $14x$ **2.** $\frac{1}{4}z$ **3.** $6k - 8$ **4.** $\frac{1}{2}y + 1$ **5.** $-.27x + .90$ **6.** $32x + 29$
7. 8 **8.** 17 **9.** 5 **10.** 7 **11.** 6 **12.** 3 **13.** 9 **14.** 14
15. 5 **16.** 3 **17.** 5 **18.** 14 **19.** 12 **20.** 1 **21.** −11 **22.** 5.6
23. $x \geq 2$ (number line: −1 0 1 2 3 4) **24.** $y > 6$ (number line: 4 5 6 7 8)
25. $a \geq 7.5$ (number line: 6 7 7.5 8 9) **26.** $z > -13$ (number line: −15 −13 −11)
27. 3520 **28.** 45 red, 15 green **29.** 87.5% **30.** 6 mph
31. $6\pi \approx 18.4$ cubic inches **32.** 9 boxes

CHAPTER 3

Exercises 3.1

1. 5^4 **3.** x^6 **5.** $(-2)^3 = -2^3$ **7.** -4^2 **9.** -7^5 **11.** $(-x)^2 = x^2$
13. Base is 4; exponent is 3; 64. **15.** Base is 2; exponent is 4; 16.
17. Base is −4; exponent is 2; 16. **19.** Base is 2; exponent is 5; −32. **21.** Base is −2; exponent is 2; −4.
23. Base is −2, exponent is 3; 8. **25.** 5^3 **27.** 2^7 **29.** 3^{10} **31.** $(\frac{1}{4})^5$ **33.** 3^{12}
35. 6^{10} **37.** x^8 **39.** z^8 **41.** x^9 **43.** 10^2 **45.** 14^2 **47.** 10^4
49. $1^3 = 1$ **51.** $(-1)^4 = 1$ **53.** $(-1)^3 = -1$ **55.** 3^4 **57.** $(xy)^5$ **59.** $(pqr)^5$
61. $25x^2y^2$ **63.** $16u^2v^2$ **65.** $8p^3q^3$ **67.** $125u^3v^3$ **69.** $81a^4b^4c^4$
71. $100{,}000x^5y^5$ **73.** $3^1 = 3$ **75.** 2^3 **77.** $14^1 = 14$ **79.** 10^2 **81.** -4^3
83. y^2 **85.** b^5 **87.** a^7 **89.** b^3 **91.** $4^1 = 4$ **93.** 7^5 **95.** 5^4
97. $x^1 = x$ **99.** y^7

Exercises 3.2

1. 1 **3.** −1 **5.** 1 **7.** $\frac{1}{9}$ **9.** $\frac{1}{81}$ **11.** $\frac{1}{125}$ **13.** $\frac{4}{9}$ **15.** $\frac{25}{49}$
17. $\frac{64}{125}$ **19.** $\frac{8}{125}$ **21.** $\frac{1}{16}$ **23.** $\frac{16}{625}$ **25.** $\frac{9}{4}$ **27.** $\frac{16}{9}$ **29.** $\frac{125}{64}$
31. $\frac{8}{125}$ **33.** 2 **35.** 4 **37.** 64 **39.** 15,625 **41.** 1 **43.** 1 **45.** $\frac{1}{9}$
47. $\frac{1}{25}$ **49.** $\frac{1}{121}$ **51.** $\frac{1}{x^4}$ **53.** $\frac{1}{a^4}$ **55.** $\frac{1}{12^{14}}$ **57.** $\frac{1}{7^{-19}} = 7^{19}$

59. $8^{-2} = \frac{1}{8^2}$ **61.** $6^{-3} = \frac{1}{6^3}$ **63.** $5^{-3} = \frac{1}{5^3}$ **65.** 2^{35} **67.** 5^{16} **69.** x^8
71. y^{14} **73.** z^{12} **75.** $\frac{1}{b^8}$ **77.** c^8 **79.** p^6 **81.** a^{20} **83.** y^4
85. $w^{-6} = \frac{1}{w^6}$ **87.** 17 **89.** 7 **91.** 0 **93.** $\frac{1}{2}$ **95.** $13^1 = 13$ **97.** 1
99. 1 **101.** $\frac{625}{81}$ **103.** $\frac{8}{7}$ **105.** w^3 **107.** a^4 **109.** x^4 **111.** q^4
113. $2w^3$ **115.** $5x^4$

Exercises 3.3

1. Binomial **3.** Monomial **5.** Binomial **7.** Monomial **9.** Binomial
11. Trinomial **13.** $x^2 + 2x + 7$; 1 **15.** $x^3 + 3x^2 + x + 7$; 1 **17.** $2x^2 - 5x + 1$; 2
19. $w^2 + 3$; 1 **21.** $y^2 + 6y$; 1 **23.** $x^3 - x^2 + 1$; 1 **25.** $-x^2 + 7$; -1 **27.** $2y^4 - 4$; 2
29. $3w^5 + w + 4$; 3 **31.** $3x^3 - x$ **33.** $-x^3 + 5x - 2$ **35.** $3y^3 + y^2 + 3y + 5$
37. $3y^3 - y^2 + y - 5$ **39.** $6x^2y - xy$ **41.** $2x^2y - 5xy$ **43.** $y^3 + y^2 + 2y + 5$
45. $-y^3 + y^2 + 3y$ **47.** $2y^2 + 2$ **49.** $2y^2 - y + 1$ **51.** Quadratic, $p(1) = 4$
53. Quadratic, $q(2) = -1$ **55.** Cubic, $h(1) = -1$ **57.** Cubic, $h(-1) = 3$
59. Quadratic, $p(-1) = 8$ **61.** 4 **63.** 4 **65.** 7 **67.** 5 **69.** 2

Exercises 3.4

1. $xy + xz$ **3.** $2r^2s + 3rs^2$ **5.** $xy^3 - 3xy^2$ **7.** $-2x^3 + 2xy^2$
9. $147m^2n - 21mn^2 + 21m^2n^2$ **11.** $48x^3y^3 + 136x^2y^4$ **13.** $21x^3y - 14x^2y^2 + 7xy^3$
15. $-2m^3n - 3m^2n^3 + m^3n^2$ **17.** $4x^2 - 3x^3y - \frac{1}{5}xy$ **19.** $2.4y^3 - 4y^2 + 2y$
21. $x^2 + 5x + 6$ **23.** $z^2 - 7z + 10$ **25.** $x^2 + 3x + 2$ **27.** $y^3 - 5y + 2$
29. $w^4 - w^3 - 41w^2 + 6w$ **31.** $p^4 - 11p^2 + 6p$ **33.** $x^3 + 2x^2 + 2x + 1$
35. $w^3 - 2w^2 + 2w - 1$ **37.** $x^4 + 5x^3 + 7x^2 + 12x$ **39.** $y^3 - y^2 - 7y + 10$
41. $x^3 + 8x^2 + 13x + 6$ **43.** $y^3 + 5y^2 + y - 15$ **45.** $w^3 - 5w^2 + 8w - 4$
47. $2x^4 - 15x^3 - 3x^2 - 40x$ **49.** $x^5 - x^2$ **51.** $x^4 - 4x^2 + 5x - 4$
53. $2a^4 + a^2 + a + 1$ **55.** $z^4 + 2z^2 - z + 2$ **57.** $x^4 + 5x - 6$ **59.** $9x^4 - x^2 - 2x - 1$

Exercises 3.5

1. $x^2 + 5x + 6$ **3.** $y^2 + 2y - 8$ **5.** $p^2 - p - 6$ **7.** $x^2 + 6x + 8$
9. $w^2 + 3w - 10$ **11.** $v^2 - 2v - 8$ **13.** $x^2 - 5x + 6$ **15.** $w^2 - 3w + 2$
17. $v^2 - 9v + 18$ **19.** $x^2 + 10x + 21$ **21.** $y^2 - 17y + 72$ **23.** $2a^2 - 7a - 15$
25. $11y^2 - 26y + 8$ **27.** $35x^2 + 59x + 14$ **29.** $40x^2 + 13x - 42$ **31.** $-2x^2 + 11x - 15$
33. $6x^2 - x - 15$ **35.** $10x^2 - 3x - 1$ **37.** $18x^2 + 30x + 8$ **39.** $x^2 + xy - 2y^2$
41. $y^2 + yz + \frac{2}{9}z^2$ **43.** $x^2 - 16$ **45.** $z^2 - 1$ **47.** $x^2 - 9$ **49.** $z^2 - 100$
51. $u^2 - 64$ **53.** $4x^2 - 9$ **55.** $1 - 64m^2$ **57.** $49x^2 - 9y^2$ **59.** $x^2 - \frac{1}{4}$
61. $\frac{1}{9}x^2 - \frac{4}{25}$ **63.** $9m^2n^2 - 49$ **65.** $.04 - x^2$ **67.** $x^2 + 8x + 16$ **69.** $y^2 + 14y + 49$
71. $w^2 + 24w + 144$ **73.** $y^2 - 4y + 4$ **75.** $x^2 - 16x + 64$ **77.** $w^2 - 12w + 36$
79. $x^2 - 20x + 100$ **81.** $4a^2 + 12a + 9$ **83.** $4y^2 - 20y + 25$ **85.** $\frac{1}{4}x^2 + x + 1$
87. $\frac{1}{9}y^2 - \frac{2}{3}y + 1$ **89.** $.25a^2 + .20a + .04$

Exercises 3.6

1. $6x + 2$ **3.** $5w + 2$ **5.** $2x^2 + x$ **7.** $7x^2 + 5x$ **9.** $4c^3 + 2c$ **11.** $4x^3 + 2x$
13. $3y^2 - 2y$ **15.** $4y$ with $R = -4y^4$ **17.** $3x^2 + 2x + 1$ **19.** $3y^2 + 2y - 1$
21. $4z^2 - 2$ with $R = 25z^2$ **23.** $10x + 5$ with $R = 5x$ **25.** $5y + 3$ with $R = 15y$
27. $6y^2 - 3y$ with $R = 24y$ **29.** $12z^2 - 8z + 6$ with $R = 48z^2$ **31.** $4x^3 + 2x^2$ with $R = -6x + 18$

33. $3a^4 + 6a$ with $R = 16a^3 - 4a^5$ **35.** $3x + 8$ **37.** $2x + 5$ **39.** $4x + 6$
41. $2x + 1$ **43.** $2x^2 + x + 1$ **45.** $2x^2 - 3x + 2$ with $R = 2$ **47.** $x^2 + 4$ with $R = 1$
49. $x^2 + x + 2$ with $R = 1$

Chapter 3 Test

1. Base is 5; exponent is 3; value is 125.
2. Base is 3; exponent is 4; value is 81.
3. Base is 2; exponent is 5; value is -32.
4. Base is 3; exponent is 2; value is -9.
5. Base is $4xy$; exponent is 3; value is $64x^3y^3$.
6. Base is $3ab$; exponent is 2; value is $9a^2b^2$.
7. 216 **8.** 1 **9.** 0 **10.** 4 **11.** -4 **12.** 16 **13.** $\frac{17}{72}$
14. 2744 **15.** 17 **16.** $\frac{5}{2}$ **17.** $\frac{1}{9}$ **18.** $\frac{1}{8}$ **19.** 125 **20.** $\frac{1}{9}$
21. -8 **22.** $2x^5 + x^3 + 3x - 1$; 5 **23.** $-y^3 + y^2 - y + 1$; 3 **24.** $3a^3 + 6a^2 + 3a$; 3
25. $-2b^5 + 2b^4 - 6b^2$; 5 **26.** $a^2 + 14a + 49$; 2 **27.** $x^2 - 8x + 16$; 2 **28.** $y^2 - 9$; 2
29. $y^2 - 121$; 2 **30.** $x^2 + 11x + 30$; 2 **31.** $3w^2 + 13w - 10$; 2 **32.** 4 **33.** 0
34. 8 **35.** $11x^2 + 7x$; $R = -33$ **36.** $4y + 2$; $R = 25$ **37.** $-a^4 + a^2 + a + 1$
38. $b^2 - 2b + 5$; $R = 10$ **39.** $x^2 + x + 1$ **40.** $3x^2 - x - 1$; $R = 2$

CHAPTER 4

Exercises 4.1

1. 1, 2, 4, 8 **3.** 1, 43 **5.** 1, 5, 7, 35 **7.** 1, 2, 4, 7, 14, 28
9. 1, 2, 3, 5, 6, 10, 15, 30 **11.** Prime **13.** No; $26 = 2 \cdot 13$ **15.** No; $27 = 3 \cdot 9$
17. Prime **19.** Prime **21.** $2 \cdot 3^2$ **23.** $2 \cdot 5^2$ **25.** $2^2 \cdot 19$ **27.** $2^3 \cdot 5^2$
29. $2^2 \cdot 3 \cdot 5^2$ **31.** 10 **33.** 1 **35.** 7 **37.** 15 **39.** 21

Exercises 4.2

1. x^4 **3.** xy **5.** $3a^2$ **7.** $6ab$ **9.** $18a^2b^2c^2$ **11.** $7(x + 2)$
13. $4(x - 2y)$ **15.** $x(y + z)$ **17.** $2a(3b - 10c)$ **19.** $5x(2x + 3y)$ **21.** $6w(3w - 4y)$
23. $3m^2n(mp - n)$ **25.** $3a(a - 2b - 2b^2)$ **27.** $6x(2x^2 - xy - 2y^2)$ **29.** $4xy(x^2y + 2 - y^2)$
31. $5x^3(6x^2 - 1 + 3x^5)$ **33.** $xy^2(y^3 - 3y - 6)$ **35.** $30s^2t^3(t^2 + 2st - 5s^2)$
37. $pqr(pr + qr + pq)$ **39.** $3m^2n(4m - 5n^2 - 2n^3)$

Exercises 4.3

1. $(x + 4)(x - 4)$ **3.** $(y + 9)(y - 9)$ **5.** $(5 + b)(5 - b)$ **7.** $(6y + 7)(6y - 7)$
9. $(z + 5)(z - 5)$ **11.** $(x + 8)(x - 8)$ **13.** $(5y + 12)(5y - 12)$ **15.** $(2x + 13)(2x - 13)$
17. $(x + yz)(x - yz)$ **19.** $(7t + 9r)(7t - 9r)$ **21.** $5(x + 2)(x - 2)$ **23.** $3(y + 5)(y - 5)$
25. $4(a + 10)(a - 10)$ **27.** $xy^2(x + 11)(x - 11)$ **29.** $7x(y + 7)(y - 7)$
31. $a^2(x + b)(x - b)$ **33.** $3t(t + 3r)(t - 3r)$ **35.** $20t^2(t + 2)(t - 2)$
37. $2(z^2 + 9)(z + 3)(z - 3)$ **39.** $3(m - n + 5)(m - n - 5)$

Exercises 4.4

1. No **3.** No **5.** Yes **7.** Yes **9.** Yes **11.** $(x - 1)^2$ **13.** $(x + 2)^2$
15. $(x + 7)^2$ **17.** $(x + 3)^2$ **19.** $(m - 11)^2$ **21.** $(x + 9y)^2$ **23.** $(2y + 4)^2$
25. $(3m + 8)^2$ **27.** $(7x - 6)^2$ **29.** $(a^2 - 8)^2$ **31.** $4(m + 15)^2$ **33.** $3(2m - 5)^2$
35. $7(a^2 + 6)^2$ **37.** $5(3z + 8)^2$ **39.** $4(5y + 3z)^2$

Exercises 4.5

1. $(x + 1)(x + 4)$ **3.** $(x + 2)(x + 7)$ **5.** $(z + 1)(z + 2)$ **7.** $(x - 2)(x - 3)$
9. $(z - 1)(z - 4)$ **11.** $(t - 3)(t - 8)$ **13.** $(y + 2)(y - 7)$ **15.** $(m + 5)(m - 8)$
17. $(y + 3)(y - 8)$ **19.** $(x - 3)(x + 7)$ **21.** $(x - 3)(x + 10)$ **23.** $(x - 3)(x + 5)$
25. $(z + 5)(z + 6)$ **27.** $(x - 6)(x - 9)$ **29.** $(x - 4)(x + 9)$ **31.** $2(x - 2)(x + 1)$
33. $3(z - 3)(z - 5)$ **35.** $7(y + 5)(y + 6)$ **37.** $5(x - 5)(x + 8)$ **39.** $10(t - 3)(t + 7)$
41. $a(a + 4)(a - 5)$ **43.** $x(x - 3)(x + 13)$ **45.** $2y(y - 1)(y - 4)$
47. $4x(y - 5)(y - 5) = 4x(y - 5)^2$ **49.** $3xy(z - 4)(z + 6)$ **51.** $(t + 7s)(t - 5s)$
53. $(y + 7z)(y + 8z)$ **55.** $(z + 4y)(z - 9y)$ **57.** $(r - 3s)(r - 27s)$ **59.** $5(x + 3y)(x + 7y)$
61. $xy(x + y)(x + 7y)$

Exercises 4.6

1. x **3.** $13x$ **5.** $-18x$ **7.** $-11x$ **9.** $31x$ **11.** $(x + 5)(2x + 1)$
13. $(x + 7)(2x + 1)$ **15.** $(x + 3)(5x + 1)$ **17.** $(x + 11)(7x - 3)$ **19.** $(z - 11)(5z + 2)$
21. $(a + 2)(3a + 2)$ **23.** $(3t + 2)(2t + 3)$ **25.** $(2t - 5)(3t + 7)$ **27.** $(x + 5)(2x - 3)$
29. $(5x + 1)(x + 5)$ **31.** $(y - 2)(7y + 4)$ **33.** $(a - 2)(5a - 2)$ **35.** $(3y - 8)(2y + 7)$
37. $(5y - 2)(5y + 4)$ **39.** $(5z - 2)(2z + 3)$ **41.** $3(x + 5)(3x - 4)$ **43.** $2(x - 3)(2x - 1)$
45. $3(z + 2)(2z + 1)$ **47.** $5(y + 3)(3y - 2)$ **49.** $2y(y - 2)(7y + 4)$
51. $2xy(2x - 1)(2x + 7)$ **53.** $3pq(4p - 3)(2p + 5)$ **55.** $3x^2(4x + 5)(3x - 2)$
57. $(4x + 5y)(4x + 3y)$ **59.** $2y(4y - z)(3y + 4z)$

Exercises 4.7

1. $\{1, 3\}$ **3.** $\{0, 7\}$ **5.** $\{\frac{1}{2}, -\frac{2}{3}\}$ **7.** $\{\frac{5}{3}, -4\}$ **9.** $\{0, -2, \frac{1}{2}\}$ **11.** $\{-3, 3\}$
13. $\{-10, 10\}$ **15.** $\{4\}$ **17.** $\{-6\}$ **19.** $\{5\}$ **21.** $\{-3, 4\}$ **23.** $\{-2, -5\}$
25. $\{-2, 6\}$ **27.** $\{-7, 6\}$ **29.** $\{-2, -14\}$ **31.** $\{\frac{1}{5}, \frac{1}{2}\}$ **33.** $\{-\frac{2}{3}, 3\}$
35. $\{-\frac{5}{2}, 3\}$ **37.** $\{-\frac{5}{2}, \frac{3}{4}\}$ **39.** $\{\frac{2}{3}\}$ **41.** $\{\frac{3}{4}, -\frac{3}{4}\}$ **43.** $\{3, 5\}$ **45.** $\{4\}$
47. $\{-\frac{1}{6}, 1\}$ **49.** $\{-\frac{5}{4}, \frac{3}{4}\}$ **51.** $\{-\frac{5}{2}, 3\}$ **53.** $\{-\frac{8}{3}, 3\}$ **55.** $\{-8, 0, 8\}$
57. $\{0, 1, 2\}$ **59.** $\{-\frac{3}{2}, 0, \frac{1}{3}\}$

Exercises 4.8

1. $L = 11$, $W = 5$ **3.** $L = 15$, $W = 5$ **5.** $b = 16$, $h = 8$ **7.** $b = 11$, $h = 4$ **9.** 8, 9 or −7, −6
11. 11, 12 or −2, −1 **13. (a)** 2 seconds **(b)** 6 seconds **15. (a)** 2 seconds **(b)** 7 seconds
17. (a) $\frac{1}{2}$ second **(b)** 4.5 seconds **19.** 1, 3 or 17, 19 **21.** −7, −5, −3
23. 4, 6, or −4, −2 **25.** 7, 8, 9 **27.** 4, 6, 8, 10

Chapter 4 Test

1. 1, 3, 9, 27 **2.** 1, 29 **3.** 3 **4.** $12xy$ **5.** $2x^2(x^2 - 3y)$ **6.** $3x^2y^2(3y + 4x)$
7. $(x - 5)(x + 5)$ **8.** $(3x - 4)(3x + 4)$ **9.** $(y + 7)^2$ **10.** $(y - 10)^2$ **11.** $4(x + 4)^2$
12. $(x^2 + 6)^2$ **13.** $(x - 7)(x - 4)$ **14.** $(x + 1)(2x + 5)$ **15.** $(3z - 10)(z - 2)$
16. $w(5w + 2)(w - 4)$ **17.** $\{-7, 7\}$ **18.** $\{-5\}$ **19.** $\{+7, -3, 0\}$ **20.** $\{-6, -\frac{2}{3}\}$
21. 7 and 8; −2 and −1 **22.** 5, 7, and 9

CHAPTER 5

Exercises 5.1

1. 0 **3.** 0 **5.** 0 **7.** Defined for all values of z **9.** 3 **11.** −1 **13.** 5
15. 2 and 4 **17.** −1 **19.** −5 and −2 **21.** $\frac{2}{3}$ **23.** $\frac{3}{7}$ **25.** $\frac{y}{4}$

27. $-\frac{5}{x^2}$ **29.** $\frac{1}{2}$ **31.** -3 **33.** $\frac{3}{4}$ **35.** 3 **37.** $\frac{1}{x+1}$ **39.** $\frac{1}{y-4}$
41. $\frac{p-2}{p+2}$ **43.** $\frac{x+3}{x+4}$ **45.** -1 **47.** $\frac{x+8}{x+5}$ **49.** $\frac{y+6}{y-6}$

Exercises 5.2

1. $\frac{9}{5}$ **3.** $\frac{6}{y}$ **5.** $6wz$ **7.** $\frac{8z}{9(x+y)}$ **9.** $\frac{(p+q)^2}{6r}$ **11.** $\frac{c}{2(a+b)}$
13. $\frac{(x+1)(x-7)}{(x-1)}$ **15.** $\frac{12}{7}$ **17.** $\frac{10}{9}$ **19.** $\frac{x+3}{x-2}$ **21.** $\frac{y-1}{y}$ **23.** $\frac{(x+2)}{(x+7)}$
25. $\frac{z-3}{z+1}$ **27.** $6y^2$ **29.** $\frac{3}{2x^2}$ **31.** $\frac{1}{3(y+1)^2}$ **33.** $\frac{1}{3b(b-4)^2}$ **35.** 6
37. $\frac{9}{10}$ **39.** $\frac{5(a+2)}{3(a-5)}$ **41.** $\frac{x+1}{x+3}$ **43.** $-\frac{2}{5}$ **45.** $\frac{(w+1)}{(w-5)}$ **47.** $\frac{2x}{x-1}$
49. $\frac{2(z+1)}{7}$

Exercises 5.3

1. $\frac{8}{x}$ **3.** $\frac{3}{b}$ **5.** $\frac{5}{z}$ **7.** $\frac{7}{2x-1}$ **9.** $\frac{11}{y+4}$ **11.** 3 **13.** $\frac{a^2+10}{5a}$
15. $\frac{5x-4}{2x}$ **17.** $\frac{b^2+7}{7b}$ **19.** $\frac{5x+2}{x(x+2)}$ **21.** $\frac{5(a-2)}{a(a+5)}$ **23.** $\frac{2(3z+7)}{z(z+3)}$
25. $\frac{-(p+4)}{p(p+1)}$ **27.** $\frac{2(4x-1)}{(x-1)(x+2)}$ **29.** $\frac{-3}{(a+2)(a+4)}$ **31.** $\frac{x^2-x+6}{(x-2)(x+2)}$
33. $\frac{-10y^2+29y+14}{(y-3)(y+3)}$ **35.** $\frac{z^2-3z-13}{(z-6)(z+6)}$ **37.** $\frac{2(3a-2)}{(a+1)(a-1)}$ **39.** $\frac{2(3x+7)}{(x-1)(x+1)}$
41. $\frac{2z^2+z+4}{(z-4)(z+4)}$ **43.** $\frac{2(2p-1)}{(p-1)(p+1)}$ **45.** $\frac{2x^2+1}{(x-1)(x+1)(x+2)}$
47. $\frac{5(a+1)}{(a-3)(a+2)(a-3)}$ **49.** $\frac{-4x^2+15x-1}{5(x-1)(x+1)}$

Exercises 5.4

1. $\frac{35}{9}$ **3.** $\frac{y^2}{x}$ **5.** $\frac{w^2}{2z}$ **7.** $\frac{a+1}{3a^2}$ **9.** $\frac{3c}{c-5}$ **11.** $\frac{a}{b}$ **13.** $\frac{12x^2+2}{6x^2-3}$
15. $\frac{1}{(a-1)(a+1)}$ **17.** $\frac{1}{(b-2)(b+3)}$ **19.** $(x-6)(x-1)$ **21.** $\frac{1}{b+a}$
23. $\frac{y+1}{y+x}$ **25.** -1 **27.** $\frac{1}{(x+1)(x+4)}$ **29.** $\frac{(a-4)(a+1)}{(a-5)(a+4)}$

Exercises 5.5

1. -5 **3.** $\frac{14}{3}$ **5.** $\frac{3}{4}$ **7.** $\frac{28}{3}$ **9.** $\frac{5}{3}$ **11.** 2 **13.** 6 **15.** -1
17. 4 is extraneous. **19.** 3 **21.** 2 is extraneous. **23.** -2 is extraneous.
25. 11 is extraneous. **27.** 2 **29.** -3 **31.** $-\frac{2}{3}$, 3 **33.** -1, 5 **35.** -1, $\frac{1}{2}$
37. 0, -5 **39.** -1, 4 **41.** -12 **43.** 4 **45.** -6, -1 **47.** $\frac{1}{2}$, -6
49. 1; -5 is extraneous.

Exercises 5.6

1. $\frac{4}{7}$ **3.** $\frac{6}{18} = \frac{1}{3}$ **5.** $\frac{35}{7} = \frac{5}{1}$ **7.** $\frac{8}{36} = \frac{2}{9}$ **9.** $\frac{20}{120} = \frac{1}{6}$ **11.** $\frac{60}{300} = \frac{1}{5}$ **13.** $\frac{5}{21}$
15. $\frac{1}{6}$ **17.** 2 **19.** 7 **21.** 5 **23.** 25 **25.** $\frac{4}{3}$ **27.** $\frac{77}{50}$ **29.** ± 6
31. ± 8 **33.** -4 and 5 **35.** -7 and 4 **37.** -4 and 2 **39.** ± 2 **41.** $\frac{10}{9}$
43. 250 miles **45.** $181\frac{9}{11} \simeq 182$ pounds **47.** **(a)** 2.5 ounces **(b)** 70 ounces **49.** 2025

Exercises 5.7

1. 2 **3.** $\frac{3}{7}$ **5.** 4 or $\frac{1}{2}$ **7.** 9 mph **9.** 110 mph **11.** $27,500
13. 14 years **15.** 15 years **17.** $5\frac{5}{11}$ ohms **19.** 24 ohms **21.** **(a)** $3\frac{3}{4}$ hours **(b)** $\frac{4}{15}$ **(c)** $\frac{5}{8}$
23. 24 minutes **25.** **(a)** 2.4 hours **(b)** Slower computer, $\frac{2}{5}$ **(c)** Faster computer, $\frac{3}{5}$

Chapter 5 Test

1. 5 **2.** 2, 3 **3.** $\frac{15}{28}$ **4.** $\dfrac{5}{3x}$ **5.** $\dfrac{1}{y-2}$ **6.** $\dfrac{z+7}{z-4}$ **7.** $\dfrac{1}{10x^2}$
8. $\dfrac{x}{2(x+3)}$ **9.** $\frac{14}{3}$ **10.** $\dfrac{z^3}{4}$ **11.** $3x(x-3)^2$ **12.** $\dfrac{y+3}{y+1}$ **13.** $\dfrac{x^2+6}{2x}$
14. $\dfrac{10-3y^2}{2y}$ **15.** $\dfrac{w}{w-1}$ **16.** $\dfrac{(x-6)(x-2)}{(x-4)(x+2)}$ **17.** $\dfrac{x(3x+1)}{3(7x-2)}$ **18.** $\dfrac{y+2}{y-2}$
19. 9 **20.** -2 **21.** 1, -3 is extraneous. **22.** -2, -1 is extraneous. **23.** $\frac{1}{6}$
24. Means: x and $x - 2$; extremes: 7 and 5 **25.** 75 miles **26.** 9600 **27.** 4 ohms
28. 15 minutes

CHAPTER 6

Exercises 6.1

1–7

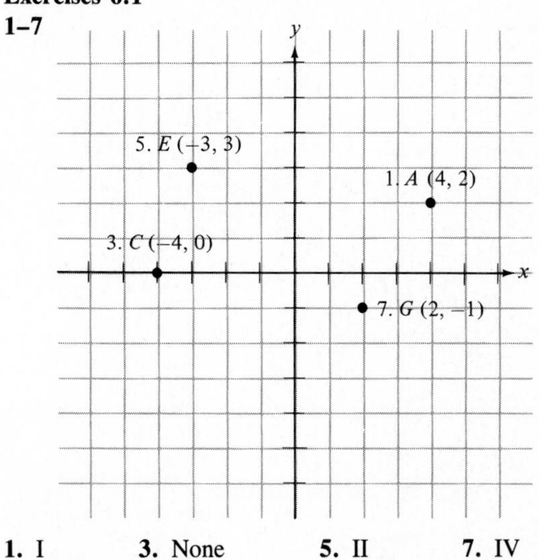

9–25

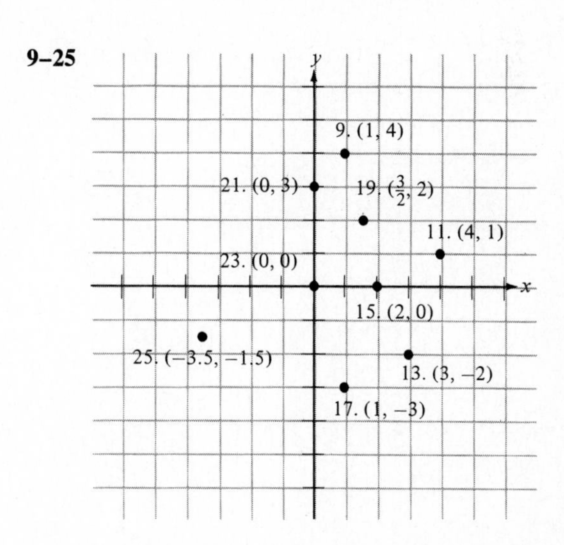

1. I **3.** None **5.** II **7.** IV

Exercises 6.2

1. (5, 3), $(-1, 9)$ **3.** $(0, -3)$, $(2, -\frac{5}{3})$, (6, 1) **5.** None **7.** (1, 5), (0, 3)
9. $(\frac{8}{3}, 0)$, $(0, -4)$, $(2, -1)$ **11.** $(\frac{1}{3}, 0)$, $(3, -2)$, $(0, \frac{1}{4})$ **13.** (0, 0), (0, 0), $(1, -4)$

15. (0, −1), (−3, 0), (3, −2) **17.** (4, −1), (5, −1), (9, −1) **19.** (−2, 1), (−2, 6), (−2, 27)
21. (3, −2), (5, −8), (1, 4) **23.** (4, 5), (2, −3), (2, −3)

25.

x	2	4	−4	0
y	4	8	−8	0

27.

x	2	4	−6	−2
y	1	2	−3	−1

29.

x	0	$\frac{1}{3}$	1	2
y	−2	0	4	10

31.

x	0	9	3	4
y	9	0	6	5

33.

x	0	5	4	10
y	2	0	$\frac{2}{5}$	−2

35.

x	0	1	−1	2
y	0	7	−7	14

37.

x	0	1	−1	2
y	0	−5	5	−10

39.

x	0	1	−1	2
y	1	−2	4	−5

41.

x	0	1	−1	2
y	−2	2	−6	6

43.

x	0	1	−1	2
y	−10	−5	−15	0

45. $y = 4x$ **47.** $x + y = 12$ **49.** $\frac{1}{3}x + \frac{1}{5}y = 1$ **51.** \$1000 **53.** \$400

Exercises 6.3

1. (2, 0), (0, 6), (1, −3)

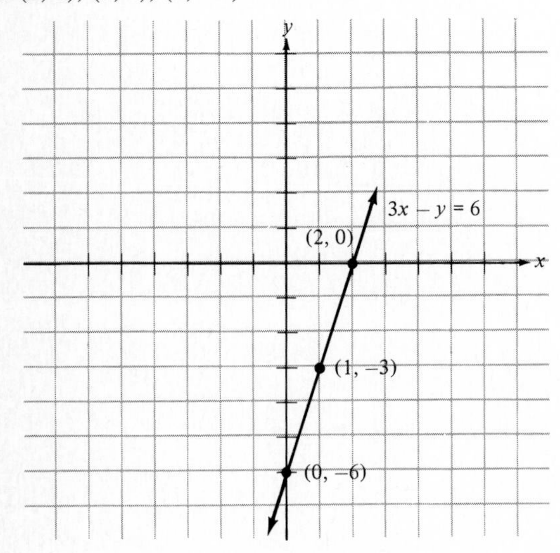

3. (3, 0), (0, −2), (6, 2)

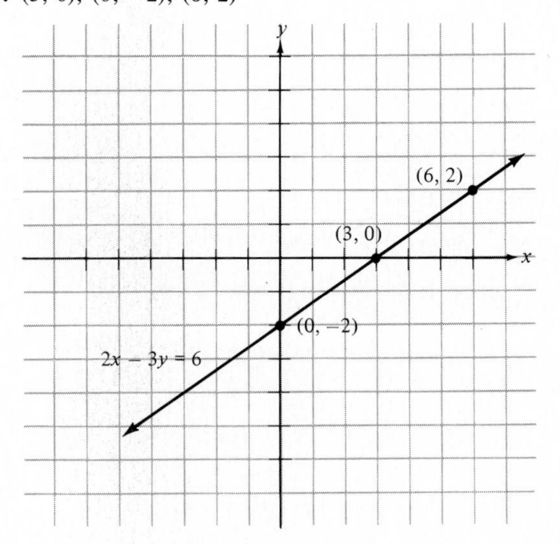

5. $(-2, 0)$, $(0, -6)$, $(-1, -3)$

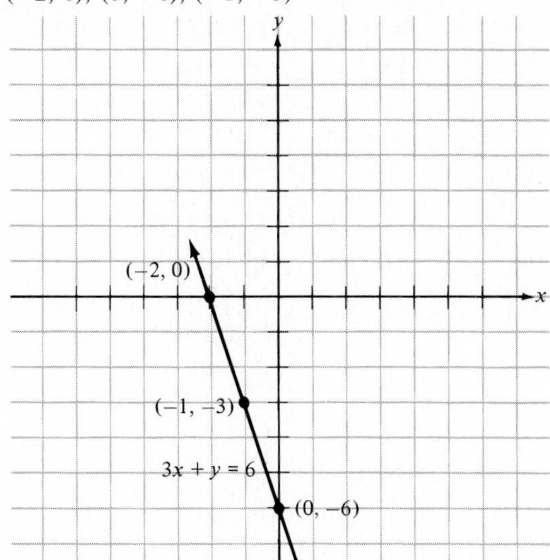

7. $(0, 0)$, $(1, 2)$, $(2, 4)$

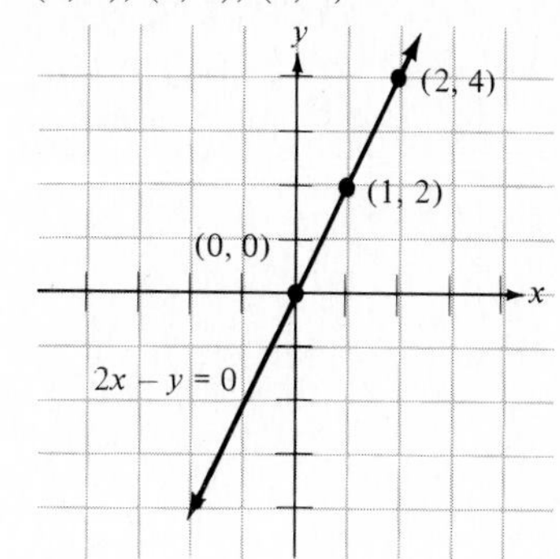

9.

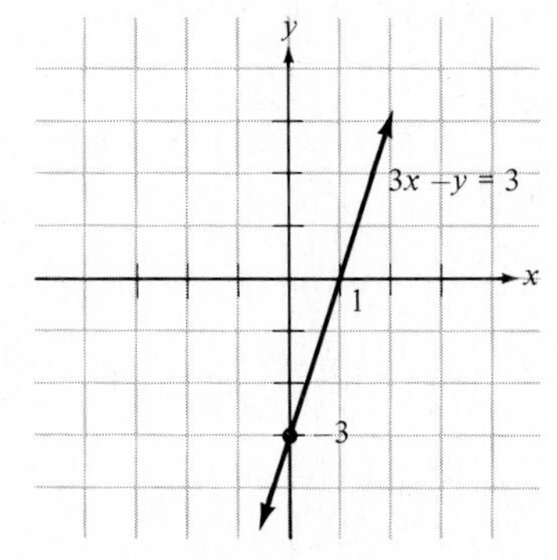

11.

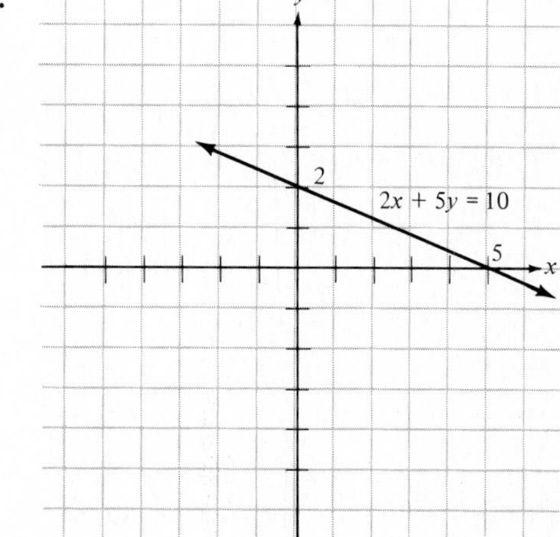

13.

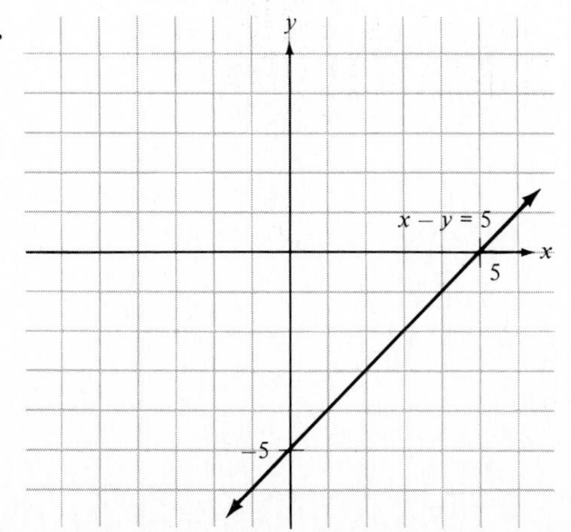

15.

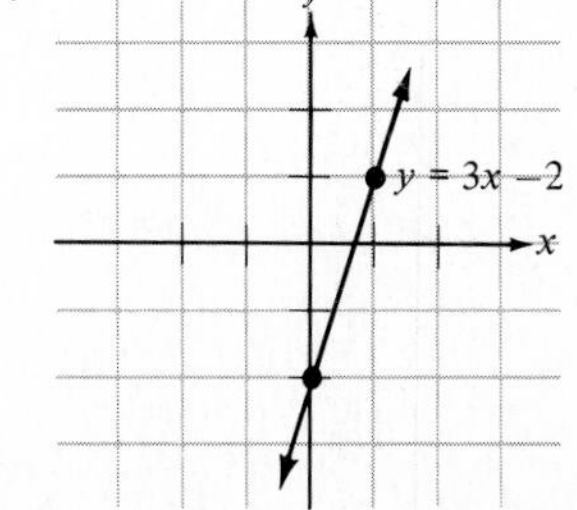

17.

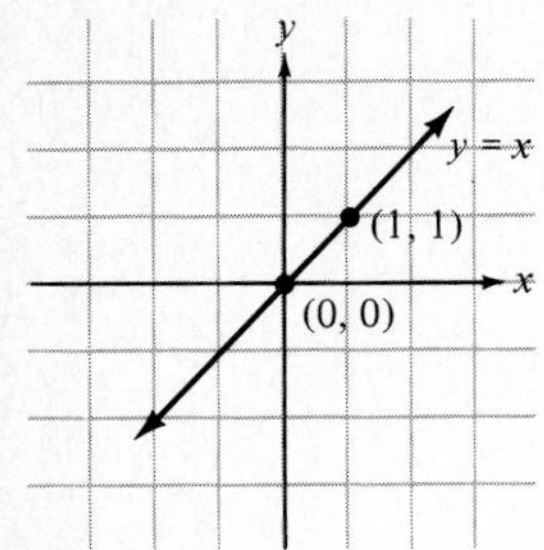

19.

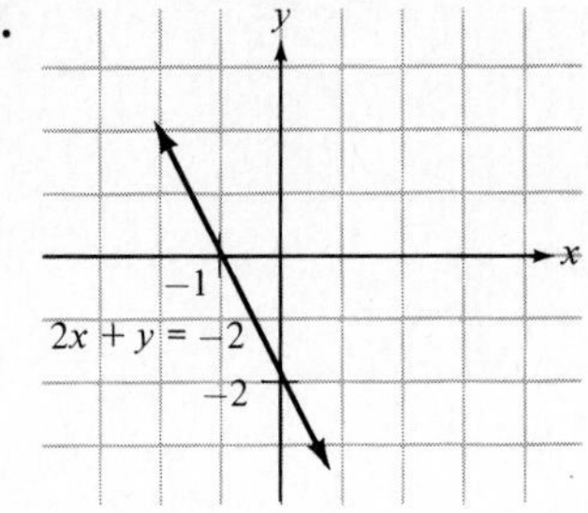

21.

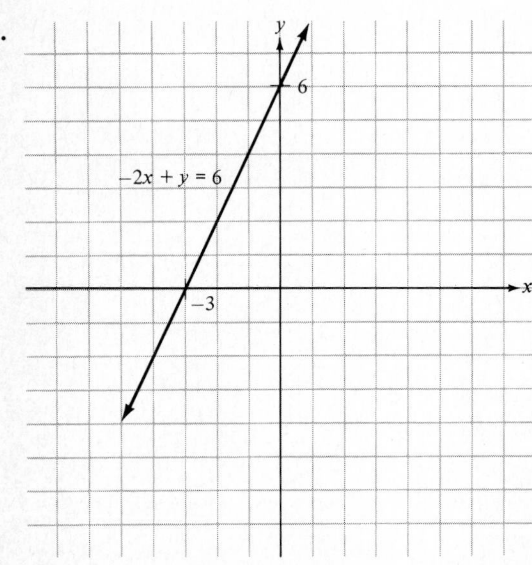

23.

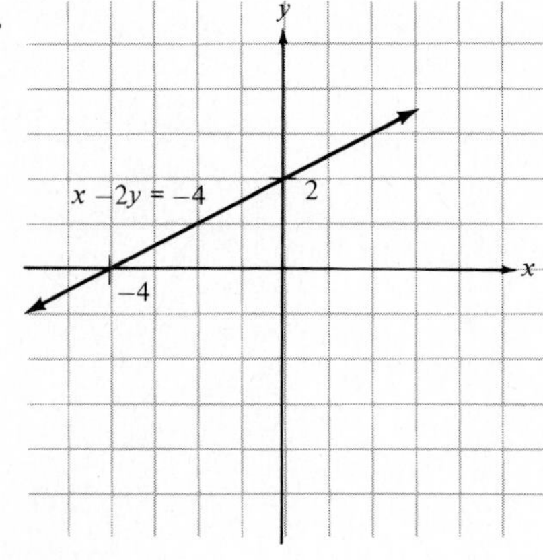

25.

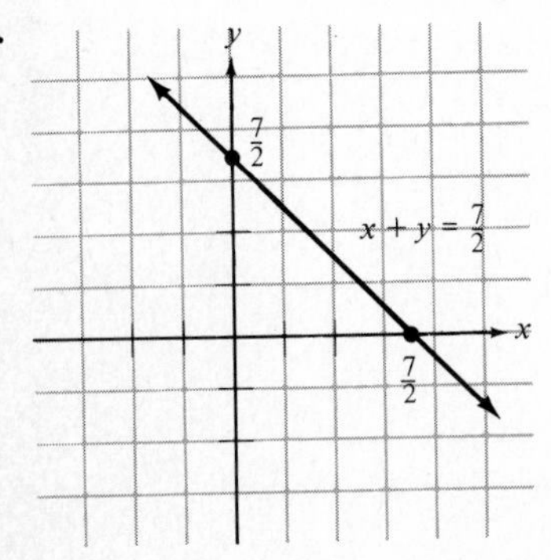

27.

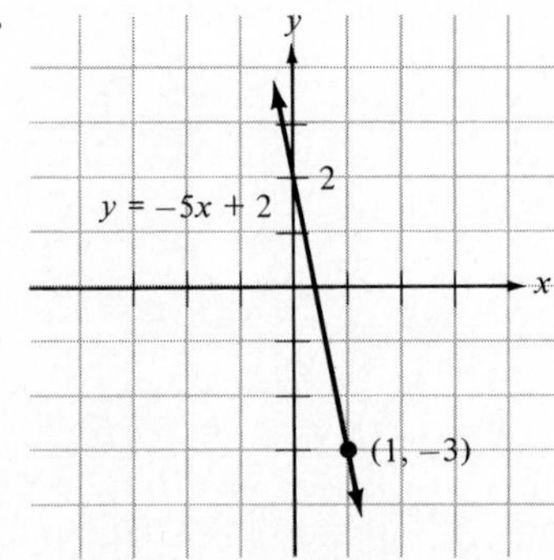

29.

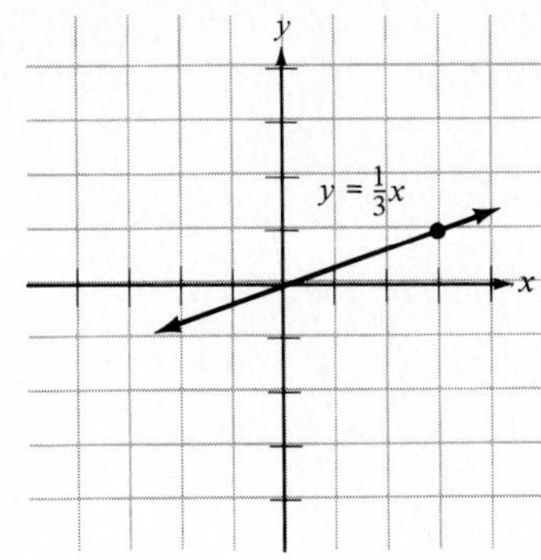

31.

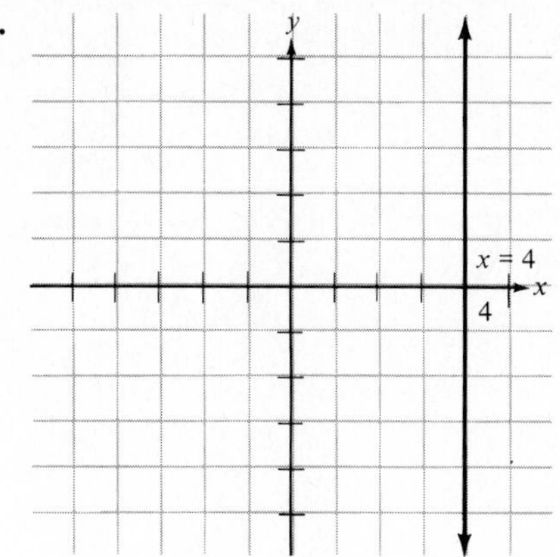

33.

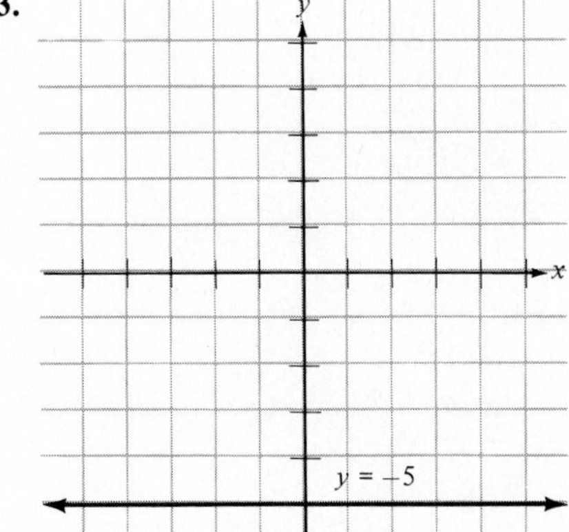

35.

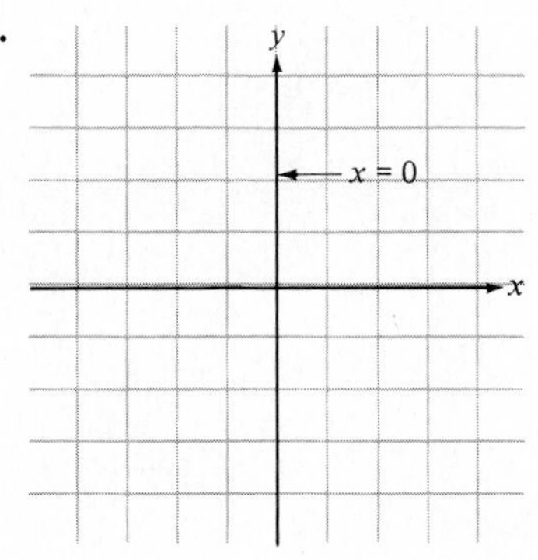

37.

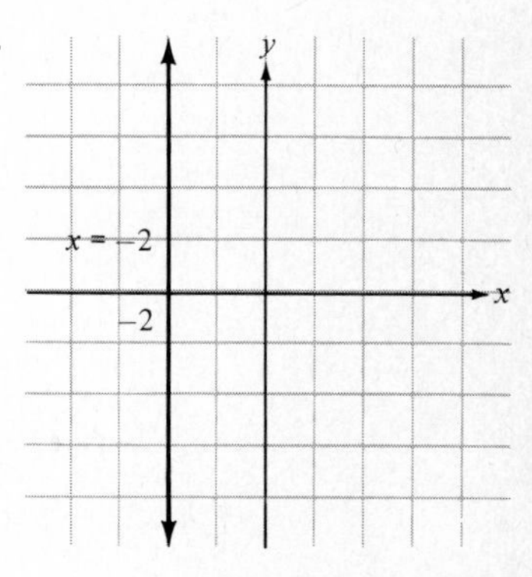

39. (a)

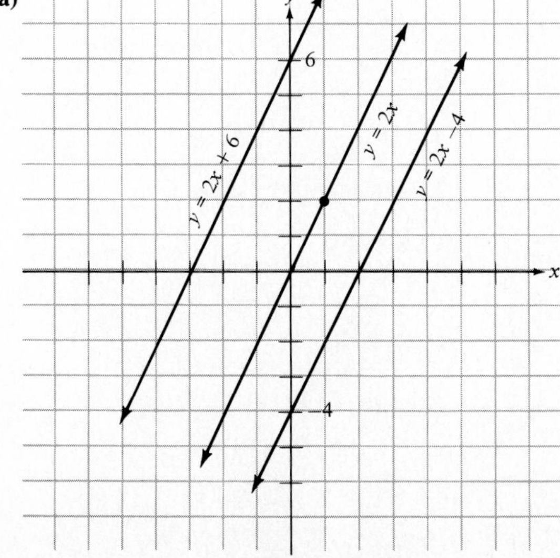

(b) No

(c) $y = 2x$

x-intercept is 0.
y-intercept is 0.

$y = 2x + 6$

x-intercept is -3.
y-intercept is 6.

$y = 2x - 4$

x-intercept is 2.
y-intercept is -4.

41. **(a)**

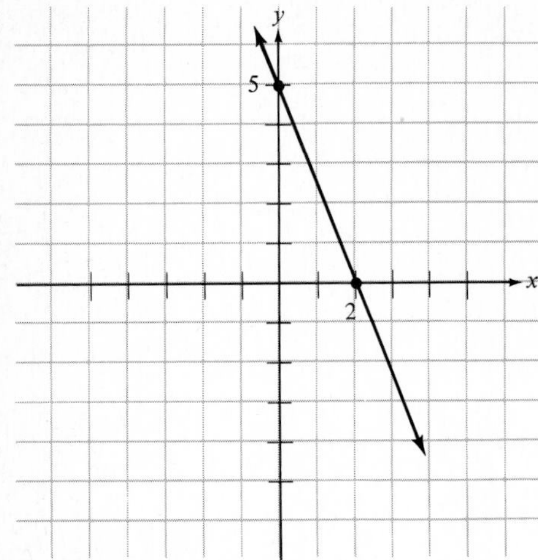

(b) \$5 **(c)** 2

Exercise 6.4

1.

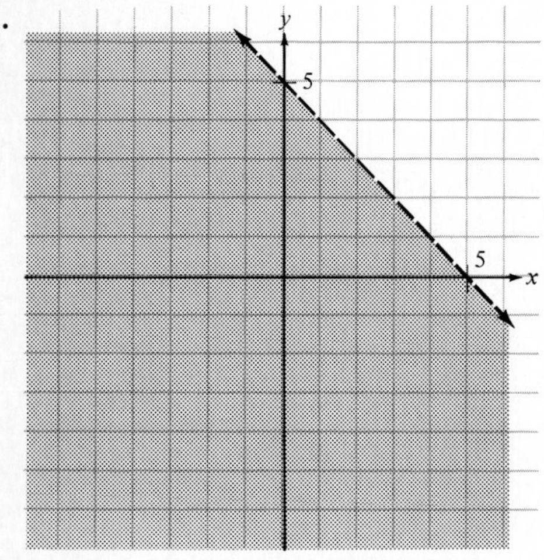

3.

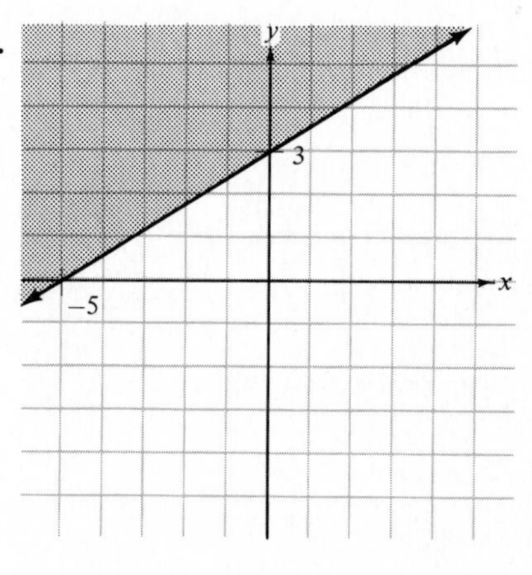

5.

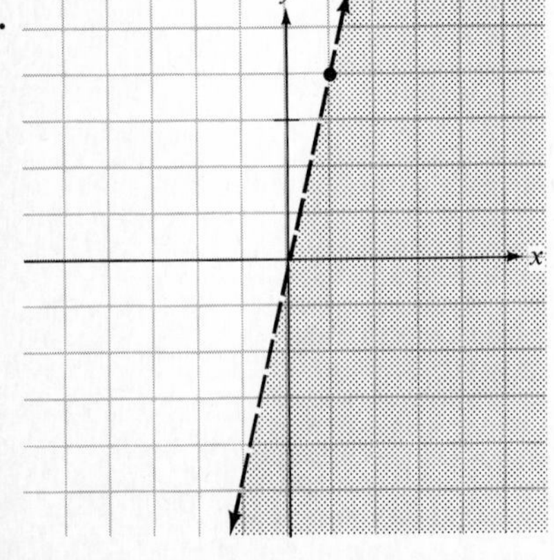

7.

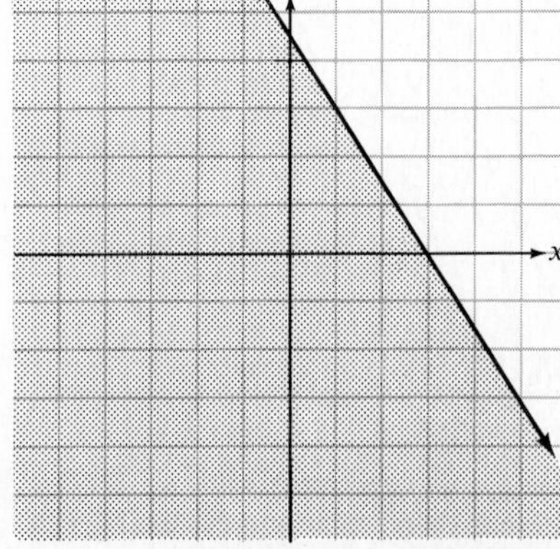

9.

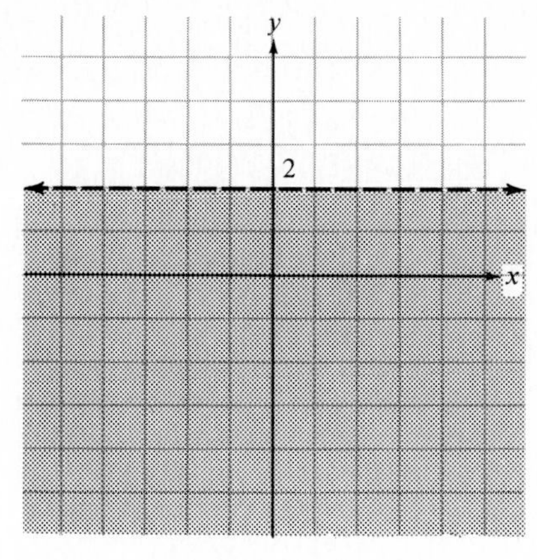

11.

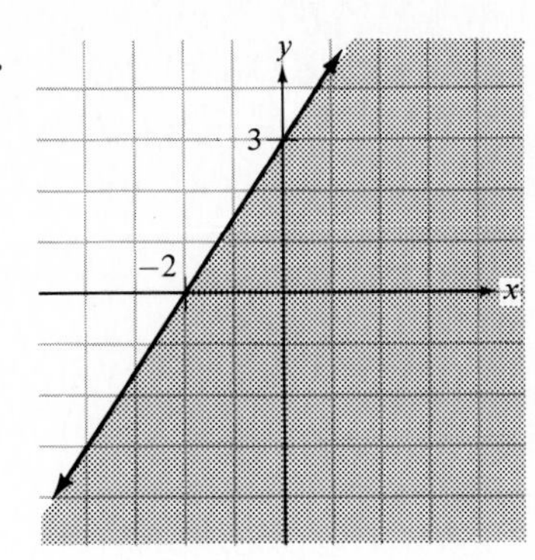

13.

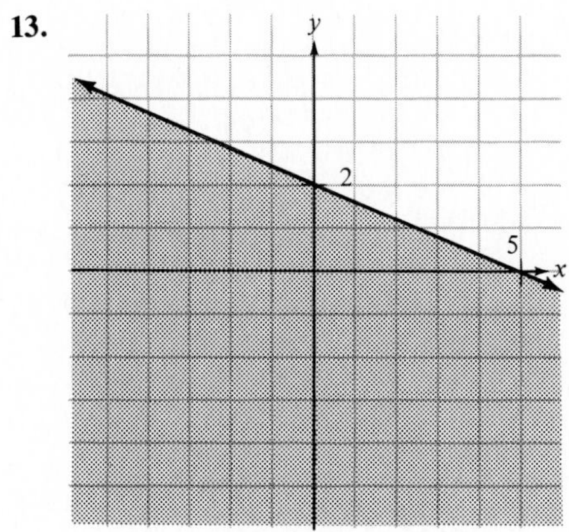

15.

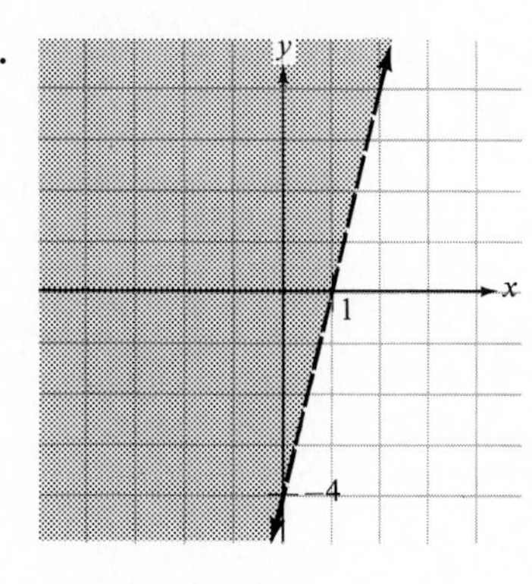

17.

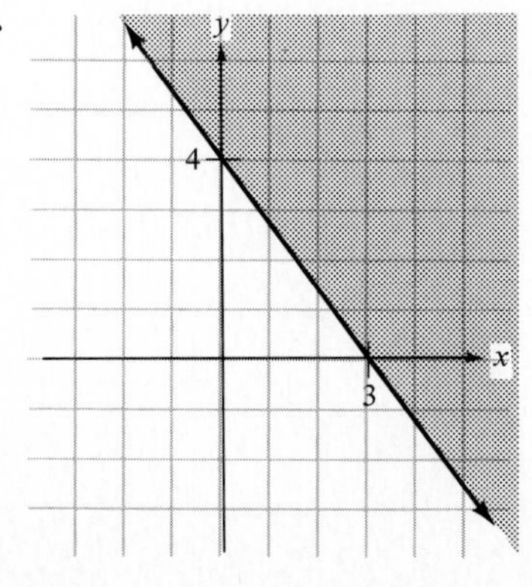

19.

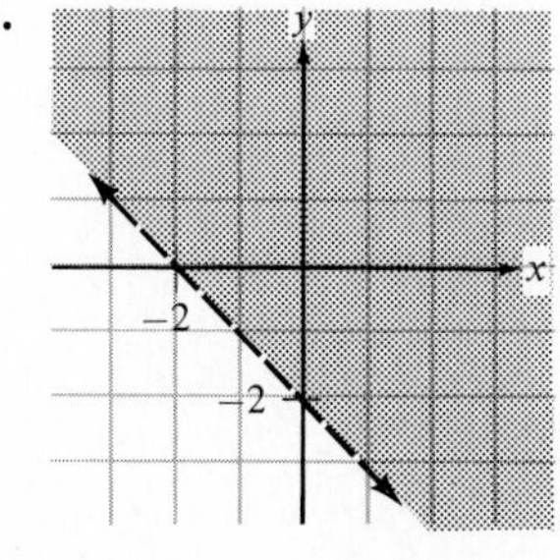

21.

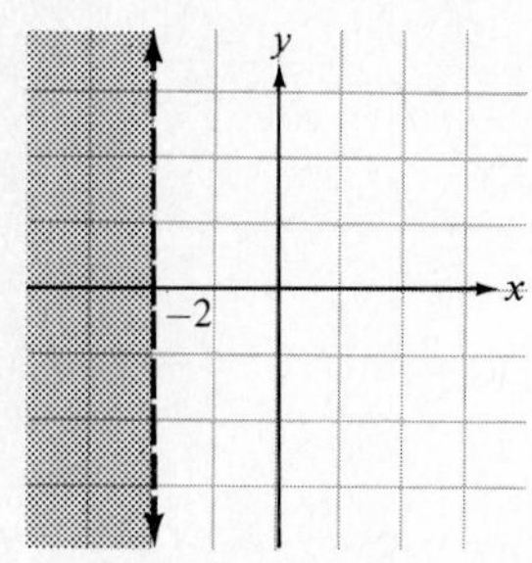

23.

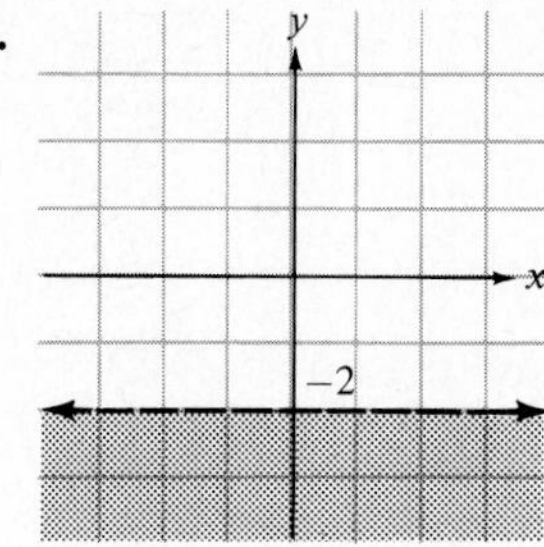

25.

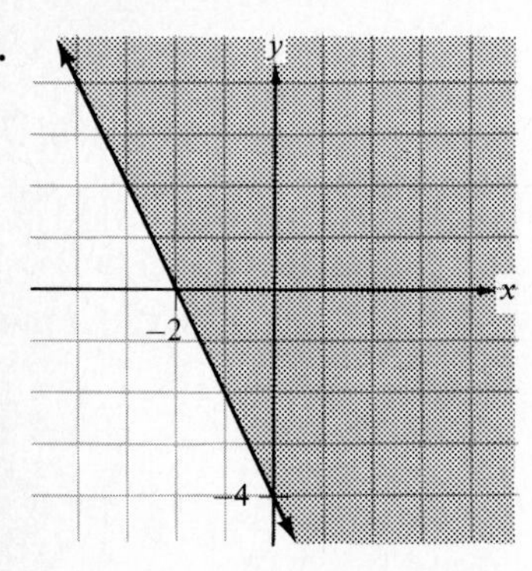

27.

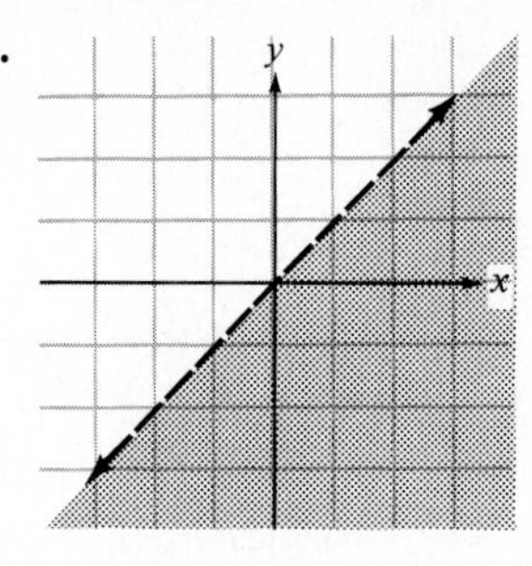

29.

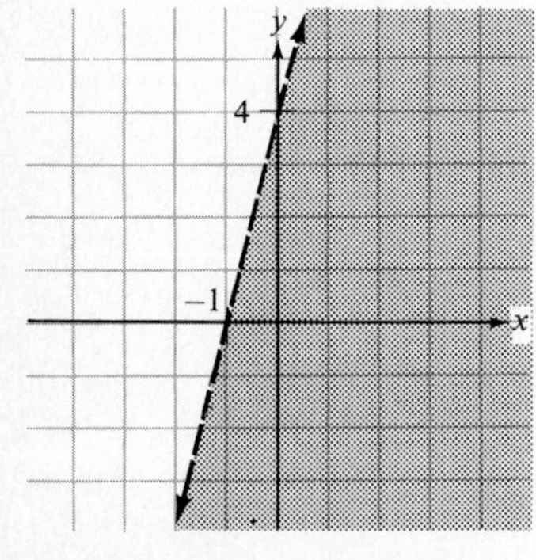

31.

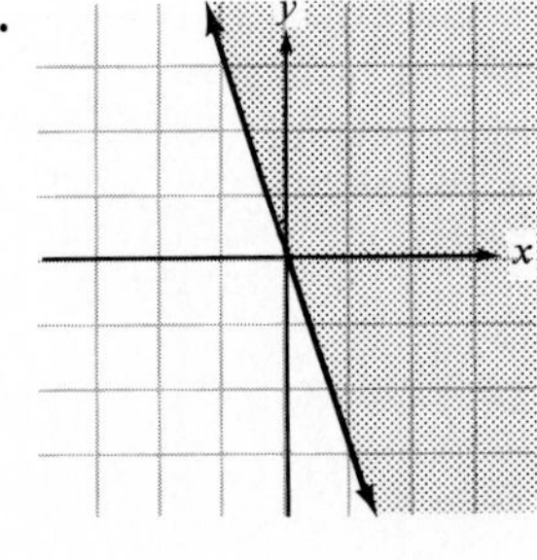

33.

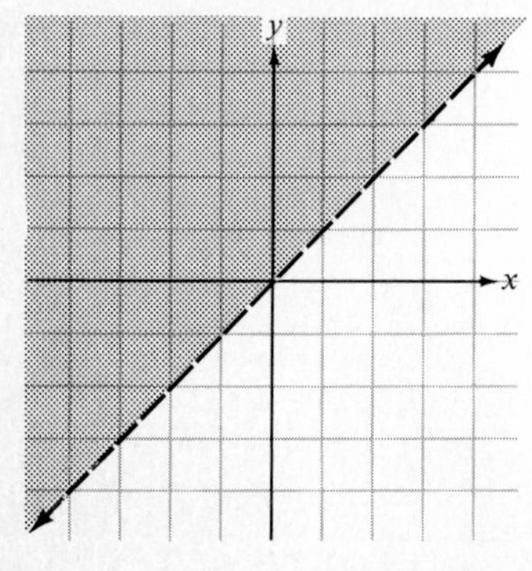

35.

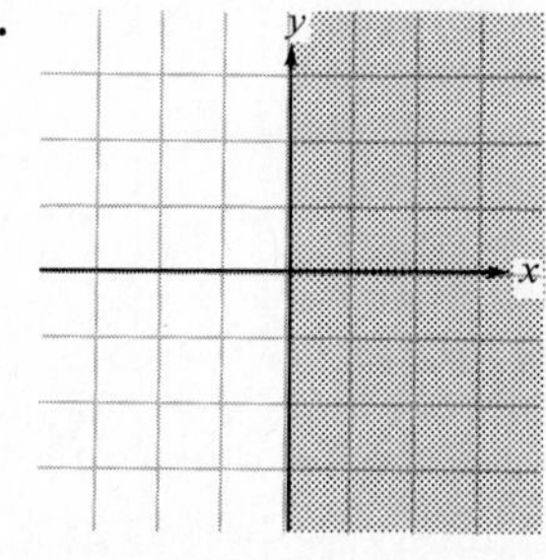

37.

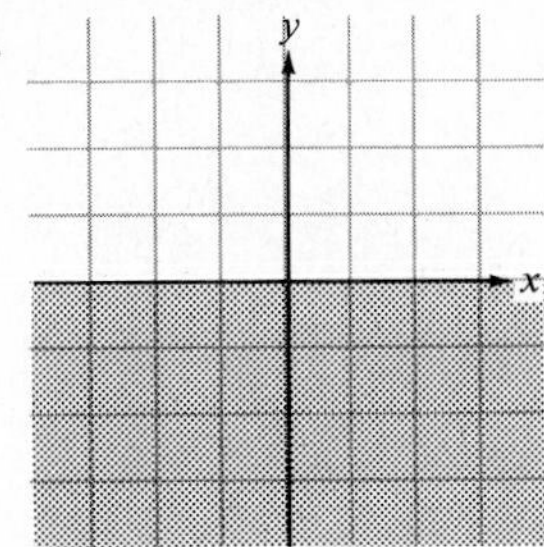

39.

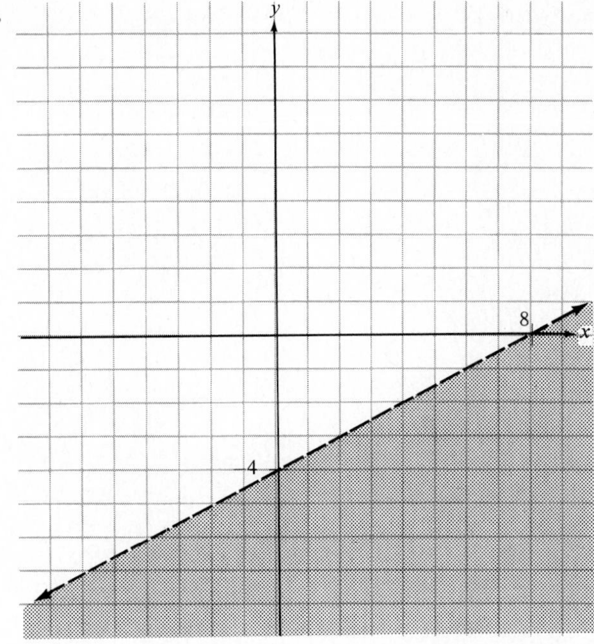

Exercises 6.5

1. (4, 1) **3.** (−2, 4) **5.** (4, 0) **7.** (0, −2) **9.** (3, 1) **11.** (−3, 2)
13. (−4, 5) **15.** (2, −1) **17.** (−2, 3) **19.** (4, 6) **21.** (6, 5) **23.** (−4, 3)
25. (−1, 2) **27.** (2, −3) **29.** Inconsistent **31.** Dependent **33.** Inconsistent
35. Dependent

Exercises 6.6

1. (4, 3) **3.** (3, −7) **5.** (−4, 1) **7.** $(\frac{7}{2}, 5)$ **9.** (6, −7)
11. $(\frac{1}{5}, 8)$ **13.** (2, −6) **15.** (3, 5) **17.** $(\frac{1}{3}, \frac{1}{2})$ **19.** (2, 1) **21.** (−7, 3)
23. $(1, \frac{1}{3})$ **25.** (0, −8) **27.** (5, 2) **29.** (3, 3) **31.** (9, 7) **33.** (6, 3)
35. $(\frac{7}{2}, 5)$ **37.** (−8, 9) **39.** (4, −2) **41.** (−2, 5) **43.** Dependent
45. Inconsistent **47.** Inconsistent **49.** Dependent

Exercises 6.7

1. (7, 14) **3.** (5, 13) **5.** (−8, −4) **7.** (17, 5) **9.** (−7, −9) **11.** (2, 5)
13. $(\frac{3}{4}, 1)$ **15.** (2, 4) **17.** (5, −2) **19.** (3, 1) **21.** (3, −6) **23.** (6, 3)
25. (8, 2) **27.** (3, 0) **29.** (−3, 5) **31.** (3, 2) **33.** (−1, −3) **35.** (−4, 10)
37. Inconsistent **39.** Dependent **41.** Dependent

Exercises 6.8

1. 87 and 127 **3.** 26 and 86 **5.** Mother receives \$6150; daughter receives \$1850.
7. 350 circuit breaker fuse boxes; 100 plug-type fuse boxes
9. 400 cars with the driver only; 1000 cars with 2 or more occupants **11.** 140 cones; 80 sundaes
13. Shaving commercial 10 times; shampoo commercial 20 times **15.** 240 regular size; 60 legal size
17. 300 small and 700 large teddy bears **19.** 120 photo packages, 40 class pictures
21. 8 liters of 80% alcohol solution and 12 liters of 30% solution

23. 24 grams 65% alloy, 36 grams 15% alloy **25.** 20 pounds at \$3 and 30 lbs at \$2
27. jet liner 450 mph, jetstream 50 mph.
29. Express passenger train speed is 65 mph, freight train 25 mph.
31. 55 mph for limousine; 45 mph for buses

Chapter 6 Test

1. $A(3, 1)$, $B(-2, 3)$, $C(-3, -4)$, $D(4, -2)$ **2.** (2, 2), (6, 12)
3. (3, 0), (0, −12), (1, −8), (4, 4)

4.

x	0	3	−1	2
y	−4	2	−6	0

5.

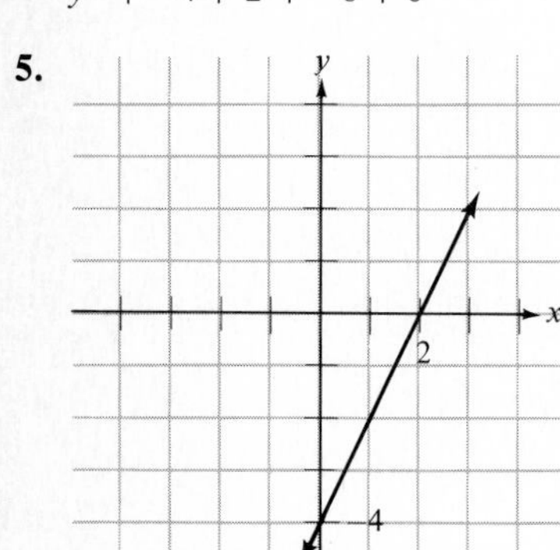

6.

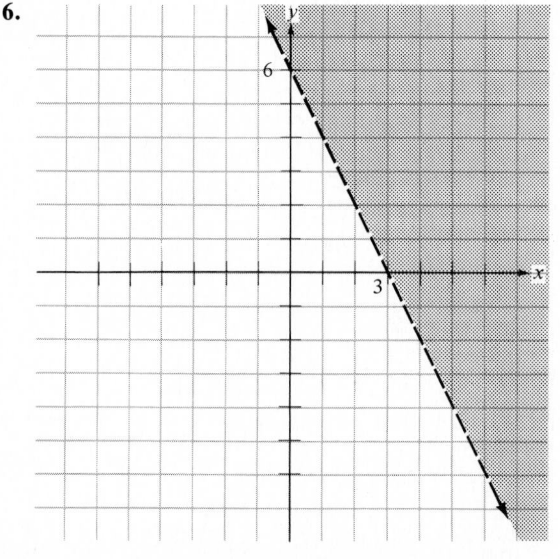

7. $x + y \leq 8$ **8.** (3, −1) **9.** (1, 2) **10.** **(a)** Inconsistent **(b)** Dependent
11. (6, 4) **12.** (−2, 10) **13.** (3, 2) **14.** (2, 4) **15.** (−1, −6)
16. (−5, −4) **17.** (2, 0) **18.** 40 and 115
19. 500 liters 50% solution, 1500 liters 10% solution **20.** Aircraft carrier 15 mph, helicopter 135 mph

CHAPTER 7

Exercises 7.1

1. 6 **3.** −6 **5.** −11 **7.** 13 **9.** 16 **11.** $\frac{1}{2}$ **13.** $\frac{2}{3}$
15. 1 **17.** 20 **19.** 32 **21.** 5.477 **23.** −8.246 **25.** 8.367
27. −8.062 **29.** −1 **31.** 4 **33.** 5 **35.** $\frac{2}{3}$ **37.** $\frac{2}{3}$ **39.** 3
41. Not real **43.** −3 **45.** −3 **47.** Not real **49.** −3 **51.** Not real
53. y **55.** $6y$ **57.** xy **59.** $2xy$ **61.** $(x + y)$ **63.** $2x$ **65.** $27y$
67. x^2 **69.** $\sqrt{(16 + 9)} = \sqrt{25} = 5$; $\sqrt{16} + \sqrt{9} = 4 + 3 = 7$; notice $5 \neq 7$
71. $\sqrt{(25 - 9)} = \sqrt{16} = 4$; $\sqrt{25} - \sqrt{9} = 5 - 3 = 2$; notice $4 \neq 2$
73. $\sqrt{(225 + 64)} = \sqrt{289} = 17$; $\sqrt{225} + \sqrt{64} = 15 + 8 = 23$; notice $17 \neq 23$
75. $\sqrt{(289 - 225)} = \sqrt{64} = 8$; $\sqrt{289} - \sqrt{225} = 17 - 15 = 2$; notice $8 \neq 2$; conclusion: $\sqrt{x^2 + y^2} \neq \sqrt{x^2} + \sqrt{y^2}$, also $\sqrt{x^2 - y^2} \neq \sqrt{x^2} - \sqrt{y^2}$

Exercises 7.2

1. 6 **3.** 7 **5.** 11 **7.** $3\sqrt{2}$ **9.** $3\sqrt{3}$ **11.** $7\sqrt{2}$ **13.** $4\sqrt{6}$
15. $8\sqrt{2}$ **17.** $10\sqrt{2}$ **19.** $7y$ **21.** $9xy$ **23.** $2x\sqrt{6x}$ **25.** $5xy\sqrt{7xy}$
27. $4x^2\sqrt{10}$ **29.** $6b^2\sqrt{10b}$ **31.** $2\sqrt{30}$ **33.** $7\sqrt{15}$ **35.** $13\sqrt{15}$ **37.** $2\sqrt[3]{7}$
39. $5\sqrt[3]{2}$ **41.** $3x\sqrt[3]{3x}$ **43.** $2x^2\sqrt[3]{4}$ **45.** $xy\sqrt[3]{xy^2}$ **47.** $3x\sqrt[4]{2x}$ **49.** 28.28
51. 11.18 **53.** 16.43 **55.** 80.62

Exercises 7.3

1. $\frac{6}{7}$ **3.** $\frac{5}{4}$ **5.** 3 **7.** 6 **9.** .6 **11.** .7 **13.** 1.2 **15.** .09
17. $\frac{\sqrt{3}}{2}$ **19.** $\frac{\sqrt{11}}{12}$ **21.** $\frac{2}{5}$ **23.** 3 **25.** $\sqrt{6}$ **27.** $\sqrt{7}$ **29.** $\frac{9\sqrt{2}}{4}$
31. $\frac{3\sqrt{11}}{11}$ **33.** $\frac{7\sqrt{6}}{6}$ **35.** $\frac{5\sqrt{2}}{6}$ **37.** $\frac{3\sqrt{3}}{8}$ **39.** $\frac{7\sqrt{2}}{12}$ **41.** $\frac{\sqrt{3}}{3}$
43. $\frac{11\sqrt{3}}{6}$ **45.** $\frac{\sqrt{3}}{3}$ **47.** $\frac{10\sqrt{17}}{17}$ **49.** $\frac{3}{4}$ **51.** $\frac{7}{6}$ **53.** $\frac{5}{x}$ **55.** $\frac{4x}{z}$
57. a **59.** $\frac{a^3}{b^4}$ **61.** $\frac{b}{a}$ **63.** $5y^2\sqrt{x}$ **65.** $\frac{2\sqrt[3]{5}}{3}$ **67.** $\frac{\sqrt[3]{9}}{3}$ **69.** $\frac{\sqrt[3]{x}}{x}$
71. $a^2\sqrt[3]{a^2}$ **73.** $\frac{2\sqrt[3]{9}}{3}$ **75.** $\frac{x\sqrt{y}}{y}$ **77.** 1.889 **79.** 4.472 **81.** 1.265
83. 1.155

Exercises 7.4

1. $8\sqrt{2}$ **3.** $2\sqrt{6}$ **5.** $6\sqrt{6}$ **7.** $3\sqrt{13}$ **9.** $-8\sqrt{2x}$
11. $3\sqrt{5} + 3\sqrt{7}$ **13.** $5\sqrt{5}$ **15.** $16\sqrt{5}$ **17.** $9\sqrt{3}$ **19.** $-2\sqrt{5}$
21. $16\sqrt{3}$ **23.** $27\sqrt{10}$ **25.** $9\sqrt{3}$ **27.** $2\sqrt{2}$ **29.** $-16\sqrt{6}$
31. $-3\sqrt{10}$ **33.** $39\sqrt{3}$ **35.** $10\sqrt{y}$ **37.** $6y\sqrt{2}$ **39.** $\sqrt{2z}$
41. $8x\sqrt{2x}$ **43.** $5n\sqrt{3n}$ **45.** $11xy\sqrt{y}$ **47.** $3xy\sqrt{z}$ **49.** $13y\sqrt{y}$

Exercises 7.5

1. $4\sqrt{5} + 5$ **3.** 15 **5.** $3\sqrt{5} - 3$ **7.** 54 **9.** $7\sqrt{6}$ **11.** $14\sqrt{2} + 42$
13. $30 + 15\sqrt{2}$ **15.** $-28\sqrt{3}$ **17.** $11 + 6\sqrt{3}$ **19.** $22 - 8\sqrt{7}$ **21.** $1 + 4\sqrt{13}$
23. $11\sqrt{2} + 13\sqrt{3}$ **25.** $24 - 10\sqrt{5}$ **27.** $8 + 2\sqrt{15}$ **29.** $-25\sqrt{3}$ **31.** 14
33. 20 **35.** 2 **37.** -2 **39.** $3 - \sqrt{2}$ **41.** $-12 - 6\sqrt{5}$ **43.** $-4 + 3\sqrt{2}$
45. $2\sqrt{6} + 2\sqrt{5}$ **47.** $2\sqrt{5} - 2\sqrt{3}$ **49.** $\sqrt{2}$ **51.** $\sqrt{5} - \sqrt{3}$ **53.** $19 - 5\sqrt{15}$

Exercises 7.6

1. 4 **3.** 8 **5.** 29 **7.** No solution **9.** 11 **11.** 5 **13.** 4
15. 4 **17.** 36 **19.** 9 **21.** 23 **23.** 5 **25.** 8 **27.** 5
29. 1 solution; -2 extraneous root
31. 12 solution; 0 extraneous root **33.** 10 solution; 1 extraneous root **35.** 0 and 2 are solutions
37. $\frac{3}{2}$ **39.** $+5, -5$ **41.** $+8, -8$

Chapter 7 Test

1. 9 **2.** -12 **3.** -3 **4.** 3 **5.** -3 **6.** ± 10 **7.** **(a)** Not real
(b) -4 **(c)** 2 **(d)** Not real **8.** $10\sqrt{3}$ **9.** $5x\sqrt{2x}$ **10.** 5 **11.** $5xy\sqrt{y}$
12. $\frac{8}{10} = .8$ **13.** 5 **14.** $4\sqrt{3}$ **15.** $2\sqrt[3]{3}$ **16.** $6\sqrt{11}$ **17.** $18\sqrt{3}$
18. $6x\sqrt{5x}$ **19.** $12\sqrt{2} + 12$ **20.** 2 **21.** $7\sqrt{2} - 11$ **22.** 32 **23.** $4 + \sqrt{3}$
24. 5 **25.** 5 **26.** No solution **27.** 12 is a solution; 4 is not a solution.

CHAPTER 8

Exercises 8.1

1. ± 4 **3.** ± 9 **5.** ± 12 **7.** $\pm\sqrt{3}$ **9.** $\pm\sqrt{17}$ **11.** $\pm\sqrt{29}$
13. $\pm 2\sqrt{2}$ **15.** $\pm 2\sqrt{10}$ **17.** $\pm 5\sqrt{2}$ **19.** $\pm 10\sqrt{2}$ **21.** $-6, 4$ **23.** $-7, 1$
25. $-5, 15$ **27.** $-2, 6$ **29.** $-2, 4$ **31.** $1 \pm \sqrt{6}$ **33.** $-3 \pm \sqrt{15}$
35. $-5 \pm \sqrt{17}$ **37.** $2 \pm \sqrt{5}$ **39.** $-3 \pm \sqrt{2}$ **41.** $2 \pm \sqrt{2}$ **43.** $-2 \pm \sqrt{3}$
45. $\frac{1}{2} \pm \sqrt{2} = \frac{1 \pm 2\sqrt{2}}{2}$ **47.** $\frac{2}{3} \pm \sqrt{2} = \frac{2 \pm 3\sqrt{2}}{3}$ **49.** $-\frac{1}{5} \pm \sqrt{2} = \frac{-1 \pm 5\sqrt{2}}{5}$

Exercises 8.2

1. Yes **3.** Yes **5.** No **7.** No **9.** 25 **11.** 64 **13.** 100
15. 49 **17.** $-3 \pm \sqrt{10}$ **19.** $2 \pm \sqrt{5}$ **21.** $8 \pm 2\sqrt{17}$ **23.** $-9 \pm 3\sqrt{10}$
25. $-1, 3$ **27.** $-3 \pm 2\sqrt{5}$ **29.** $7 \pm 4\sqrt{3}$ **31.** $-21, -1$ **33.** $-3, 1$
35. $3 \pm 2\sqrt{2}$ **37.** $\frac{-1 \pm \sqrt{33}}{4}$ **39.** -1 **41.** $\frac{1 \pm \sqrt{5}}{4}$ **43.** $\frac{-1 \pm \sqrt{41}}{10}$
45. $\frac{-1 \pm \sqrt{13}}{3}$

Exercises 8.3

1. $-2, -1$ **3.** $-1 \pm \sqrt{6}$ **5.** $\frac{-3 \pm \sqrt{5}}{2}$ **7.** $2 \pm \sqrt{6}$ **9.** $5 \pm \sqrt{13}$
11. $\frac{1 \pm \sqrt{13}}{6}$ **13.** $\frac{1 \pm \sqrt{17}}{4}$ **15.** $-1, \frac{5}{2}$ **17.** $-2, -\frac{1}{2}$ **19.** $\frac{1 \pm 2\sqrt{2}}{7}$
21. $-\frac{1}{2}, 2$ **23.** 1 **25.** $\frac{-5 \pm \sqrt{17}}{2}$ **27.** $-1, 3$ **29.** $-\frac{1}{2}, \frac{2}{3}$ **31.** $\frac{-1 \pm \sqrt{29}}{2}$
33. $\frac{-1 \pm \sqrt{13}}{2}$ **35.** $-3 \pm \sqrt{14}$ **37.** $\frac{1 \pm \sqrt{17}}{4}$ **39.** $\frac{1 \pm \sqrt{73}}{12}$

Exercises 8.4

1. $3 + i$ **3.** $6 + 4i$ **5.** $-4 + 2i$ **7.** $-8 + 0i = -8$ **9.** $0 - 9i = -9i$
11. $9 + 6i$ **13.** $9 - 3i$ **15.** $5 + 3i$ **17.** $-2 + i$ **19.** $6 - i$ **21.** $-12 + 24i$
23. $7 + 6i$ **25.** $13i$ **27.** $37 - 5i$ **29.** $34 + 13i$ **31.** $1 + 2i$
33. $\frac{-3 - 4i}{25} = -\frac{3}{25} - \frac{4}{25}i$ **35.** $1 - 3i$ **37.** $\frac{12 - 5i}{13} = \frac{12}{13} - \frac{5}{13}i$
39. $\frac{5 + 27i}{13} = \frac{5}{13} + \frac{27}{13}i$ **41.** $i^1 = i, i^2 = -1, i^3 = -i, i^4 = 1, i^5 = i, i^6 = -1, i^7 = -i, i^8 = 1$

Exercises 8.5

1. $2, 7$ **3.** $-4, -2$ **5.** $-3, \frac{1}{2}$ **7.** $-\frac{5}{2}, 3$ **9.** ± 4 **11.** $\pm 4i$
13. $-2 \pm 3i$ **15.** $3 \pm 7i$ **17.** $-1 \pm \sqrt{10}$ **19.** $\frac{-3 \pm \sqrt{185}}{4}$ **21.** $\frac{1 \pm i\sqrt{3}}{2}$
23. $-1 \pm 2i$ **25.** $-2, -1$ **27.** $\frac{2 \pm \sqrt{2}}{2}$ **29.** $\frac{-1 \pm i\sqrt{7}}{4}$ **31.** $3 + i$

Exercises 8.6

1. Opens upward; vertex at (0, 0)

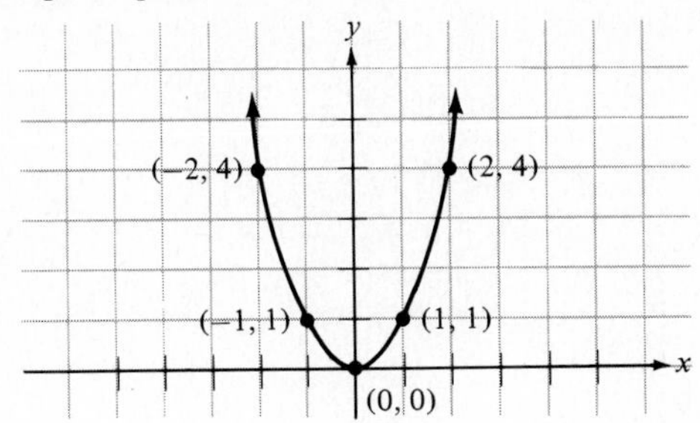

3. Opens upward; vertex at (0, 3)

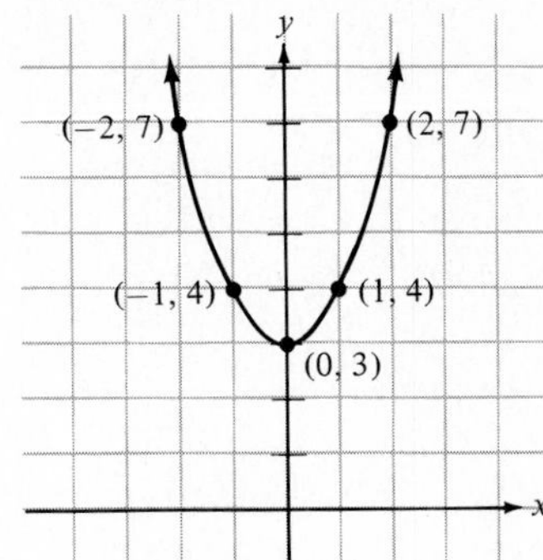

5. Opens upward; vertex at (0, −2)

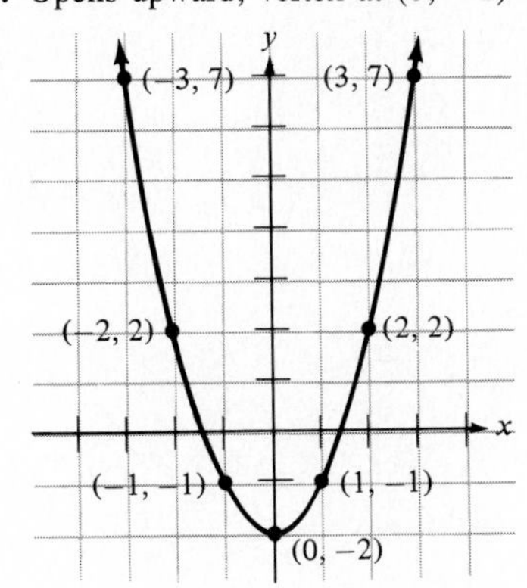

7. Opens downward; vertex at (0, 3)

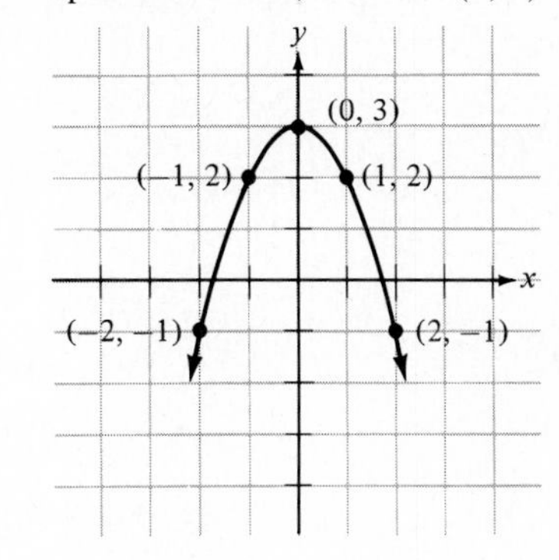

9. Opens upward; vertex at (0, 0)

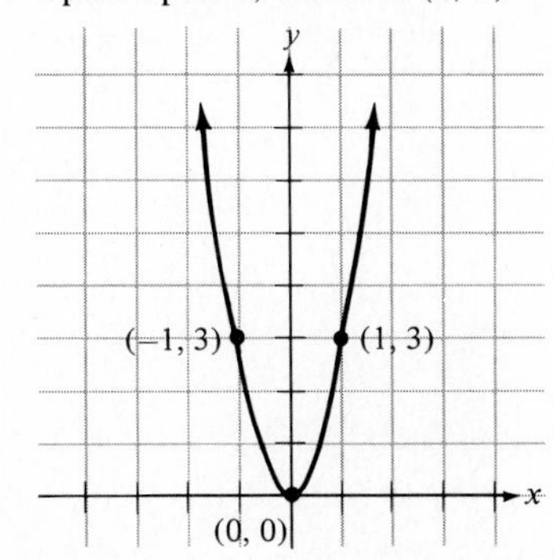

11. Opens downward; vertex at (0, 1)

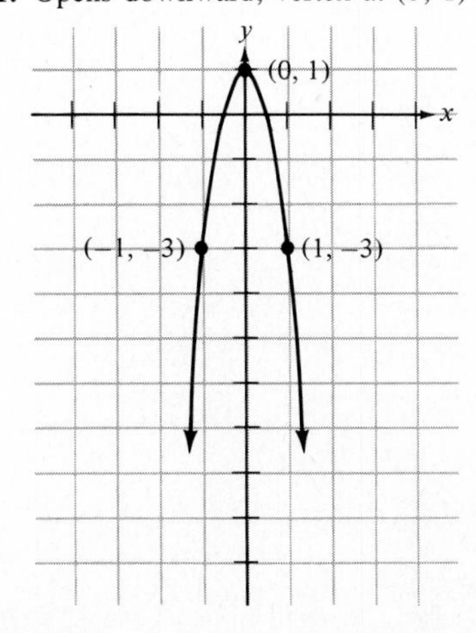

13. Opens upward; vertex at (0, 0)

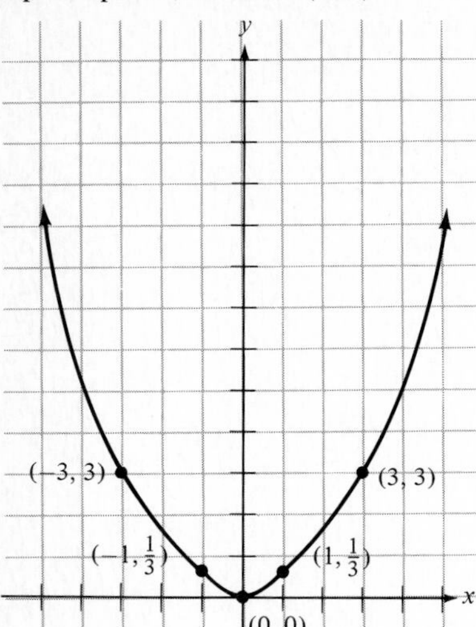

15. Opens upward; vertex at (0, −2)

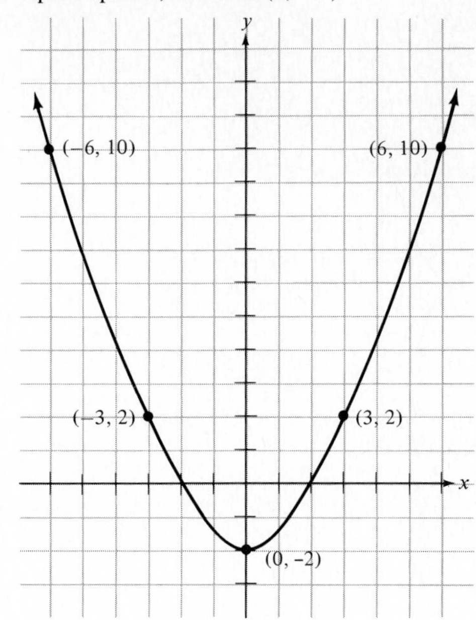

17. Vertex at (2, 0); axis of symmetry is $x = 2$

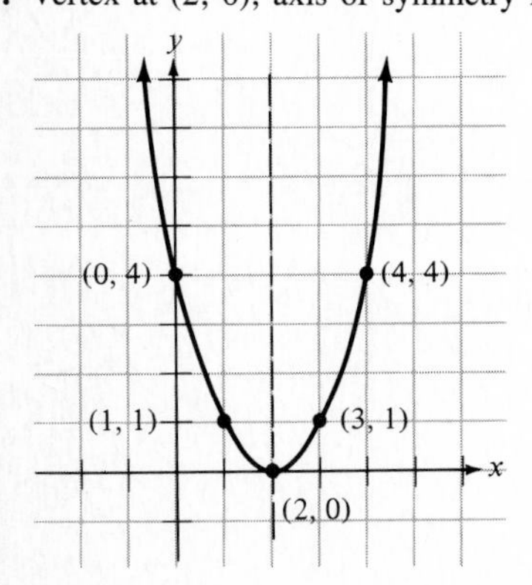

19. Vertex at (4, 0); axis of symmetry is $x = 4$.

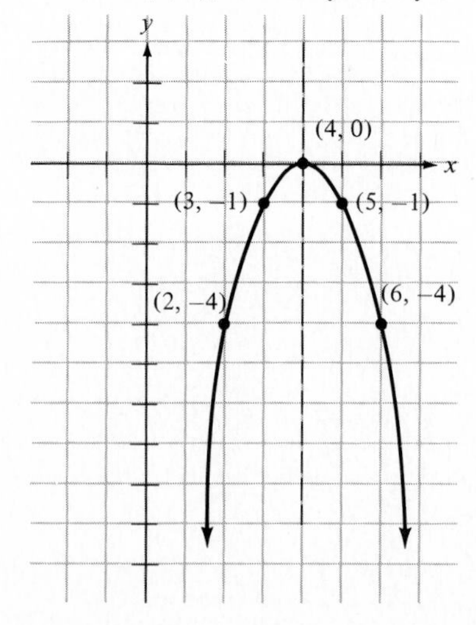

21. Vertex at (4, 2); axis of symmetry is $x = 4$.

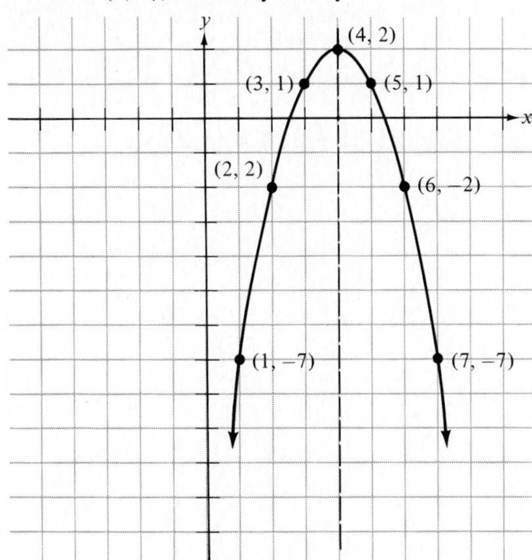

23. Vertex at $(-1, -3)$; axis of symmetry is $x = -1$.

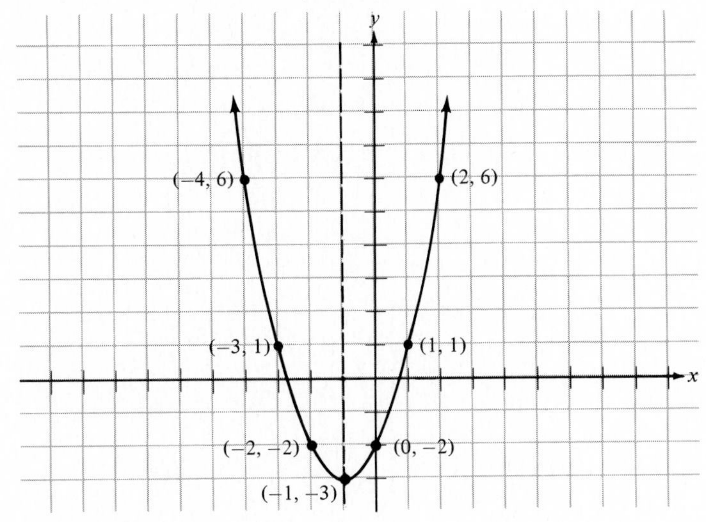

25. $y = (x - 1)^2$; vertex at (1, 0); axis of symmetry is $x = 1$.

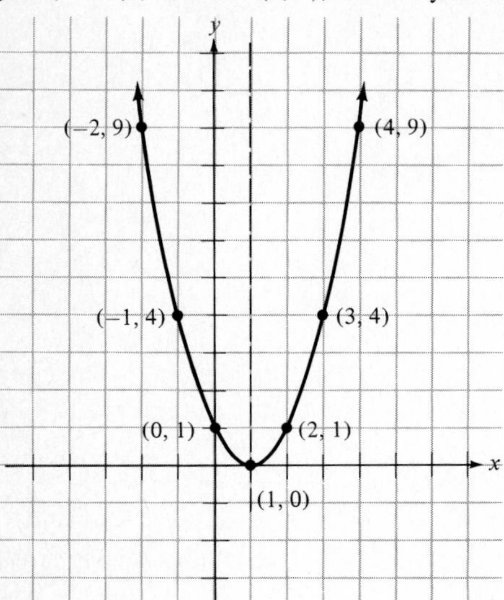

27. $y = (x - 2)^2$; vertex at (2, 0); axis of symmetry is $x = 2$.

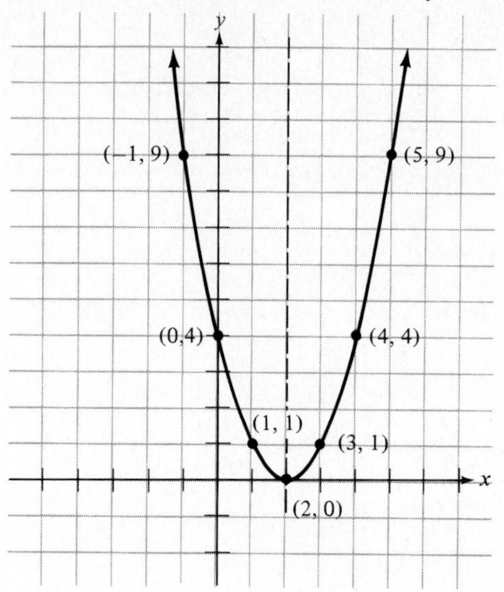

29. $y = (x - 5)^2$; vertex at (5, 0); axis of symmetry is $x = 5$.

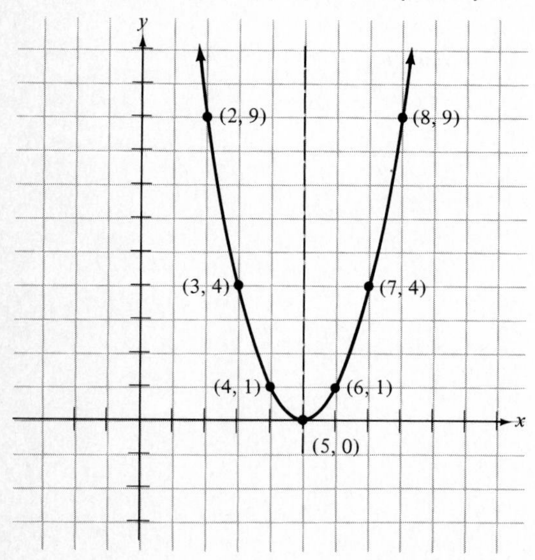

Exercises 8.7

1. 25 inches
3. $5\sqrt{5}$ inches
5. $25\sqrt{2}$ inches
7. 15 feet
9. $L = 90$ feet, $W = 60$ feet
11. $L = 13$ inches, $W = 6$ inches
13. 4.5 seconds
15. 50 seconds
17. $90\sqrt{2} \approx 126$ feet
19. $4\sqrt{29} \approx 21.6$ feet
21. $\sqrt{5} \approx 2.24$ miles (it will be close!)
23. $w = \dfrac{-1 + \sqrt{5}}{2}$
25. $\dfrac{5(\sqrt{5} + 1)}{2}$
27. $-4 + 4\sqrt{5} = 4(-1 + \sqrt{5})$
29. $.618 \approx .6 = \frac{3}{5}$

Chapter 8 Test

1. ± 7
2. $\pm 4\sqrt{5}$
3. $-8, 4$
4. $-8, 2$
5. No
6. Yes
7. $-11, 1$
8. $2, 6$
9. $-2, -\frac{1}{2}$
10. $\dfrac{3 \pm \sqrt{89}}{10}$
11. $8 + 6i$
12. $2 - 4i$
13. $-3 + 21i$
14. $8 + 6i$
15. $13 - i$
16. $4 + 19i$
17. $\dfrac{-4 + 8i}{5} = -\dfrac{4}{5} + \dfrac{8}{5}i$
18. $\dfrac{11 - 27i}{25} = \dfrac{11}{25} - \dfrac{27}{25}i$
19. $-1 \pm 2i$
20. $\dfrac{-6 \pm 2i\sqrt{33}}{6} = -1 \pm \dfrac{i\sqrt{33}}{3}$

21. Vertex at (0, 2); axis of symmetry is $x = 0$.

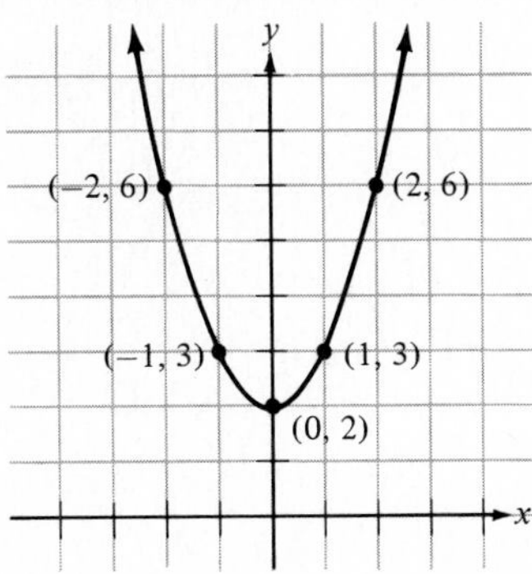

22. Vertex at (0, 1); axis of symmetry is $x = 0$.

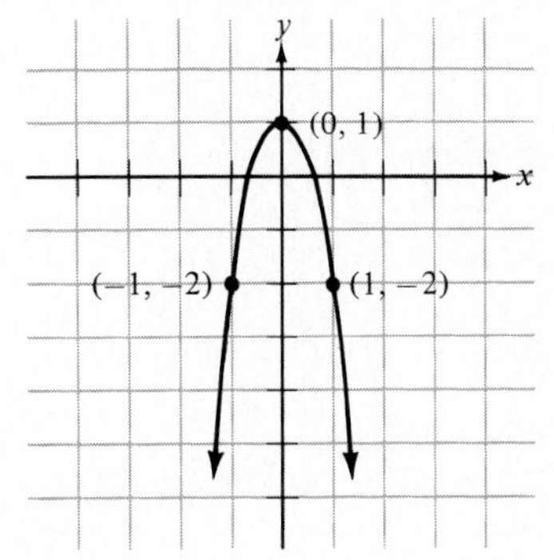

23. Vertex at (0, −4); axis of symmetry is $x = 0$.

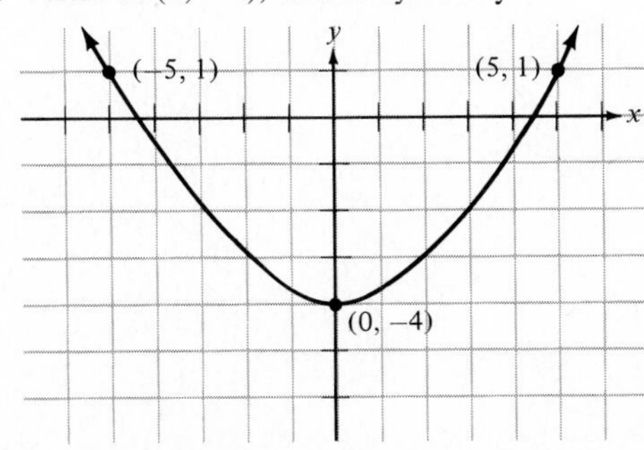

24. Vertex at (−1, 4); axis of symmetry is $x = -1$.

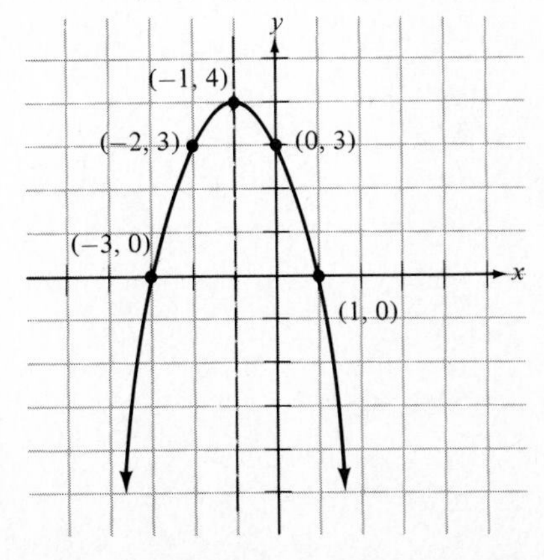

25. $4\sqrt{13} \approx 14.4$ feet **26.** **(a)** $5\sqrt{2} \approx 7$ inches **(b)** $20\sqrt{2} \approx 28$ inches **(c)** 50 square inches **27.** $2\sqrt{109} \approx 208$ feet

CHAPTER 9

Exercises 9.1

1. Yes; $D = \{0, 1, 2, 3\}$; $R = \{0, 3, 6, 9\}$ **3.** Yes; $D = \{-1, 0, 1\}$; $R = \{3, 5, 7\}$
5. Yes; D is the set of real numbers; R is the set of all real numbers.
7. Yes; $D = \{-3, -2, -1, 0, 1, 2, 3\}$; $R = \{0, 3, 12, 27\}$ **9.** No **11.** No
13. Yes; $D = \{-3, -2, -1, 0, 1, 2, 3\}$; $R = \{-27, -8, -1, 0, 1, 8, 27\}$ **15.** Yes; D all reals, $R = \{4\}$
17. $\{x \mid x \text{ is real and } x \neq 0\}$ **19.** $\{x \mid x \text{ is real and } x \geq 0\}$ **21.** $\{x \mid x \text{ is real and } x \neq 1, x \neq 2\}$
23. $\{x \mid x \text{ is real}\}$ **25.** $\{x \mid x \text{ is real and } x \geq \frac{1}{2}\}$ **27.** $\{x \mid x \text{ is real and } x \neq \pm 1\}$
29. $\{x \mid x \text{ is real and } x > 2\}$ **31.** $\{x \mid x \text{ is real and } x \neq \pm 1\}$ **33.** $\{x \mid x \text{ is real and } x \neq 5\}$
35. **(a)** $f(0) = 1$ **(b)** $f(3) = 7$ **(c)** $f(-2) = -3$ **37.** **(a)** $g(0) = 3$ **(b)** $g(\sqrt{2}) = 5$ **(c)** $g(-2) = 7$
39. **(a)** $h(0) = 2$ **(b)** $h(-3) = 1$ **(c)** $h(5) = 3$ **41.** **(a)** $F(0) = \frac{1}{4}$ **(b)** $F(2) = \dfrac{3}{-2}$ **(c)** $F(3) = -5$
43. **(a)** $G(0) = 5$ **(b)** $G(3) = 4$ **(c)** $G(-3) = 4$ **45.** **(a)** $H(0) = 5$ **(b)** $H(-1) = 5$ **(c)** $H(20) = 5$
47. Not a function **49.** Function **51.** Function **53.** Function

Exercises 9.2

1. $\frac{3}{2}$ **3.** 1 **5.** $\dfrac{-4}{8} = -\dfrac{1}{2}$ **7.** 0 **9.** Not defined **11.** $\frac{4}{3}$
13. -1 **15.** 2 **17.** -3 **19.** $\dfrac{b}{-a}$ **21.** $-\frac{1}{2}$ **23.** $\frac{2}{3}$ **25.** 0
27. False **29.** False **31.** True

Exercises 9.3

1. $y = 4x - 7$ **3.** $y = 3x + 3$ **5.** $y = 5x - 7$ **7.** $y = -x + 2$
9. $y = 2x + 13$ **11.** $y = \frac{5}{2}x - 1$ **13.** $y = -\frac{1}{3}x + 6$ **15.** $y = 9x + 4$ **17.** $y = 3$
19. $y = 7x + 2$ **21.** $y = 11x + 3$ **23.** $y = -2x + 5$ **25.** $y = -5x + 7$
27. $y = -\frac{5}{3}x + 8$ **29.** $y = -6$ **31.** $m = 2$ **33.** $m = 3$ **35.** $m = 4$
37. $m = \frac{2}{5}$ **39.** $m = 0$ **41.** $y = 2x - 1$ **43.** $y = 4x - 6$ **45.** $y = -\frac{2}{3}x + 2$
47. $y = 7$ **49.** $x = -4$ **51.** $y = -7$

Exercises 9.4

1. 27 **3.** 66 **5.** 77 **7.** 10 **9.** 10 **11.** 20 **13.** 27
15. 60 **17.** 9 **19.** 144 **21.** 6 **23.** 25 **25.** 4 **27.** 5 **29.** $\frac{100}{9}$
31. 4 **33.** 18 **35.** 1 **37.** 15.75 miles **39.** 2100 pounds **41.** 10
43. 16 **45.** 150 pounds

Exercises 9.5

1. 10 **3.** 5 **5.** 5 **7.** 125 **9.** 64 **11.** 16 **13.** 1 **15.** 9
17. 8 **19.** 4 **21.** $\frac{1}{4}$ **23.** $\frac{1}{6}$ **25.** $\frac{1}{9}$ **27.** $\frac{1}{8}$ **29.** $\frac{1}{27}$ **31.** $\frac{1}{4}$
33. 36 **35.** 81 **37.** 81 **39.** $x^{13/15}$ **41.** $z^{7/4}$ **43.** $x^{17/12}$ **45.** 6
47. 2 **49.** $(xy)^{2/3}$ **51.** $(yz)^{5/3}$ **53.** 25 **55.** 12 **57.** 8 **59.** W
61. y^3 **63.** $x^{9/4} = (\sqrt[4]{x})^9$ **65.** 3 **67.** 49 **69.** $a^{1/3} = \sqrt[3]{a}$ **71.** x^2
73. $\sqrt{6}$ **75.** 3 **77.** $\dfrac{x^6z^3}{y^5}$ **79.** $\dfrac{b^8c^6}{a^5}$

Exercises 9.6

1.

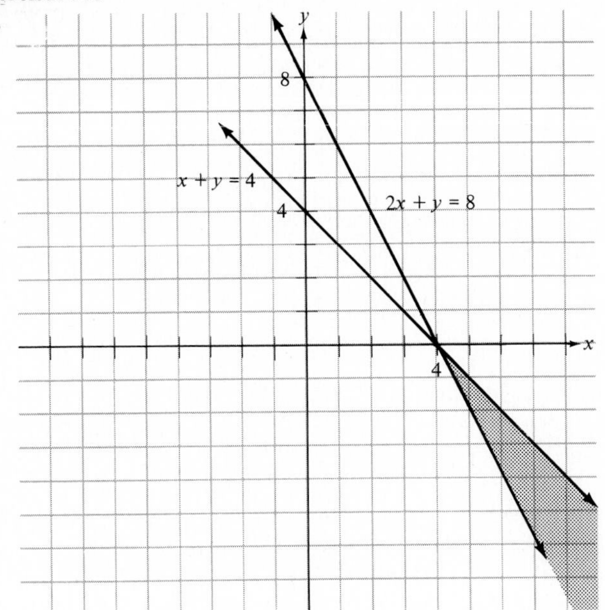

3.

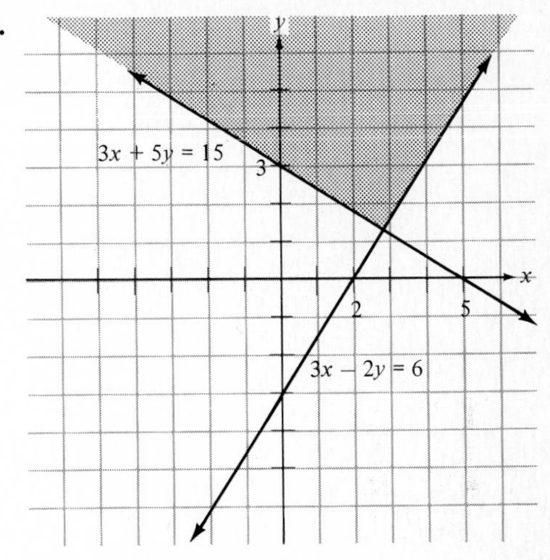

5.

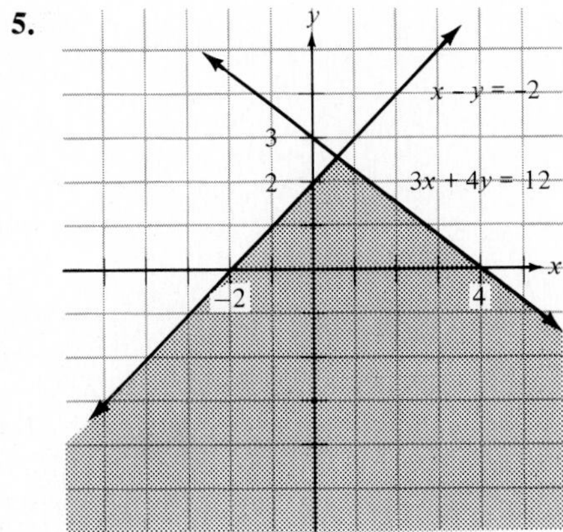

7.

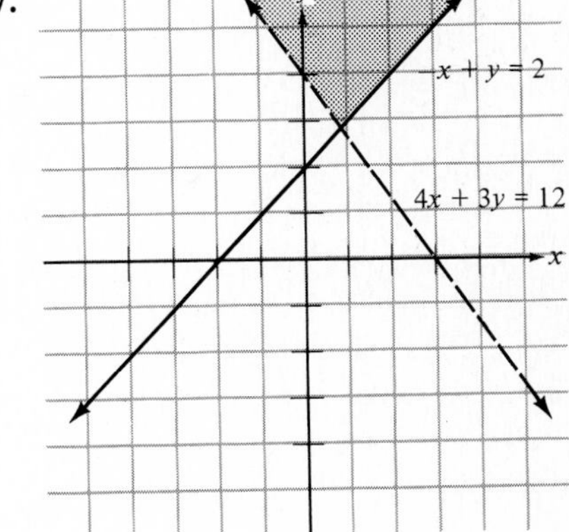

9.

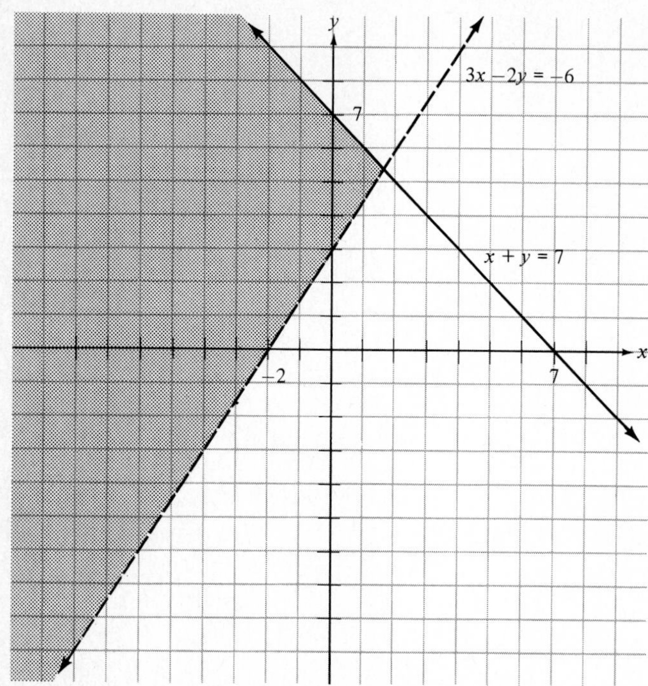

y
x
$3x - 2y = -6$
7
$x + y = 7$
−2
7

11.

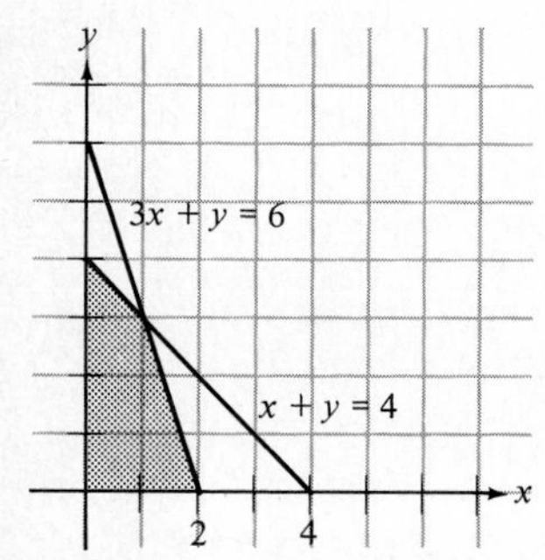

y
x
$3x + y = 6$
$x + y = 4$
2
4

13.

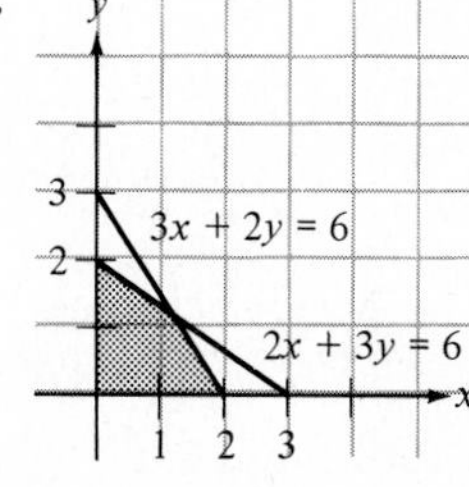

y
x
3
2
$3x + 2y = 6$
$2x + 3y = 6$
1
2
3

15.

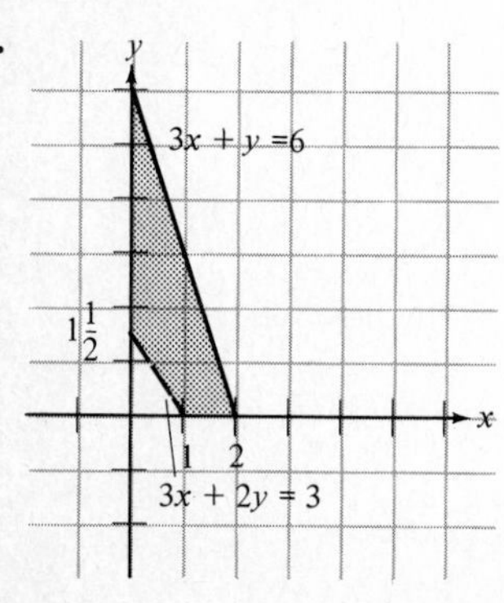

y
x
$3x + y = 6$
$1\frac{1}{2}$
1
2
$3x + 2y = 3$

17.

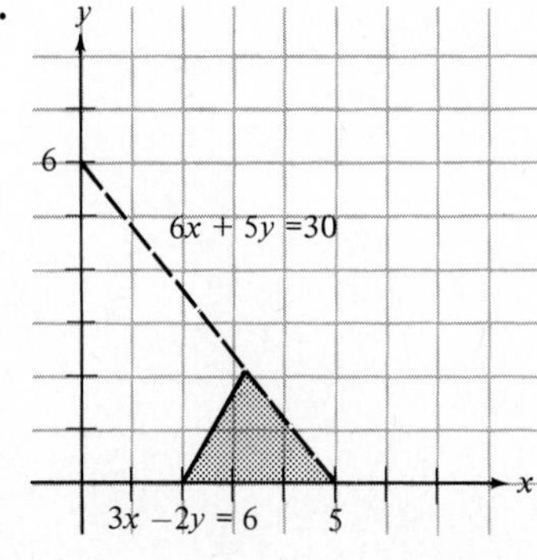

y
x
6
$6x + 5y = 30$
$3x - 2y = 6$
5

19.

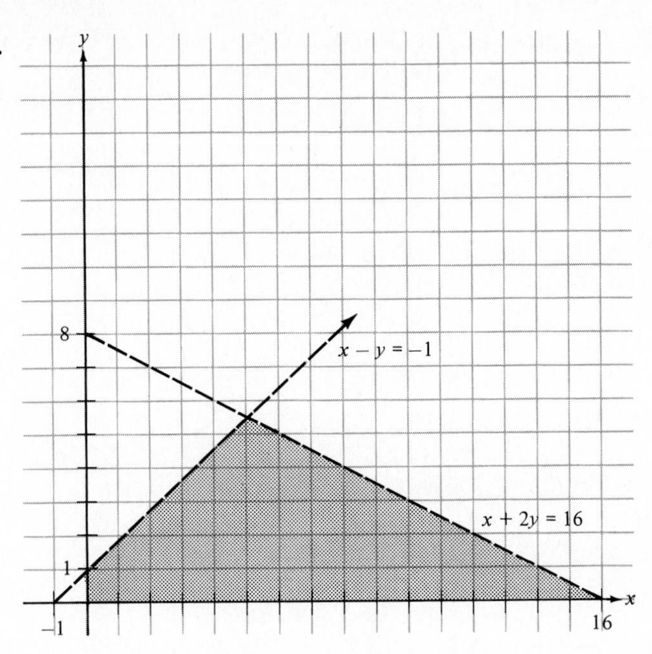

21.

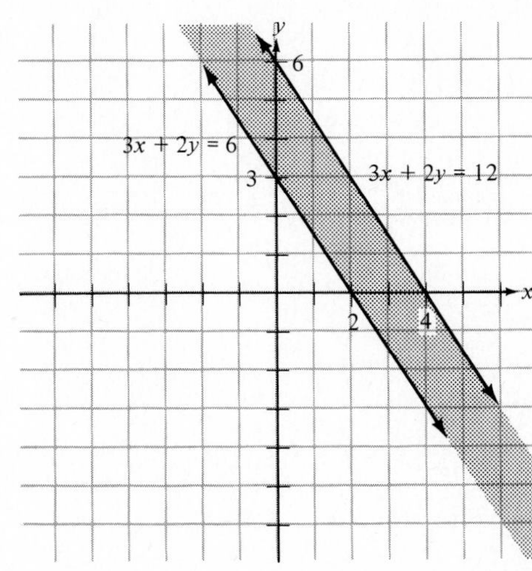

23.

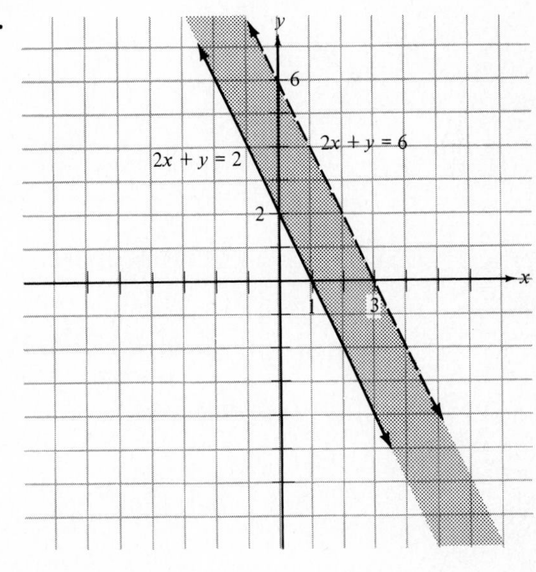

25.

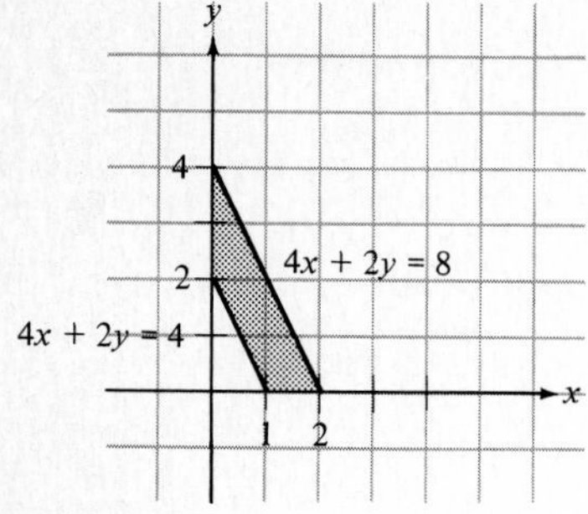

Chapter 9 Test

1. A function; $D = \{-2, 0, 2\}$; $R = \{-16, 0, 16\}$ **2.** Not a function
3. $\{x \mid x \text{ is real and } x \neq 4\}$ **4.** $\{x \mid x \text{ is real}\}$ **5.** $\{x \mid x \text{ is real and } x \geq -2\}$
6. $\{x \mid x \text{ is real and } x \neq -1, x \neq 6\}$ **7. (a)** $f(0) = 1$ **(b)** $f(2) = 11$ **(c)** $f(-2) = -9$
8. (a) $g(0) = 1$ **(b)** $g(5) = 4$ **(c)** $g(8) = 5$ **9. (a)** 3 **(b)** -3 **10.** $y = \frac{2}{3}x + 1$
11. $y = -7x + 4$ **12.** $y = 2x + 5$ **13.** 10 **14.** $\frac{40}{3}$ **15.** 24
16. 700 pounds per square foot **17.** 8 **18.** $\frac{1}{9}$ **19.** 49 **20.** 4 **21.** x
22. 25 **23.** $\dfrac{x^8z^2}{y^9}$ **24.**

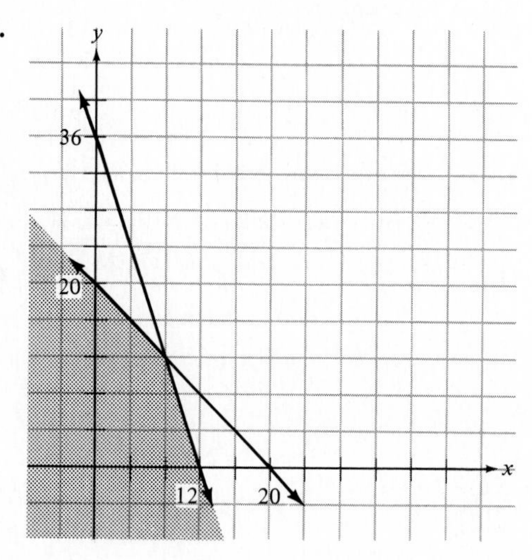

INDEX